THE RELUCTANT FARMER

THE
RELUCTANT
FARMER

THE RISE OF
AGRICULTURAL
EXTENSION
TO 1914

ROY V. SCOTT

UNIVERSITY OF
ILLINOIS PRESS

URBANA
CHICAGO
LONDON

To my parents,
Illinois farmers
who have known
the realities
of rural life.

CONTENTS

PREFACE

SEVERAL YEARS AGO after taking his degree, one of my first graduate students joined the Peace Corps and was sent to Latin America. In Peru his superiors discovered that the young man had written his master's thesis on the subject of the farmers' institutes in Mississippi. That experience presumably made him an expert on agricultural education, and he was assigned duties in that area.

The episode suggested to me that a survey of the rise of agricultural extension education in the United States might be of some value. Conditions in the United States prior to 1914 were dramatically different from those prevailing in Peru a half century later, but still significant parallels existed. There can be little doubt that increased agricultural productivity is a necessary prerequisite to industrial growth and a rising standard of living, regardless of time or place, or that agricultural extension in some form has a role in improving agricultural productivity. Consequently, I would hope that this study might be of some importance not only to historians, agricultural extension personnel, and others interested in rural change over time in the United States but also to those persons concerned with the problems of underdeveloped agricultural economies wherever they exist.

This monograph is the story of the search for a teaching device effective with rural adults. The search was a lengthy one, extending from at least the 1780s to 1914. Public figures and other prominent men of the Revolutionary generation established agricultural societies and sought by example to improve the farming practices of their neighbors. The farm press, local agricultural clubs and societies, country fairs, and pioneer college instruction in agriculture appeared before the Civil War. In the course of that struggle, the Morrill Act provided for an agricultural college in each state, and a quarter of a century later the Hatch Act led to the establishment of agricultural experiment stations. These agencies and institutions served useful purposes, but all of them found it difficult, if not impossible, to convince farmers of the value of science. Needed was an extension system capable of proving to farmers that "book farming" was not a joke and that agricultural science, properly applied, would produce a better life for them and their families. Accordingly, rural leaders, agricultural educators, and various business interests developed farmers' institutes, reading courses in agriculture, movable schools, and other methods calculated to *tell* agriculturists how to improve their operations. Farmers listened but they were little moved. It remained for Seaman A. Knapp and a few innovative men in the North to point the way through demonstration, a method that *showed* farmers ways to better and more productive agriculture.

The passage of the Smith-Lever Act in 1914 marked the end of the long search for that effective teaching technique, and I have made no effort to discuss results or developments after that date. In the main, the bibliography contains only those materials and items actually used in writing the study; it excludes many others that were consulted.

Several other writers have dealt with aspects of agricultural extension in the United States. Alfred C. True, for many years head of the Office of Experiment Stations, wrote on the general subject some four decades ago, but his work tends to be encyclopedic in character and lacks interpretation. Russell Lord's *The Agrarian Revival* is a popular, journalistic account. More useful are volumes on specific phases of the topic covered in this book. Joseph C. Bailey has written an excellent biography of Seaman A. Knapp. Gladys

Baker's *The County Agent* deals primarily with developments since 1914 and gives only scant attention to origins of extension. In *The County Agent and the Farm Bureau,* M. C. Burritt describes the early development of the farm bureau and the roots of the special relationship that it enjoyed with the extension service.

The writing of this monograph would have been impossible without the assistance and encouragement of many individuals, most of whom I cannot mention here. I must confirm my debt to the late Fred A. Shannon, who directed my graduate studies and who was largely responsible for whatever virtues my first book may have had. Ralph W. Hidy of the Graduate School of Business Administration at Harvard University has made many suggestions over a period of several years; his wife, Muriel, has been a source of much inspiration. Professors W. Magruder Drake of the University of Southwestern Louisiana and Warren F. Kuehl of the University of Akron gave important help at an early stage of the project. The American Philosophical Society and the Mississippi State University Development Foundation provided travel grants and other aid. Numerous librarians, especially those at Mississippi State, Michigan State, Clemson, and Cornell universities, have been most cooperative and helpful. Mrs. Kathleen Young, in charge of the interlibrary loan office at Mississippi State, and Miss Willie D. Halsell deserve particular notice; both function with outstanding competence and cheerfulness. The editors of several scholarly journals, most notably *Business History Review*, have allowed me to use materials that appeared in those periodicals. Miss Sherry L. Morrison typed the manuscript. Finally, I must again thank my wife, Jane Brayford Scott, who made all of the sacrifices that only a historian's wife knows. Of course, I am solely responsible for any errors in fact or interpretation that may remain.

R. V. S.

THE RELUCTANT FARMER

I. THE ROOTS
OF AGRICULTURAL EXTENSION

"FARMERS, LIKE OTHER PEOPLE, hesitate to believe and act on theories, or even facts, until they see with their own eyes the proof of them in material form," wrote Cornell University's Isaac P. Roberts in 1889. The same year, speaking before the Association of American Agricultural Colleges and Experiment Stations, J. S. Newman of Alabama emphasized the need for land-grant schools to bring agricultural science more to the attention of farmers. "We must carry it home to them," he said. "How we are to do this is the most important question for us to consider."[1] Professor Roberts had described a significant characteristic of farmers everywhere; his southern associate posed in a concise way a major problem in rural education and improvement. It was not solved in the United States until 1914.

Modern cooperative agricultural extension education is a product of evolution and experimentation; its development represented a victory of change and progress over traditionalism and apathy. For more than a century before the historic Smith-Lever Act became

---

[1] Association of American Agricultural Colleges and Experiment Stations, *Proceedings*, 1889 (Washington, 1889), pp. 29, 31. Cited hereafter as AAACES, with year.

3

law, agricultural leaders labored energetically if largely in vain to induce the rural population to improve its farming methods. Individual innovators were always present, although their impact is difficult to determine.[2] A few prominent men made especially notable contributions. Agricultural societies, fairs, farm papers, land-grant colleges, and experiment stations appeared, but none met adequately the needs of agriculture. Essentially, such individuals and agencies were unable to reach and influence in an effective way the great body of ordinary farmers. To do so was the great task of agricultural extension.

Much of the progress that did occur in agriculture before and during the nineteenth century was the result of efforts by a relatively few prominent and wealthy men. George Washington corresponded with other agricultural innovators, both at home and abroad, he tried newer methods on his own property, and he sought to use the office of the presidency to improve agriculture. Jefferson's experiments with the plow and his other efforts toward agricultural change are well known.[3]

A multitude of lesser figures, in the aggregate, made an even greater contribution. In the Revolutionary generation were such men as Samuel H. Black of Delaware and Thomas Pinckney of South Carolina. A medical doctor born in 1742, Black used his LaGrange farm as an experiment station to test the value of fertilizers and lime. In time he acquired a reputation as a speaker on agricultural topics, using the lecture platform to disseminate his ideas among farmers. Pinckney had large land holdings in South Carolina, where he experimented with rice and other crops. He, too, corresponded with leading agriculturists, and he imported various kinds of cattle from

[2] Allan G. Bogue, "Pioneer Farmers and Innovation," *Iowa Journal of History,* LVI (January, 1958), pp. 1–36; Gould P. Colman, "Innovation and Diffusion in Agriculture," *Agricultural History,* XLII (July, 1968), pp. 173–187.

[3] Ralph M. Brown, "Agricultural Science and Education in Virginia before 1860," *William and Mary College Quarterly History Magazine,* XIX (April, 1939), pp. 198–199. Jefferson's contributions are discussed in Everett E. Edwards, "Jefferson and Agriculture," U.S. Department of Agriculture, *Agricultural History Series* 7 (Washington, 1943).

Europe. His son served as the first president of the Pendleton Farmer's Society.[4]

A generation later came John Taylor of Caroline County, Virginia. In 1813 he published his *Arator* in which he urged the use of manures to restore fertility, called for the growing of corn and clover instead of tobacco, and advocated contour plowing to control erosion. In fact, Taylor was interested in any technique that promised to improve agriculture, and his was a powerful voice calling for change.

Among his contemporaries were several other innovators worthy of mention. Thomas Mann Randolph was an exponent of contour plowing; Stephen McCormick experimented with improved farm tools; Fielding Lewis was widely known for his use of lime to rejuvenate soils; and Philip Tabb, John Singleton, William Merriweather, and W. C. Nicholas used advanced techniques that could not have failed to influence neighbors.

Towering over a next wave of southern agrarian reformers was Edmund Ruffin of Virginia. An actual farmer, he approached his problems intelligently and logically. From his experience he produced the theories advanced in his famous book on calcareous manures, a volume that in 1895 the United States Department of Agriculture praised as the best of its type in the English language. Later, for ten years Ruffin published the *Farmer's Register* and served as an officer of the Virginia State Board of Agriculture and of the state agricultural society.[5]

During the two or three decades before the Civil War, the South and particularly the Southeast was especially well supplied with agrarian innovators. In fact, one writer has contended that more

[4] C. M. Allmond, ed., "The Agricultural Memorandums of Samuel H. Black: 1815–1820," *Agricultural History*, XXXII (January, 1958), pp. 56–58; Clyde E. Woodall and George H. Aull, "The Pendleton Farmer's Society," *Agricultural History*, XXXI (April, 1957), pp. 36–37.

[5] Avery O. Craven, "The Agricultural Reformers in the Ante-Bellum South," *American Historical Review*, XXXIII (January, 1928), pp. 305–307, 309–311. Ruffin's colorful career is covered in detail in Avery Craven, *Edmund Ruffin, Southerner: A Study of Secession* (New York, 1932).

"scientific agricultural practices prevailed in the Lower South in the two decades before the Civil War than in the four or five decades which followed," a judgment that seems to be well founded.[6] Presumably, the decline in soil fertility in the older areas, combined with the fall in cotton prices in the 1840s and 1850s, forced some planters to develop newer methods, while others adopted them after being convinced that change was economically feasible.

Prominent among such individuals were David Dickson of Georgia, Wade Hampton of South Carolina, and John C. Jenkins of Mississippi. In 1845 Dickson invested $25,000 in Hancock County land and began to apply new methods in his planting operations. He used contour plowing, shallow cultivation, and a crop rotation system, and he experimented with subsoil drainage, commercial and natural fertilizers, and soil-building crops like field peas. In contrast to many of his peers, Dickson was a shrewd businessman, living simply, minimizing overseer costs, and supervising personally the management of his property. Profits were plowed back into the enterprise, and he avoided the common pitfall of excessive credit.[7] Far better known, of course, was Hampton. On his showplace estate near Columbia, he raised purebred livestock, often exhibiting them at shows sponsored by the state agricultural society. Jenkins, a planter near Natchez, was widely known in that area as an innovator.[8]

Prominent innovators were less conspicuous in the North during the antebellum years. The small, single-family farm in Pennsylvania or Illinois failed to produce men with time and money to devote to agricultural improvement. At the same time, southern tobacco and cotton cropping led to soil exhaustion and related problems to a degree unknown in the North.

[6] James C. Bonner, "Advancing Trends in Southern Agriculture," *Agricultural History*, XXII (October, 1948), p. 248.

[7] Chester M. Destler, "David Dickson's 'System of Farming' and the Agricultural Revolution in the Deep South, 1850–1885," *Agricultural History*, XXXI (July, 1957), pp. 30–33; Willard Range, *A Century of Georgia Agriculture, 1850–1950* (Athens, 1954), p. 22.

[8] Albert G. Seal, "John Carmichael Jenkins, Scientific Planter of the Old Natchez District," *Journal of Mississippi History*, I (January, 1939), pp. 14–18; Clyde E. Woodall and William H. Faver, "Famous South Carolina Farmers," *Agricultural History*, XXXIII (July, 1959), pp. 140–141.

Still, the North produced some outstanding leaders. Foremost, perhaps, was Jesse Buel. After a career in printing and publishing in New York, Buel turned to agriculture, acquiring a farm near Albany which he promptly converted into a showplace of better methods. By experimentation he developed a variety of innovations and theories, and he soon began to transmit his ideas to farmers. In the 1820s he served as recording secretary of the New York State Board of Agriculture, editing and apparently writing two volumes on agriculture issued by that body. The next decade found him prominent in the establishment of the New York State Agricultural Society, and in 1834 he induced the society to launch the *Cultivator*, a farm journal that under his management became the most popular in the nation. Five years later he published his *Farmer's Companion*, a volume that contained thoughts formulated over two decades in the forefront of the movement for agricultural betterment.[9]

The Civil War and its results dramatically altered the place of the North and the South in the movement for agricultural change. Southern defeat and the trials of reconstruction removed or impoverished many of the planter class who might otherwise have been innovators. The sharecropper and crop lien systems retarded progress in many ways. On the other hand, the war did much to mechanize northern farms, and if farmers there remained largely unimpressed with the value of science, they seemed more alert than their southern counterparts.

The march of settlement beyond the Missouri especially produced a number of notable innovators. Men soon learned that new methods were essential to survival, so they became experimenters by necessity. Moreover, prairie and plains settlement, more than eastern frontiers, involved business groups and other relatively large operators with capital in sufficient quantities to allow them to try new ideas.

Among those individuals who showed the way to others was Hardy W. Campbell. Born in 1859, he became a homesteader in Brown County, Dakota Territory, in 1879. During the next fifteen years, he formulated his own ideas and systematized the work of others until

[9] Harry J. Carman, "Jesse Buel, Early Nineteenth Century Agricultural Reformer," *Agricultural History*, XVII (January, 1943), pp. 1–11.

he had produced what came to be known as the Campbell system of dry farming. Later he was employed by various western railroads to manage farms maintained by them as part of their colonization programs.[10]

A few of the bonanza farmers in the Red River Valley were more than wheat growers. James B. Powers, for example, was something of an innovator on his great holdings in Minnesota and Dakota. Abandoning a complete reliance on wheat, he produced livestock, introduced new field crops and new seed varieties, and interested himself in soil conservation methods. George Grant, a Scottish silk merchant who in 1877 bought 69,000 acres in Ellis County, Kansas, from the Kansas Pacific Railroad, amazed and perhaps instructed his neighbors by bringing in blooded cattle, horses, and sheep. In Texas the Spur Ranch, an enterprise in Garza, Kent, Crosby, and Dickins counties that from 1885 to 1904 was owned by a British syndicate, conducted experiments with field crops, various grasses, and garden truck.[11]

Innovators among ordinary farmers were usually unknown beyond their immediate neighborhoods. But in 1880 Samuel T. K. Prime, an Illinois stock raiser and agricultural journalist, published a volume discussing the methods of a number of outstanding farmers in the Middle West. Each farmer described his own methods. Some of the writers were pretentious, but the book did demonstrate that a few ordinary operators were alert to change and it suggested that their successes had impressed their neighbors. Moreover, while the circulation of Prime's book is unknown, it did constitute one means by which information concerning better techniques was disseminated.[12]

[10] Mary W. M. Hargreaves, "Hardy W. Campbell (1850–1937)," *Agricultural History*, XXXII (January, 1958), pp. 62–65.

[11] Hiram M. Drache, *The Day of the Bonanza: A History of Bonanza Farming in the Red River Valley of the North* (Fargo, N. Dak., 1964), pp. 83–86, 169–170; Mary A. Rich, "Railroads and Agricultural Interests in Kansas, 1865–1915" (M.A. Thesis, University of Virginia, 1960), pp. 56–58; W. C. Holden, "Experimental Agriculture on the Spur Ranch, 1885–1909," *Southwestern Social Science Quarterly*, XIII (1932), pp. 16–18, 23.

[12] Earle D. Ross, "A Neglected Source of Corn Belt History: Prime's Model Farms," *Agricultural History*, XXIV (April, 1950), pp. 109–111.

Despite regional handicaps innovators were not completely lacking in the post-Civil War South. James Monroe Smith, a successful planter from Oglethorpe County, Georgia, was one who did not object to the label "book farmer." Long an advocate of scientific agriculture, he practically converted his property into an experiment farm. Visitors came to see the results of fertilization, careful seed selection, and diversification, and Smith even suggested that the legislature appropriate funds to provide transportation for those farmers who wanted to observe his experiments. Meanwhile, W. B. Montgomery of Starkville, Mississippi, imported blooded cattle from Europe, planted clover and other grasses, and did much to introduce dairying into his area of the South.[13]

Individual innovators played a significant role in the improvement of agriculture in the nineteenth century, but organized efforts also were important. Such agencies as agricultural societies and fairs provided forums for innovators to present new ideas to other farmers, while at the same time they encouraged a multitude of ordinary agriculturists to adopt newer concepts and techniques.

The appearance of the first agricultural societies in the United States was clearly related to developments in Europe. The idea of agricultural improvement was sweeping the British Isles and the Continent in the eighteenth century, and the formation of agricultural societies was one manifestation of that movement. In the colonies the American Philosophical Society, founded in 1743, devoted some attention to agricultural topics. Four decades later leading Americans began to establish societies devoted exclusively to agricultural improvement.

The first organization founded for that purpose was the Philadelphia Society for Promoting Agriculture, established in 1785. It was soon followed by others. The South Carolina Society for Promoting and Improving Agriculture appeared the same year. Within a decade comparable groups were launched in New Jersey, New York, Massachusetts, and Connecticut. Even beyond the Appalachians, the

---

[13] E. Merton Coulter, *James Monroe Smith; Georgia Planter, before Death and after* (Athens, 1961), pp. 17–19; John K. Bettersworth, *Mississippi: A History* (Austin, Tex., 1959), p. 347.

Kentucky Society for Promoting Useful Knowledge was formed at Lexington as early as 1787.

Such organizations had little to do with ordinary farmers. With the possible exception of the Kennebec Society of Maine, the first agricultural societies consisted of wealthy landowners, northern and southern, and merchants and professional men. Dues tended to be high and membership aristocratic. The South Carolina society, for instance, numbered among its first officers a chief justice of the United States, several members of Congress and governors of the state, and a signer of the Declaration of Independence. Essentially, these groups served only to enlighten their own members and to encourage them to experiment with new ideas, many of which were imported from Europe, and to report the results. Regular meetings allowed for an exchange of information. Prizes in the form of cash or medals rewarded those who successfully introduced new methods or displayed improved animals or other farm products. Many of the societies published their transactions or produced articles that went into other publications, thereby allowing the organizations to reach an audience larger and less selective than their narrow membership.[14]

More important were local agricultural societies that flourished wondrously in the middle decades of the nineteenth century. Parenthood of that type of organization is given to Elkanah Watson, a wealthy landowner and speculator in New York who purchased a Massachusetts farm in 1807 and set out to use better methods on his property. In 1811 he was the leading figure in the establishment of the Berkshire Agricultural Society at Pittsfield. The idea quickly became popular, and soon county societies appeared throughout New England, the Middle Atlantic states, and as far west as Ken-

[14] Alfred C. True, *A History of Agricultural Education in the United States, 1785–1923* (Washington, 1929), pp. 7–14; Stevenson W. Fletcher, *Pennsylvania Agriculture and Country Life, 1640–1840* (Harrisburg, 1950), pp. 345–350; Curtis P. Nettels, *The Emergence of a National Economy, 1775–1815* (Vol. II of *The Economic History of the United States*, Henry David and others, eds., New York, 1962), p. 245; "The Massachusetts Society for Promoting Agriculture," *Garden and Forest*, V (August 3, 1892), p. 371; Kenyon L. Butterfield, "Farmers' Social Organizations," in Liberty Hyde Bailey, ed., *Cyclopedia of American Agriculture* (4 vols., New York, 1909–1910), IV, 291.

tucky and Illinois. In the first flush of enthusiasm, a number of eastern states appropriated money to aid the county groups in their work. By 1819 Watson claimed that there were at least 100 such societies in the United States. The number continued to grow in the early 1820s, but rural depression, combined with a case of excess expectations, caused the movement to lose much of its vigor in the late 1820s and 1830s.[15]

A dramatic revival came in the two decades before 1860; in fact, the years between 1850 and 1870 have been called the "golden age" of agricultural societies in the United States. Authorities in 1852 estimated that there were 300 societies in thirty-one states and five territories, but according to the Commissioner of Patents the number had increased to more than 900 by 1858. In Ohio, a state that was reasonably typical, the expansion of the movement was as follows:

YEARS	NUMBER OF SOCIETIES ESTABLISHED
1818–27	5
1828–37	10
1838–47	24
1848–57	49

For the nation as a whole, even the Civil War decade saw further growth, despite the adverse affect of the struggle on agricultural societies in the South. Alabama, for example, had at least 30 societies in the 1850s, but that number was never again reached after 1865. Still there were some 1,330 agricultural societies in the United States in the late 1860s.[16]

[15] Hugh M. Flick, "Elkanah Watson's Activities on Behalf of Agriculture," *Agricultural History*, XXI (October, 1947), pp. 193–198. Agricultural societies in different states are discussed in such studies as Cornelius O. Cathey, *Agricultural Developments in North Carolina, 1783–1860* (Chapel Hill, 1956), pp. 74–81; John H. Moore, *Agriculture in Ante-Bellum Mississippi* (New York, 1958), pp. 88–91; James C. Bonner, *A History of Georgia Agriculture, 1732–1860* (Athens, 1964), pp. 110–119; Carl R. Woodward, *The Development of Agriculture in New Jersey, 1640–1880* (New Brunswick, 1927), pp. 166–178.

[16] True, *History of Agricultural Education*, pp. 12–14, 23; Ohio State Board of Agriculture, *A Brief History of the State Board of Agriculture, the State Fair,*

Typical perhaps of the early county and local agricultural societies was one established at Edwardsville, Illinois, in 1822. Given the name Madison County Agricultural Society, the organization met annually until it disappeared later in the decade. Its primary function was the arranging of shows at which produce of various types was displayed and prizes awarded. For example, at one show, the society offered prizes for the best essay on pork-curing methods, the best sample of malt liquor of not less than thirty gallons, the greatest number of wolf scalps, and the best piece of linsey-woolsey.[17] In older settled regions, of course, the societies devoted their attention to crops, machinery, and farm animals, although household products were often displayed.

Local or county agricultural societies led directly to the establishment of state agricultural societies and later to the creation of state boards and departments of agriculture. The agricultural society formed in South Carolina in 1785 was in a sense a "state" organization; later as the local society movement gained momentum, there was a tendency to convert private state organizations to public ones or, more commonly, to create entirely new public agencies. In 1819 New York established its state board of agriculture; similar bodies appeared in New Hampshire and Pennsylvania in 1820 and 1823. Across the country the California state agricultural society was formed in 1854. Nine years later a newly established board of agriculture assumed the most important functions of the older organization. By 1910, all but one state had public agencies that devoted at least a portion of their time to the improvement of agriculture.[18]

District and Agricultural Societies and Farmers' Institutes in Ohio (Columbus, Ohio, 1899), pp. 9–24; U.S. Commissioner of Patents, Report, 1858 (Washington, 1859), part II, pp. 90–91; U.S. Commissioner of Agriculture, Report, 1867 (Washington, 1868), pp. 364–403; Weymouth T. Jordan, "Agricultural Societies in Ante-Bellum Alabama," Alabama Review, IV (October, 1951), p. 244.

[17] Willard C. Flagg, "Report on the Agricultural, Horticultural, and Other Industrial Interests of the County of Madison," Illinois State Agricultural Society, Transactions, VIII (1869–70), pp. 209–211.

[18] Percy W. Bidwell and John I. Falconer, History of Agriculture in the Northern United States, 1620–1860 (Washington, 1925), p. 189; Henry H. Metcalf, New Hampshire Agriculture. Personal and Farm Sketches (Concord, N. H., 1897), p. 17; Claude B. Hutchison, ed., California Agriculture (Berkeley, 1946), p. 428;

The state agricultural agencies had numerous duties, some rather poorly defined, and their emphasis changed over time. At the outset they usually had some supervisory or advisory relationship to the county groups, in several instances managing the state aid that was extended to the local associations. In time the state organizations assumed more definite educational and regulatory functions. Probably their most common task was the management of the state fairs that gradually emerged.[19]

While the pioneer agricultural societies were general in nature, aiming at the improvement of all kinds of farming, specialized groups soon made their appearance. The first horticultural society was founded in New York in 1818; it managed to survive until 1837. Longer-lived was the Pennsylvania Horticultural Society, organized in Philadelphia in 1827. In 1848 there appeared the American Pomological Society, an organization that was closely modeled after groups in England and Belgium. Soon there was a dramatic flowering of societies for all kinds of farmers, dairymen, cattle feeders, and others. By 1910 there were 217 specialized societies, not including purely local groups. Among that number were 59 associations for poultry raisers and 4 for beekeepers.[20]

In a sense, the agricultural society movement culminated with the creation of the United States Agricultural Society. George Washington had proposed the establishment of a national society, but noth-

"State Organizations for Agriculture," in Bailey, *Cyclopedia of American Agriculture*, IV, 328; Edward Wiest, *Agricultural Organization in the United States* (Lexington, Ky., 1923), pp. 291–300.

[19] The evolution of the state agricultural agencies may be seen in I. D. Graham, "The Kansas State Board of Agriculture: Some High Lights of History," Kansas State Historical Society, *Collections*, XVII (1928), pp. 788–813; Benjamin M. Davis, "Agricultural Education: State Organizations for Agriculture and Farmers' Institutes," *Elementary School Teacher*, XI (November, 1910), pp. 136–145; Moore, *Agriculture in Ante-Bellum Mississippi*, pp. 88–90; Merrill E. Jarchow, *The Earth Brought Forth: A History of Minnesota Agriculture to 1885* (St. Paul, 1949), pp. 248, 250, 253, 256; Cathey, *Agricultural Developments in North Carolina*, pp. 81–83.

[20] Massachusetts Horticultural Society, *History of the Massachusetts Horticultural Society, 1829–1878* (Boston, 1880), pp. 44–45; "Agricultural Societies," in U.S. Commissioner of Agriculture, *Report*, 1866, p. 526; "American Agricultural Societies," in Bailey, *Cyclopedia of American Agriculture*, IV, 341–344.

ing was done. Fifty years later in 1841 Solon Robinson and others met at Washington and established an organization for the primary purpose of directing the grant of Hugh Smithson toward the betterment of agriculture. The establishment of the Smithsonian Institution eliminated the society's reason to be, and it seems to have evaporated.

The United States Agricultural Society was formed June 14, 1851. The Massachusetts State Board of Agriculture took the lead in calling for such an agency; at the organizational meeting in Washington 153 delegates from twenty-seven states appeared. Among them were such notables as Daniel Webster, Stephen A. Douglas, and Millard Fillmore. Probably the leading figure was Marshall P. Wilder of Massachusetts who served as president for six years.

The organization functioned with considerable success until the Civil War. Annual meetings attracted delegates from practically every state and territory who met to discuss matters of mutual interest and to hear papers on agricultural topics. In addition the society sponsored national exhibitions and field trials in Massachusetts and in other states as far west as Kentucky and Illinois. A variety of publications carried information to many farmers who could not attend the annual meetings and national exhibitions. Like lesser groups of the same type, the society did much to stimulate interest in purebred stock, improved farm machinery, and better cropping methods. But its most important role lay in its impact on Congress. Its resolutions reflected the views of a cross section of the nation's more articulate farmers, and Congress was hesitant to flout them. Consequently, the society contributed to the rise of sentiment that would utimately result in the passage of the Morrill Land-Grant Act and the measure creating the United States Department of Agriculture.[21]

Whether a society was national or local in scope, in the majority of cases the holding of a fair constituted its primary function. Fairs and exhibitions, in fact, became the principal means by which societies of all types were able to reach the mass of farmers.

[21] "Agricultural Societies," pp. 514–516; Lyman Carrier, "The United States Agricultural Society, 1852–1860," *Agricultural History*, XI (October, 1937), pp. 279–281.

As early as 1796 Washington urged that fairs be held. Nothing came of his proposal, but in 1804 the Commissioner of Patents suggested that an organized display of livestock and other domestic products might facilitate the expansion of markets. Proceeding on that assumption, authorities in Washington arranged a three-day affair that opened October 17, 1804. At a second fair held the next spring, premiums amounting to $100 were given for the best animals sold during the gathering. This effort was discontinued after a third fair in November, 1805. Four years later the Columbian Agricultural Society held at Georgetown a meeting that one historian has called the first agricultural fair in the United States. Those in attendance included the President and his wife and other members of society in Washington. Meanwhile, in 1808 the annual sheep shearings began on the estate of George Washington Parke Custis, social gatherings that also sought to promote the sheep industry.

Still, credit for launching agricultural fairs in the United States goes to Elkanah Watson. In 1807, the same year that he bought his farm in Massachusetts, he acquired a pair of Merino sheep and displayed them in the Pittsfield town square. That experience, combined with his acquaintance with stock shows in England, encouraged him to establish a local show for ordinary farmers. The first was held at Pittsfield in October, 1810, when more than two dozen farmers joined him in exhibiting their stock. The next year premiums were offered to those displaying stock, and the modern agricultural fair was born.

The early success of the Berkshire Agricultural Society and Watson's promotional efforts in creating other societies produced a rapid multiplication of fairs. In 1820 Watson could report the existence of societies in almost every county in New England. New York State was well covered, and societies had appeared in the Middle West and in such southern states as Virginia, Maryland, and Kentucky. The arranging of a fair was in the majority of cases a society's first order of business.[22]

[22] "Agricultural Societies," pp. 516–517; Butterfield, "Farmers' Social Organizations," pp. 291–292; Ella G. Agnew, "Agricultural Fairs—Yesterday and Today," *Southern Planter*, XVI (August 15, 1929), p. 5; Flick, "Elkanah Watson's Activi-

The agricultural fair movement followed hard on the heels of the frontier as it retreated westward. For example, the first fair in Maine was held in 1819; only a little over a year later a statewide exposition opened at Hallowell. The pioneer state fair in New York came in 1841, and Wisconsin launched its first state fair at Janesville a decade later. In Ohio local fairs appeared as early as the 1820s; the state fair came in 1850. Far to the west, Arizona joined the ranks of states with a fair in 1912.

By the 1850s and 1860s agricultural fairs were beginning to take on the geographical distribution that would continue. The states of the Old Northwest were clearly the leaders in numbers of fairs and of societies that sponsored them. The Middle Atlantic states trailed, and New England was third. The South and the West were far behind, with the adverse affect of the Civil War clearly apparent in southern figures for 1868.[23]

Agricultural fairs, according to most authorities, enjoyed their golden period between 1850 and 1870, with some variation depending upon geography and other factors. During those decades few other agencies existed to instruct the farmer, and certainly the societies and their fairs were the only means by which large numbers of ordinary farmers might be contacted directly. During those years, also, the fairs were more educational in nature, with less of the horse racing and carnival atmosphere that later would be the dismay of farmers. Indeed, exhibitions of livestock and new machinery, displays of home-grown produce, and more or less informative talks by

---

ties," pp. 193–195; Bidwell and Falconer, *History of Agriculture*, p. 187; Fred Kniffen, "The American Agricultural Fair: The Pattern," Association of American Geographers, *Annals*, XXXIX (December, 1949), p. 266.

[23] Clarence A. Day, *A History of Maine Agriculture, 1604–1860* (Orono, Maine, 1954), pp. 232–234; John D. Howe, "The New York State Agricultural Society—Its History and Objects," New York State Department of Farms and Markets, *Bulletin* 161 (Albany, 1924), p. 10; "Development of Wisconsin Fairs," *Wisconsin Agriculturist and Farmer*, LVIII (August 17, 1929), p. 3; C. W. Burkett, *History of Ohio Agriculture* (Concord, N.H., 1900), p. 191; Ohio State Board of Agriculture, *Brief History*, p. 7; *Hoard's Dairyman*, XLIV (November 22, 1912), p. 512; Wayne C. Neely, *The Agricultural Fair* (New York, 1935), pp. 83–88.

a variety of speakers gave the fairs of those decades many of the characteristics of later-day farmers' institutes.[24]

In addition to the ordinary fairs the last decades of the nineteenth century witnessed a number of exhibition extravaganzas that were partly educational in character. For instance, when the corn crop in the middle Missouri Valley proved to be unusually good in the fall of 1887, Sioux City businessmen resolved to sponsor a festival. The Sioux City Corn Palace Exposition Company was formed, a giant building constructed, and the first exposition opened in October. Cornelius Vanderbilt, Chauncey Depew, and Grover Cleveland attended the affair; so did thousands of farmers who learned something of corn raising. The event proved to be an annual one for a few years; the last one was held in 1891. Similar in purpose and nature was the Blue Grass Palace, an annual exhibition held in southwest Iowa from 1889 to 1892.[25]

More famous was the National Corn Exposition. A function of the National Corn Association, the first exposition was held in Chicago in 1908. Later affairs were held in Omaha, Columbus, Dallas, and Columbia, South Carolina. At the Dallas exposition, in February, 1914, thirty-seven land-grant colleges and experiment stations and the United States Department of Agriculture had exhibits of corn and other crops and offered visitors suggestions concerning better growing methods. In 1909 more than 100,000 people visited the show.[26]

While agricultural societies of various types and their fairs reached considerable numbers of farmers, their impact was less than that of agricultural periodicals. In fact, most authorities maintain that prior to 1914 it was through the medium of an expanding farm press that

[24] There are a number of excellent accounts of agricultural fairs in addition to those already cited. Included are Earle D. Ross, "The Evolution of the Agricultural Fair in the Northwest," *Iowa Journal of History and Politics*, XXIV (July, 1926), pp. 445–480, and William W. Rogers, "The Alabama State Fair, 1865–1900," *Alabama Review*, XI (April, 1958), pp. 100–116.

[25] John E. Briggs, "The Sioux City Corn Palaces," *Palimpsest*, III (October, 1922), pp. 313–326; Bruce E. Mahan, "The Blue-Grass Palace," *ibid.*, pp. 327–335.

[26] *American Fertilizer*, XXXIX (September 6, 1913), p. 30; *ibid.*, XL (January 10, 1914), p. 61; *Hoard's Dairyman*, XXXIX (January 15, 1909), p. 1386.

the majority of farmers came into contact with ideas originating outside their own communities.

A case can be made for the contention that agricultural journalism predated even the gentlemen's organizations like the Philadelphia Society. Jared Elliot's *Essays upon Field Husbandry in New England* appeared in 1748 as a periodical, and a half a dozen numbers were issued by 1759. Later, the Massachusetts Society for Promoting Agriculture established the *Massachusetts Agricultural Repository and Journal,* a publication that continued for thirty-four years but was issued only semiannually or less frequently.[27]

A short-lived pioneer was *The Agricultural Museum.* Edited by David Wiley, the secretary of the Columbian Agricultural Society, the paper was the organ of that organization. The first issue was published at Georgetown, July 4, 1810, but the paper disappeared after only two years.

American agricultural journalism is usually dated from the establishment in Baltimore in 1819 of the *American Farmer.* Edited by John S. Skinner, the paper survived until 1834. During its existence Skinner filled the paper with agricultural information of all kinds, drawing on his own experiences and travels as well as on information obtained from others, both in the United States and abroad. Subjects covered the entire range of farm topics, including new crops, improved tillage methods, fertilization, breeding of livestock, and drainage.[28]

From these early beginnings farm journals increased rapidly in number if not always in quality. They first became common in lower

[27] Richard Bardolph, "Agricultural Education in Illinois to 1870: The Press" (Ph.D. Thesis, University of Illinois, 1944), pp. 143–144. Professor Bardolph's outstanding dissertation was published under the title, *Agricultural Literature and the Early Illinois Farmer* (Urbana, Ill., 1948). An excellent discussion of the origins of agricultural journalism is George F. Lemmer, "Early Agricultural Editors and Their Farm Philosophies," *Agricultural History,* XXXI (October, 1957), pp. 3–22.

[28] Claribel R. Barnett, "*The Agricultural Museum:* An Early American Agricultural Periodical," *Agricultural History,* II (April, 1928), pp. 99–101; L. C. Gray, *History of Agriculture in the Southern United States to 1860* (2 vols., Washington, 1933), II, 788; Harold T. Pinkett, "The *American Farmer,* A Pioneer Agricultural Journal," *Agricultural History,* XXIV (July, 1950), pp. 146–148.

18

New England, the Middle and South Atlantic states, and in Ohio; but in a remarkably short time they were being published in the Mississippi Valley. Among the pioneers were the *Southern Agriculturist* (Charleston, 1822), the *Plough Boy* (Albany, 1819), the *New England Farmer* (Boston, 1822), the *Genesee Farmer* (Rochester, 1831), and the *Farmer's Cabinet* (Philadelphia, 1836). The first farm paper in Illinois was the *Western Plough Boy*, a semimonthly published in Edwardsville in 1831. It put forth twenty-five issues before succumbing. Nine years later there appeared the paper that in time would be the bible for midwestern farmers, the *Union Agriculturist and Western Prairie Farmer*. The first farm paper in Maine was the *Kennebec Farmer*, edited by Ezekiel Holmes. Originated in 1833, it soon changed its name to *Maine Farmer* and survived for almost a century. John Sherwood's *Farmer's Advocate*, launched in 1838, was the pioneer farm paper in North Carolina, but the *American Cotton Planter* (Montgomery, 1853) and the *Southern Cultivator* (Augusta, 1843) proved to be the leading papers in the South.[29]

In the century after 1810 some 3,600 farm periodicals appeared in the United States and Canada. The casualty rate was high, and the majority issued only a few numbers. In fact, there probably were no more than 30 in existence in the United States at any one time before 1840 and not over 60 between then and 1870. In 1913 one authority estimated that there were about 450 farm papers in circulation, including the various specialized journals.[30]

Circulation figures were by no means accurate, but in a general

[29] Bardolph, "Agricultural Education in Illinois," p. 149; Gray, *History of Agriculture*, II, 788; Bidwell and Falconer, *History of Agriculture*, p. 316; Flagg, "Agricultural, Horticultural, and Other Industrial Interests," p. 242; Day, *Maine Agriculture*, p. 240; Wesley H. Wallace, "North Carolina's Agricultural Journals, 1838–1861: A Crusading Press," *North Carolina Historical Review*, XXXVI (July, 1959), p. 276; Weymouth T. Jordan, "Noah B. Cloud and the *American Cotton Planter*," *Agricultural History*, XXXI (October, 1957), p. 44.

[30] Stephen C. Stuntz, "List of the Agricultural Periodicals of the United States and Canada Published during the Century July, 1810, to July, 1910," U.S. Department of Agriculture, *Miscellaneous Publication* 398 (Washington, 1941); Bardolph, "Agricultural Education in Illinois," p. 143; Pennsylvania Farmers' Annual Normal Institute, *Proceedings*, 1913 (Harrisburg, 1913), p. 141.

way they did show that a growing number of farmers subscribed to agricultural papers. As early as 1840 total circulation in the northern states was set at 100,000. Thirteen years later thirty-three papers in the North reported a circulation of 234,000, while ten in the South went to almost 33,000 addresses. Even these figures indicated that only a small percentage of the total farm population read agriculture papers. In fact, Luther Tucker, a farm journal editor who was in a position to know, claimed that in 1850 something less than one farmer in two thousand was a subscriber.[31]

A dramatic expansion in circulation came in the post–Civil War period, and the growth continued into the twentieth century. In 1913 a survey showed that almost every farmer in the Northeast subscribed to one or more papers, and the eastern Middle West was at least as well covered. Growth in circulation in the South was slower and came later; President J. C. Hardy of Mississippi Agricultural and Mechanical College admitted in 1902 that southern farmers read less than their northern contemporaries. But even in the South, by the early years of the twentieth century farm papers were significant as a force in rural change. By 1903 twenty-five papers in the region claimed a circulation of 278,000, and that figure climbed to 636,000 by 1911.[32]

Sources of information that appeared in farm journals varied over time. In earlier years the periodicals cooperated with agricultural societies, printing their news and transactions. Probably the farm papers made their greatest contribution after the agricultural experiment stations came into existence. By carrying in their columns distilled versions of station bulletins, the papers made the findings of scientific experimentation available to a larger number of farmers in language that the ordinary reader could understand. Moreover, many periodicals published information taken from English and

[31] Bidwell and Falconer, *History of Agriculture*, p. 194; Gray, *History of Agriculture*, II, 788; Lemmer, "Early Agricultural Editors," p. 3.

[32] Oscar A. Beck, "The Agricultural Press and Southern Rural Development, 1900–1940" (Ph.D. Thesis, George Peabody College, 1952), p. 28; Stevenson W. Fletcher, *Pennsylvania Agriculture and Country Life* (Harrisburg, 1955), p. 439; AAACES, *Proceedings*, 1902, p. 70.

other foreign papers, thereby exposing their readers to the latest ideas from abroad and otherwise broadening their horizons.

From whatever source the data came, farm papers were a successful medium for disseminating information over a wide area. When a subscriber read something of interest he discussed it with his neighbors, thereby allowing the paper to enlighten many. In 1851 Edmund Ruffin maintained that the progress registered by agriculture in the preceding thirty years was due mainly to the rural press. That, perhaps, was an exaggeration, but almost seventy-five years later the United States Department of Agriculture found that of a group of farmers interviewed over 40 percent considered farm papers their most valuable source of information.[33]

Rural sentiment concerning farm journals was by no means always favorable. In numerous cases, in fact, criticism was warranted. Some papers resorted to obvious puffing and peddled products or ideas that were unsound. In other instances they devoted too much attention to topics of little interest to farmers, such as the growing of flowers or the beautification of the rural home. Others seemed to make a practice of presenting long discussions of showplace farms, accounts of which made the ordinary farmer more discontented with his lot. Even the best papers were guilty of giving advice that was not applicable except in specific instances. Finally, at least one modern writer has charged that most farm editors followed, rather than led, the process of change in the countryside.[34]

A growing number of county papers and metropolitan dailies

[33] Bidwell and Falconer, *History of Agriculture*, p. 194; *American Agriculturist*, XXXVII (June, 1878), p. 219; *Cultivator and Country Gentleman*, XLIV (April 26, 1881), p. 350; Fletcher, *Pennsylvania Agriculture and Country Life*, p. 439; A. L. Demaree, "The Farm Journals, Their Editors, and Their Public, 1830–1860," *Agricultural History*, XV (October, 1941), p. 188; *Wallaces' Farmer*, XXXIV (December 10, 1909), p. 1587; *ibid.*, XXXIII (November 13, 1908), p. 1389; C. Beaman Smith and K. H. Atwood, "The Relation of Agricultural Extension Agencies to Farm Practices," U.S. Bureau of Plant Industry, *Circular* 117 (Washington, 1913), pp. 18, 22.

[34] Demaree, "Farm Journals," p. 187; Fletcher, *Pennsylvania Agriculture and Country Life*, p. 439; *Wallaces' Farmer*, XXV (October 5, 1900), p. 973; John T. Schlebecker, "Dairy Journalism: Studies in Successful Farm Journalism," *Agricultural History*, XXXI (October, 1957), p. 23.

carried agricultural information to their readers. By 1840 a few editors were devoting attention to agriculture, and their coverage improved after the Civil War in terms of both quality and quantity. In Wisconsin, for example, W. D. Hoard used the *Jefferson County News* to keep farmers informed more than a decade before he established his *Hoard's Dairyman*. The *Monroe Sentinel* and the *Whitewater Register* were two other local papers in the same state that carried more than political news and community affairs. Among the daily newspapers the *New York Tribune* and the *New York Sun* were reported to be especially informative, and during the first decades of the twentieth century C. P. J. Mooney of the Memphis *Commercial Appeal* was a powerful voice for agricultural improvement in the middle South.[35]

While some men developed societies, fairs, and the rural press as means of educating the rural population, others were thinking in terms of formal education for farmers. After all, they said, if men in other professions were trained in the classroom, there was no reason why farmers could not be similarly prepared. Consequently, professorships of agriculture appeared, there were efforts to establish private agricultural schools, and finally the demand arose that each state create a public institution in which agriculture would be given the treatment that it deserved.

A number of the older, classical colleges either undertook to offer some instruction in agriculture or were pushed into doing so. In 1792 Columbia College established a professorship of natural history, chemistry, and agriculture. Even before the University of North Carolina opened its doors its trustees talked bravely of agricultural education, but it was sixty years before the university es-

---

[35] Bardolph, "Agricultural Education in Illinois," pp. 69, 73, 91; Allan G. Bogue, *From Prairie to Corn Belt: Farming on the Illinois and Iowa Prairies in the Nineteenth Century* (Chicago, 1963), pp. 195–196; "Fifty Years of Cooperative Extension in Wisconsin, 1912–1962," Wisconsin Agricultural Extension Service, *Circular* 602 (Madison, 1962), pp. 16–17; *American Agriculturist*, XXVII (June, 1868), p. 211; James W. Silver, "C. P. J. Mooney of the Memphis *Commercial Appeal*, Crusader for Diversification," *Agricultural History*, XVII (April, 1943), pp. 81, 84.

tablished its School for the Application of Science to the Arts, a division that gave no complete course in agriculture but did attempt to demonstrate to students the application of chemistry to farming. In the 1830s and the 1840s the gifts of Benjamin Bussey and Abbot Lawrence to Harvard University were expected to lead to instruction in agriculture, but little was done. In 1850 Brown University and the University of Mississippi established professorships in agriculture. The latter institution filled the post with the appointment of Benjamin L. C. Wailes. Even earlier, in 1842 Union University, a Baptist school at Murfreesboro, Tennessee, created a professorship in agriculture; South Carolina College was not far behind.[36]

In the 1850s Philip St. George Cocke offered $20,000 to the University of Virginia to establish a professorship in agriculture, but his generosity was rejected when he demanded the right to name the professor. In 1853 a chair of agricultural chemistry was established at the University of North Carolina,[37] and the next year William Terrell of Hancock County gave $20,000 to the University of Georgia to enable that school to name Daniel Lee of New York as its professor of agriculture. The University of Iowa, in its first circular dated September 1, 1855, promised that its chemistry department would devote attention to the problems of agriculture, and its catalogue for 1864–65 listed a course entitled "Analysis of Soils and Manures, and Agricultural Chemistry." Other established colleges that claimed to offer instruction in agriculture in the 1850s included the Michigan State Normal School at Ypsilanti and the University of Michigan. At the latter school the first professor of

[36] Bidwell and Falconer, *History of Agriculture*, p. 194; Lindsey O. Armstrong, "The Development of Agricultural Education in North Carolina" (M.A. Thesis, North Carolina State College, n.d.), pp. 4–8, 20–21; Paul W. Gates, *The Farmer's Age: Agriculture, 1815–1860* (Vol. III of *The Economic History of the United States*, Henry David and others, eds., New York, 1960), p. 361; Walter C. Bronson, *The History of Brown University* (Providence, R.I., 1914), pp. 286–287; Bettersworth, *Mississippi*, p. 240; Gray, *History of Agriculture*, II, 791.

[37] *Ibid.*, p. 792. Thirty years earlier, both Jefferson and Madison had urged that agriculture be taught at the University of Virginia. See H. G. Good, "Early Attempts to Teach Agriculture in Old Virginia," *Virginia Magazine of History and Biography*, XLVIII (October, 1940), pp. 342–345.

agriculture was Charles Fox, a minister at Grosse Isle, who in 1854 published a *Text-Book of Agriculture*, reputed to be the pioneer volume of its type to appear in the West.[38]

It perhaps could be argued that the first systematic resident instruction in agriculture came at Eleazer Wheelock's Indian school, but most writers would place the distinction elsewhere. The Gardiner Lyceum, "America's first agricultural school," was established in Maine in 1822. The school was the brainchild of Robert Hallowell Gardiner, who had convinced himself and a few others that farmers, surveyors, and millwrights needed to know the principles underlying their businesses. The institution opened in 1823, and by November twenty students were enrolled in the three-year program. The next year the school added a winter course for those who could not attend for the longer period. The school prospered for a few years, but in the 1830s supporters lost interest and the institution declined, like a multitude of others of the time.[39]

A few other enthusiasts were attempting to establish private colleges in which agriculture might receive primary attention. Such was largely the objective of Stephen Van Rensselaer when in 1824 he established at Troy, New York, the institution that bears his name. The primary emphasis of the school, however, proved to be in engineering. Meanwhile, an agricultural school existed at Derby, Connecticut, for some years after 1824.[40]

Several privately supported agricultural colleges appeared in the 1840s, but none of them was long-lived. In 1846 Daniel Lee, then editor of the *Genesee Farmer*, joined Rawson Harmon of Wheatland, New York, to launch an agricultural college on Harmon's 200-acre farm. According to the school's prospectus the promoters

[38] John T. Wheeler, *Two Hundred Years of Agricultural Education in Georgia* (Danville, Ill., 1948), pp. 105–106; *American Agriculturist*, XVII (June, 1858), p. 167; Barton Morgan, *A History of the Extension Service of Iowa State College* (Ames, Iowa, 1934), pp. 5–6; AAACES, *Proceedings*, 1909, p. 41.

[39] *Ibid.*, 1907, p. 51; Neil E. Stevens, "America's First Agricultural School," *Scientific Monthly*, XIII (December, 1921), pp. 531–536; Day, *Maine Agriculture*, p. 251.

[40] Gates, *Farmer's Age*, pp. 360–361; Bidwell and Falconer, *History of Agriculture*, p. 194.

expected to teach chemistry, geology, and other natural sciences relating to agriculture and to explore and present information on the physiology of farm animals and proper rations for them. Similar schools appeared or were planned at Poughkeepsie, New York; Germantown, Pennsylvania; and Newburgh, New York. At the latter school James Darrach, a graduate of Yale but a "holder of his own plow," was employed as instructor.[41]

Privately established institutions were soon followed by public or semipublic ones. By 1858 Farmers' College at Cincinnati had been in operation for ten years, while in Cleveland the Ohio Agricultural College offered instruction in subjects pertaining to agriculture. At Havana, New York, the cornerstone of the People's College was laid in 1858 with Mark Hopkins and Horace Greeley as principal speakers. Two years later the New York State Agricultural College at Ovid opened its doors to twenty-seven students. In the middle South the Tennessee legislature chartered Franklin College in 1846, reportedly the first state-sponsored agricultural college in the United States, only to see it converted into a classical college in the 1850s. In 1861 the legislature in Illinois sought to aid the Illinois Agricultural College at Irvington by bestowing upon it a residual of the state's seminary land grant. The General Court of Massachusetts incorporated agricultural colleges in 1848 and 1856, but neither progressed beyond that stage.[42]

Pennsylvania, Michigan, Iowa, and Maryland were the states that

[41] *American Agriculturist*, V. (February, 1846), p. 69; *ibid.*, (April, 1846), p. 109; *ibid.*, (May, 1846), p. 167; Gates, *Farmer's Age*, p. 360.

[42] Diedrich Willers, *The New York State Agricultural College at Ovid, N.Y. and Higher Agricultural Education* (Geneva, N.Y., 1907), pp. 5–11; Earle D. Ross, *Democracy's College: The Land-Grant Movement in the Formative Stage* (Ames, Iowa, 1942), pp. 24–28; Edward D. Eddy, *Colleges for Our Land and Time: The Land-Grant Idea in American Education* (New York, 1957), pp. 15–16; Stanley G. Watts, "Knowledge for the Tennessee Farmer: Agricultural Extension in Tennessee Prior to 1914" (M.A. Thesis, Mississippi State University, 1967), pp. 45–46; Burt E. Powell, *The Movement for Industrial Education and the Establishment of the University* (Urbana, Ill., 1918), p. 159; Allan Nevins, *Illinois* (New York, 1917), p. 29; George W. Smith, "The Old Illinois Agricultural College," Illinois State Historical Society, *Journal*, V (January, 1913), pp. 476–478; Lilley B. Caswell, *Brief History of the Massachusetts Agricultural College* (Springfield, Mass., 1917), pp. 1–4.

pushed ahead most vigorously. The Michigan constitution of 1850 authorized the establishment of an agricultural college, and five years later the legislature acted favorably. Meanwhile, in Pennsylvania a group led by Evan Pugh induced the legislature to establish the Farmers' High School, an institution that became a college in 1862. As early as 1848 the General Assembly in Iowa asked Congress for a land grant for an agricultural college, but it was ten years later that the legislature created the Iowa State Agricultural College and Farm. A college in Maryland opened in 1859 on a 400-acre tract near Washington.[43]

So far as ordinary farmers were concerned, all of these efforts availed little. Like an early attempt to teach agriculture at the University of Missouri, they died "like a seed on a rock."[44] Lack of interest among farmers, hostility among academic men, and shortage of funds were handicaps that could not be overcome.

But if nothing else the pioneer efforts pointed to one major necessity—more adequate financial support. To meet that demand, proponents seized upon earlier precedents and accepted the concept of federal assistance in the form of grants of public land.

Historians have disagreed vigorously concerning the parenthood of the Morrill Land-Grant Act. A number of writers from Illinois gave credit to Jonathan Baldwin Turner, but Justin S. Morrill and others have their supporters. Probably the most accurate view is that of Earle D. Ross who concluded that no one individual can be considered to be the "father" of the act; instead, it was the inevitable result of a long struggle for industrial education, carried forward by a large number of energetic and dedicated men.[45]

[43] Eddy, *Colleges for Our Land and Time*, pp. 16–21; Pennsylvania Agricultural College, *The Agricultural College of Pennsylvania* (Philadelphia, 1862), pp. 11–16; *American Agriculturist*, XVII (June, 1858), pp. 166–167; George H. Callcott, *A History of the University of Maryland* (Baltimore, 1966), pp. 142–145.

[44] William C. Etheridge, "The College of Agriculture," in Jonas Villes, ed., *The University of Missouri: A Centennial History* (Columbia, Mo., 1939), p. 296.

[45] H. K. Smith to A. C. True, October 27, 1895, Office of Experiment Stations Records (National Archives, Washington); Mary Turner Carriel, *The Life of Jonathan Baldwin Turner* (Urbana, Ill., 1961), pp. 130–131; Donald R. Brown, "The Educational Contributions of Jonathan Baldwin Turner" (M.A. Thesis, University of Illinois, 1954), pp. 71, 157–158; William B. Parker, *The Life and Public Service*

In any event Justin S. Morrill introduced his famous bill on December 17, 1857, and saw it accepted in the House by a vote of 105 to 100 on April 22, 1858. At the next session the Senate passed its own version, but President Buchanan vetoed a compromise bill on February 24, 1859. Abraham Lincoln was more favorably inclined toward the concept of industrial education; more important, no doubt, was the absence of the southern congressmen and senators who had so vigorously opposed the growth of federal power. The ultimate result was the approval on July 2, 1862, of the Morrill Land-Grant Act that would in time produce a nationwide system of colleges in which agriculture would receive the attention so long denied it.[46]

Iowa was the first state to accept the gift offered by the Morrill Act, doing so only two months after Lincoln had affixed his signature. Others soon followed, and by 1870 thirty-seven states had agreed to establish separate agricultural and mechanical colleges or had bestowed the grant upon existing private colleges or state universities. Expansion continued until ultimately there were sixty-nine land-grant colleges scattered throughout the fifty states and Puerto Rico. A second Morrill Act, which came in 1890, contributed to the increase in the number of schools by requiring that the southern states make provisions by which Negroes might share in the grants. As a result, seventeen of the sixty-nine schools were institutions attended primarily by Negro students.[47]

The early years of the colleges were less than promising. In fact, for twenty years and perhaps more they could only be described as failures. With a few exceptions enrollments in agriculture were so small as to be almost nonexistent, faculties were weak and often incompetent, and even enthusiasts could not agree concerning what should be taught. In truth there was little scientific knowledge to teach.

Among other problems, the colleges suffered from a case of excess

of Justin Smith Morrill (New York, 1924), p. 262; Earle D. Ross, "The 'Father' of the Land-Grant College," *Agricultural History*, XII (April, 1938), pp. 185–186.

[46] Parker, *Morrill*, pp. 263–268; *U.S. Statutes at Large*, Vol. XII, pp. 503–505.

[47] Eddy, *Colleges for Our Land and Time*, pp. 48–51, 102–103, 257–259.

expectations. When the schools were first established, those men who were deeply interested in the institutions expected them to work a miracle in the countryside. Instead, nothing happened. Students did not appear, farmers generally ignored the schools or displayed little but contempt for them, and under the leadership of classicists the colleges were soon placing emphasis upon subjects which at best had only a distant and indirect relationship to agriculture.[48]

Basic in explaining the difficulties of the colleges was the fact that in a very real sense agricultural education was not economically feasible when land was of little value. During the first decades that the colleges existed any farmer with the necessary capital could go west and "by gently tickling the rich, virgin prairies, secure an abundant harvest without any education whatever." In essence, agricultural college graduates were forced to compete with unskilled labor. Only when land increased in value did it become feasible to apply to farming the techniques of science. The colleges, in short, "were born one whole generation before their time."[49]

Opposition by the classicists was another cause of the failure of the early land-grant colleges. The overwhelming majority of the professors of traditional subjects made no effort to hide their contempt for agriculture and for those who sought to teach it. When Isaac P. Roberts arrived at Cornell in 1874 he found a cool reception. "Because agriculture was then regarded by most of the classically educated members of the Cornell faculty as quite unworthy of a place in education beside the traditional subjects," he recalled later, "we suffered a sort of social neglect and felt ourselves in an alien atmosphere." As late as 1891, according to Alfred T. Atkeson, the president of West Virginia University "looked upon agriculture as an undesirable invasion of the holy precincts of classical knowledge."[50]

[48] AAACES, *Proceedings*, 1896, pp. 48–49; *Cultivator and Country Gentleman*, XXXIX (November 26, 1874), p. 755.

[49] Isaac P. Roberts, *Autobiography of a Farm Boy* (Albany, N.Y., 1916), p. 142; *Breeder's Gazette*, XLIV (November 18, 1903), p. 858.

[50] Roberts, *Autobiography of a Farm Boy*, p. 109; Thomas C. Atkeson and Mary M. Atkeson, *Pioneering in Agriculture: One Hundred Years of American Farming and Farm Leadership* (New York, 1937), p. 133.

Some critics, including academic men, resorted to ridicule. It was common to label the new schools as "cow colleges," but wits in Massachusetts at least displayed originality by calling their institution the "bull and squash college."[51]

Particularly distressing was the dislike or contempt for the colleges that most farmers displayed. Not only did agriculturists refuse to enroll their sons but rural spokesmen were unrestrained in voicing their hostile opinions of the institutions. A convention of farmers in Illinois in 1870 expressed the views of many when it denounced the curriculum at the Illinois Industrial University as being totally irrelevant to the needs of agriculturists. According to the malcontents the college would not "allow a boy to go there and study such agricultural or mechanical branch as he may choose without taking everything else in the curriculum. If a man has peculiar faculties for blacksmithing, in God's name let him be a blacksmith. Metaphysics, what is it? Ten pages will contain the substance of the labors of all the metaphysical fools from Aristotle down." Consequently, in Illinois as elsewhere farmers generally concluded that the land-grant schools were humbugs and that nothing of value could come from them.[52]

As late as 1906, a controversy that erupted between farmers and officials of Clemson College following a meeting on the campus suggested the void that often separated the colleges and the people they were expected to serve. Rural spokesmen were harshly critical of their reception at the school and of the institution's facilities and faculty. In fact, said the outraged farmers, the treatment afforded them "was a downright insult. Accommodations were a disgrace. The barracks were dirty, and the rooms were dusty, moldy and the beds contained bedbugs. The table fare was poor, unhealthy, unpalatable, and badly prepared." Moreover, the faculty "paid very little attention to the visitors" and "few of the professors had shaved faces."

[51] Caswell, *Massachusetts Agricultural College*, p. 23.

[52] Carriel, *Turner*, p. 207; Wilbur H. Glover, "The Agricultural College Lands in Wisconsin," *Wisconsin Magazine of History*, XXX (March, 1947), p. 262.

Stung by such charges Clemson officials replied in kind. The attacks, they said, were politically motivated. Accommodations were those that had been offered visitors for years. Admittedly there were a few bedbugs in the barracks, despite fumigation of the rooms for thirty-six hours "resulting in the killing of thousands of insects of all kinds, rats and mice, and the sparrows on the eaves of the buildings." But, asked the indignant educators, what of the farmers' homes?[53]

Another difficulty stemmed from a failure to teach practical agriculture, a shortcoming that resulted in part from confusion concerning the proper objectives of an agricultural college. Some early leaders, having no precedents to guide them, concluded that the colleges should do little except offer future farmers a general education, not greatly different from that given in the classical colleges. In this vein a curator of the University of Missouri stated that "too much in practical agriculture should not be expected, as the main purpose is to develop the social and mental nature of the students." That goal was fine, responded a farmer, "but what are they [the professors] going to do about hog cholera?" Understandably, the curator had no answer.[54]

The effort to refute the charge that the colleges presented no practical instruction was one of the factors that led to the widespread adoption of a student labor system. At the outset, in fact, student labor seemed to produce desirable results. Enrollments at those institutions employing the system were generally higher than elsewhere, and advocates were quick to point to the virtues of student labor. But soon problems arose.

Some institutions failed to strike any sort of a reasonable balance between the practice and the theory of the subject. One overly enthusiastic exponent of student labor proclaimed that "muscle must be put on the same level as brain." That questionable goal, presumably, was reached in a few cases; at least one graduate of the system

---

[53] Attacks on Clemson College by the Farmers and Some Items in Reply, 1906, Clemson University History Collection (Clemson University Library, Clemson, S.C.).

[54] Quoted in Etheridge, "College of Agriculture," p. 300.

30

later recalled that when he was at the college, agricultural instruction was "both practical and theatrical."[55]

Apparently the majority of boys were less than enthusiastic about student labor. According to the professor of agriculture at Kansas State Agricultural College in the early 1880s, his students "worked . . . as well, in most cases, as the average hired man, although bright and shining examples of an entirely opposite disposition have not been wanting." One of those many students who were of the "opposite disposition" was Perry G. Holden, later an extension pioneer in Iowa. Recalling that his first job was cleaning out a hog house with the temperature above 100 degrees, at a wage of 8½ cents an hour, Holden claimed that the student labor policy was an utter failure, causing students to resort to deception and fraud to escape it.[56]

Nor were the faculties such as to inspire confidence. Competent teachers of agriculture were rare. In essence, a college setting out to establish a course in agriculture could either obtain a scientist trained at a classical college or it could employ a farmer who in some manner had won more than a local reputation. The chances of failure were great in either case. Scholarly chemists and their counterparts in other sciences were likely to try to fit agricultural education into a classical mold. All too common were stories of teachers of agriculture who were completely alien to the eccentricities of grain harvesters or to the underside of cows. Such men aroused considerable distrust among farmers and the enthusiasts who had done so much to establish the land-grant schools. On the other hand, practical farmers were handicapped by their lack of teaching experience. Moreover, with a few notable exceptions, their knowledge was limited to their own area and experiences, a fact that their students were quick to notice.[57]

All of the professors, regardless of their abilities, were badly over-

[55] Eugene W. Hilgard, "Progress in Agriculture by Education and Government Aid," *Atlantic Monthly*, XLIX (April, 1882), p. 535; AAACES, *Proceedings*, 1897, p. 33.

[56] Kansas State Agricultural College, *Fourth Biennial Report*, 1883–84 (Topeka, 1885), p. 28; P. G. Holden Memoirs (Michigan State University Library, East Lansing).

[57] Carriel, *Turner*, p. 189; *Breeder's Gazette*, XXIII (February, 1883), p. 82.

31

worked. Most of the early agricultural colleges attempted to function with only one teacher of agriculture. He was expected to manage the college farm as well and, since students in agriculture were in short supply, he found himself teaching a variety of other subjects. At Kansas in the late 1870s, for instance, Edward M. Skelton was responsible for physiology, physical geography, household economy, and United States history, as well as his courses in agriculture. Logically enough, he found his work to be highly unsatisfactory.[58]

Enrollments especially indicated trouble in the academic groves. Although such institutions as Mississippi Agricultural and Mechanical College had sizeable numbers of students from the outset, most of the institutions failed miserably to attract scholars in agriculture. At the University of Wisconsin, for example, only one student had completed the four-year course by 1884. North Carolina counted 17 students in its agricultural course in 1887; between 1875 and 1887 the University of Minnesota averaged one student a year; and in 1882 only 2 of the 254 students at Purdue were in agriculture. In 1890 James Wilson claimed that only one student at the Iowa school would admit that he went to Ames to study farming. Some 85 to 90 sons of farmers entered the University of California in 1896, but not one enrolled in the agricultural course. Five students did so, but they came from urban homes.[59]

The lack of a body of scientific information to teach and the need to solve problems faced by ordinary farmers were other difficulties; these pointed to the need for experiment stations. The first agency of that type seems to have been established in Saxony in 1851, and soon similar institutions appeared in England and France.

In the United States some of the pioneer agricultural colleges began experimental work almost as soon as they were established. The

[58] Kansas State Agricultural College, *Biennial Report*, 1876–78, p. 14.

[59] Wilbur H. Glover, "The Agricultural College Crisis of 1885," *Wisconsin Magazine of History*, XXXII (September, 1948), p. 17; *Southern Farm Gazette*, X (July 1, 1905), n.p.; *Southern Farm Magazine*, XII (December, 1904), p. 16; University of Minnesota Board of Regents, *Fourth Biennial Report, Supplement I* (St. Paul, 1887), p. 122; *American Agriculturist*, XLI (March, 1882), p. 100; AAACES, *Proceedings*, 1898, p. 44; *Pacific Rural Press*, LII (February 20, 1897), p. 114.

Maryland law of 1856 that created the agricultural college directed the institution to conduct experiments on its model farm. The first distinct agricultural experiment station appeared in Connecticut in 1875, and California, North Carolina, New York, and New Jersey followed during the next few years. In Louisiana in 1884 a private station was established by the Louisiana Scientific and Agricultural Society, and two years later work began on university land.[60]

By the 1880s agricultural educators across the nation were convinced that a successful experimentation program required some form of federal assistance. The upshot was the introduction into Congress between 1882 and 1887 of a number of bills providing for a federal appropriation for experiment stations. Finally, proponents found powerful friends in Representative William H. Hatch of Missouri and Senator James Z. George of Mississippi. From his post as chairman of the House Committee on Agriculture Hatch especially pushed the issue, and the bill that carried his name passed easily in 1887. The measure provided for the creation within each of the colleges established under the Morrill Act of 1862 of a department to be designated as an agricultural experiment station "in order to aid in acquiring . . . useful and practical information . . . and to promote scientific investigation and experiment respecting the principles and applications of agricultural science." Each state received initially a grant of $15,000 annually, a sum that was expected to be adequate to induce legislatures to give their assent and to contribute to the support of the station.[61]

While the agricultural colleges were wrestling with what seemed to be an impossible task and while such agencies as rural societies,

[60] Byron D. Halstead, "A New Factor in American Education," *The Chautauquan*, XVI (December, 1892), p. 287; *American Agriculturist*, XXXVII (July, 1878), p. 251; *ibid.*, XLI (April, 1882), p. 149; *Cultivator and Country Gentleman*, L (July 16, 1885), p. 593; Frederick W. Williamson, *Origin and Growth of Agricultural Extension in Louisiana, 1860–1948* (Baton Rouge, 1951), pp. 24–25; AAACES, *Proceedings*, 1912, p. 82; Leland E. Call, "Agricultural Research at Kansas State Agricultural College before the Enactment of the Hatch Act," Kansas Agricultural Experiment Station, *Bulletin* 441 (Manhattan, Kans., 1961).

[61] Eddy, *Colleges for Our Land and Time*, pp. 94–97; *U.S. Statutes at Large*, Vol. XXIV, p. 440.

fairs, and farm journals were beginning to educate the countryside, there were early calls for the establishment of a more formal arrangement for the instruction of farmers. In fact, throughout the nineteenth century and even earlier, there were farseeing men who in their proposals foreshadowed the establishment of modern agricultural extension.

Both Washington and Jefferson envisioned the establishment of some arrangement by which farmers might come into direct contact with men who could give them the instruction they needed. In his last message to Congress Washington proposed the establishment of a national university with a chair of agriculture which would have among other duties the responsibility for "diffusing information to farmers." At various times Jefferson advocated the formation in Virginia of a statewide system of local agricultural societies that would have extension functions going beyond those common to such groups.[62]

The idea of an extension role for the land-grant colleges appeared years before the Morrill Act became a reality. Speaking before the Agricultural Society of Essex County, Massachusetts, in 1851, a proponent pointed out that the effect of such colleges would extend far beyond their campuses. Each graduate, he said, would be a source of inspiration, causing farmers in time to adopt the new techniques that former students were using. In addition, he believed those who taught in the schools might from time to time issue publications which he supposed would have a wide circulation among practicing farmers. The agricultural colleges would "be so many lights, which will shed their rays not only upon those who are brought into immediate contact [with them], but diffuse their beams abroad, illuminating remote places, finding their way into obscure recesses. . . ."

Perhaps less poetically, in the course of his agitation for the establishment of land-grant colleges, Jonathan Baldwin Turner hinted at their possible extension role. As early as 1851 he suggested that such

[62] Edwin Bay, *The History of the National Association of County Agricultural Agents, 1915–1960* (Springfield, Ill., 1961), p. 5; Wheeler McMillen, *Land of Plenty: The American Farm Story* (New York, 1961), pp. 95–96; Russell Lord, *The Care of the Earth: A History of Husbandry* (New York, 1962), p. 190.

institutions might have a general open house for farmers during commencement, with exhibits and lectures by the faculty and advanced students for those attending the ceremonies. Two years later, in a memorial addressed to the state legislature, the Fourth Industrial Conference held in Springfield, Illinois, suggested that an industrial university such as Turner proposed might send its staff members out into the state to diffuse knowledge. In 1865, while urging Illinois to establish a land-grant school, Turner claimed that a college of that type might well be a "dispenser of knowledge to the people."[63]

Such ideas found their way into the Morrill Act and into the state legislation establishing a number of the land-grant colleges. The act of 1862 directed the institutions founded under its terms to "promote the liberal and practical education of the industrial classes in the several pursuits and professions of life," an injunction that clearly foreshadowed some form of adult education. Meanwhile, the Iowa law establishing that state's agricultural college in 1858 provided for an institution "which shall be connected with the entire agricultural interests" of the state. Admittedly vague, that law nevertheless seemed to imply that the college would have a role that went beyond resident instruction. More specific was the authorization found in a Michigan law which provided that the "State Board of Agriculture [which administered the college] may institute winter courses of lectures, for other than students at the institution. . . ." Even more definite was the provision in the New Jersey law accepting the land-grand fund that required the officers of the college to deliver at least one lecture in each county each year.[64]

The concept of extension was also accepted at an early date by the

[63] U.S. Commissioner of Patents, *Report,* 1851, part II, pp. 34, 40; Carriel, *Turner,* pp. 113, 159.

[64] *U.S. Statutes at Large,* Vol. XII, p. 504; Morgan, *Extension Service of Iowa State College,* p. 11; Epsilon Sigma Phi, *The Spirit and Philosophy of Extension Work* (Washington, 1952), p. 4; "History of Cooperative Extension Work in Michigan, 1914–1939," Michigan State College, Extension Division, *Extension Bulletin* 229 (East Lansing, 1941), p. 5; Liberty Hyde Bailey, *Annals of Horticulture,* 1891 (New York, 1892), p. 139; American Association of Farmers' Institute Workers, *Proceedings,* 1902 (Washington, 1902), p. 38. Cited hereafter as AAFIW, with year. Carl R. Woodward and Ingrid N. Waller, *New Jersey's Agricultural Experiment Station, 1880–1930* (New Brunswick, N.J., 1932), p. 492.

United States Department of Agriculture. In fact, the act creating the department gave the agency the duty "to diffuse among the people of the United States useful information on subjects connected with agriculture. . . ." Commissioner Isaac Newton, in his first report dated January, 1863, acknowledged that responsibility when he noted that the Department of Agriculture had as one of its functions the dissemination of "useful information" on all agricultural fields. Eight years later, in his annual report of 1871, Commissioner Frederick Watts contended that there should be close contact between his department and the agricultural colleges in order that these agencies could better perform their extension functions.[65] Here, in fact, Watts was describing an arrangement that almost a half century later would bring the county-agent system into being.

But before any teaching system could be reasonably effective, farmers had to convince themselves that agricultural education and the application of agricultural science were beneficial to them. That proved to be a development in which the agrarian crusade played a major role.

[65] *U.S. Statutes at Large,* Vol. XII, p. 387; U.S. Commissioner of Agriculture, *Report,* 1862, p. 20; *ibid.,* 1871, p. 3.

II. AGRARIAN DISCONTENT
AND FARMER EDUCATION

THE WAVE OF AGRARIAN DISCONTENT that swept the United States between the 1860s and the end of the nineteenth century was much more than an economic and political movement. The Grange and the Alliance movement taught farmers that cooperatives had a place in rural America; the Populist party and its antecedents made substantial contributions to political thought and action. But often overlooked were other accomplishments. The agrarian organizations induced their members to read and to think, to consider questions intelligently, and to reject the traditionalism, emotionalism, and apathy so deeply implanted in their nature. Concurrently, farmers realized that education beyond the elementary level was not to be ridiculed but rather was to be embraced with growing enthusiasm. The rural organizations helped to strengthen the fledgling land-grant colleges and to focus their attention more surely on the needs of farmers. By raising an incessant clamor, in fact, the grangers and other groups forced the colleges to seek means by which they might improve their standing with farmers, thereby pointing those institutions toward the concept of extension.

Another contribution of the agrarian crusade was more subtle, but in the end perhaps equally significant. In the late 1860s, when grangerism was still only a plan in the mind of Oliver Hudson Kelley, many farmers tended to think of themselves as Jeffersonian agrarians, different from and superior to other people and living in a pastoral world in which the realities of the marketplace were not the dominant factors in determining their status and welfare. The long postwar depression and the agrarian movement that it produced shattered that illusion. By 1900 farmers knew beyond any doubt that they were small businessmen, using land, labor, and capital to produce commodities for the market. They also saw that, unlike other businessmen, growers of wheat, cotton, and livestock could do little to fit production to demand. But the agitation that was a part of rural discontent awakened farmers to another approach that promised at least partial relief. More efficient farming hopefully would reduce costs and enlarge net earnings. Thus it was that after the heartbreak of 1896, scientific agriculture and improved techniques aroused an interest among farmers that was largely lacking a few years earlier.

If more production were the only criterion, the late nineteenth century should have been a golden period for farmers. Influenced by the restless spirit engendered by the Civil War and aided immeasurably by the new transcontinental railroads, settlement swept westward so rapidly after 1865 that in 1890 the Census Bureau could report the disappearance of the frontier. In the course of that development, vast amounts of new land came into use; between 1860 and 1900 acreage in farms increased 106 percent and the number of farms rose 181 percent. Meanwhile, in the older settled regions, at least, land was tilled more intensively. The percentage of improved land rose from 40 to 49, while the size of the average farm fell from 199 to 146 acres.

Along with expanded acreage and more intensive cultivation went mechanization. Farmers might reject agricultural education and sneer at "book farming," but they displayed little hesitancy in the adoption of labor-saving machinery. With the exception of the South where such crops as cotton and tobacco did not lend themselves to mechanization and other areas where rough terrain and small and irregular fields prohibited the utilization of the new tools, farmers accepted enthusiastically grain binders, gang plows, and horse-drawn drills and planters. During the years from 1860 to 1900, in fact, the value of machinery climbed noticeably more sharply than did that of all farm property.

The certain result of the use of machinery and more intensive cultivation on greater acreages was sharply increased production. Taking the crop years 1859 and 1899, corn production rose 217 percent, wheat shot up 281 percent, and cotton, particularly the victim of the Civil War, exceeded the prewar figure by 79 percent. Moreover, farmers as a whole kept more livestock. The number of sheep declined but cattle and hogs rose more than 80 percent and milk cows doubled.

Unfortunately, these developments produced no bonanza for the nation's agriculturists. The patterns of farm prices were sharply downward after 1870, influenced by the deflationary monetary and fiscal policies of the government and even more by the adverse relationship of supply to demand. Gross income per farm fell from $945 in 1870 to $738 in 1880 and $669 in 1890. A slight revival in the later years of that decade produced a figure of only $674 in 1900.[1] In short, farmers in the late nineteenth century produced food and fiber in dramatically greater amounts, but the flood of farm commodities failed completely to yield larger incomes for the growers. These conditions provided the mainsprings of the agrarian crusade; here, too, was the economic environment from which arose the demand for agricultural extension.

The course of the agrarian crusade is well known and requires only a brief outline here. Although the roots of rural discontent may be found earlier, it was not until 1867 that Oliver Hudson Kelley and his associates created the Patrons of Husbandry or the Grange. The organization enjoyed little success until the early 1870s when alleged maltreatment by railroad managers, middlemen, corrupt politicians, and others sent farmers into the society by the thousands. By January,

[1] Statistics on production, acreages, and value of farm property have been taken from *U.S. Thirteenth Census: Agriculture*, pp. 67, 75, 84, and from *Historical Statistics of the United States, 1789–1945* (Washington, 1949), pp. 101, 103–104, 106, 108. Gross income per farm has been calculated from data given in Frederick Strauss and Louis H. Bean, "Gross Farm Income and Indices of Farm Production and Prices in the United States, 1869–1937," U.S. Department of Agriculture, *Technical Bulletin* 703 (Washington, 1940), p. 23.

1875, there were reported to be 21,697 local granges in the United States. Ten months later membership was set at 758,767 of which almost 52 percent was found in the Middle West. These numbers represented peak figures; a decline began in 1876 and by 1880 much of the organization's power and most of its militancy were lost.

Yet the Grange did not die. Something of a revival came in the late 1880s when in such states as Illinois the organization was an important component of the Farmers' Alliance movement. Further growth came after 1900. That year total membership was only 187,482, but fifteen years later it was 540,085. This resurgence of the Grange came not in the Middle West, where the order had once enjoyed such strength, but in New England and the Middle Atlantic states, with a lesser following in Ohio, Indiana, and elsewhere.[2]

In contrast to the old Grange the Alliance movement appeared spontaneously in various areas of the country. The Texas Farmers' Alliance arose in the 1870s, the Louisiana Farmers' Union developed early in the next decade, and the Agricultural Wheel was born in Arkansas in 1882. These organizations and shortly afterwards the North Carolina Farmers' Association merged to form the National Farmers' Alliance and Industrial Union, commonly called the Southern Alliance. Meanwhile, in Chicago in 1880, Milton George, the editor of the *Western Rural* who was familiar with earlier developments in New York, established the National Farmers' Alliance or the Northern Alliance. In due course that organization spread westward and northwestward to enlist large numbers of prairie and plains farmers in its ranks. Additional orders that constituted a part of the Alliance movement included the Farmers' Mutual Benefit Association, formed in Illinois in 1883; the Patrons of Industry, a Michigan group that appeared in 1887; a society for Negro farmers in the South; and a number of less well-known groups. Membership figures are by no means incontestable, partly because of a distressing ten-

[2] For figures on membership, see Solon J. Buck, *The Granger Movement: A Study of Agricultural Organization and Its Political, Economic and Social Manifestations, 1870–1880* (Cambridge, Mass., 1913), chart following p. 58; Carl C. Taylor, *The Farmers' Movement* (New York, 1953), p. 137; Roy V. Scott, *The Agrarian Movement in Illinois, 1880–1896* (Urbana, Ill., 1962), pp. 56–60; Kenyon L. Butterfield, *Chapters in Rural Progress* (Chicago, 1908), pp. 138–141.

dency of leaders to exaggerate the strength of their following, but it is certain that by 1890 the various groups included millions of farmers. A recent scholar has set total farm family memberships at 1,053,-000, or one out of every four rural families in the United States.[3]

Although the Farmers' Alliances were more than political organizations, by 1890 aggressive leaders were moving toward political action. After significant victories in Kansas and elsewhere in 1890, the third-party men formed the Populist party and named James B. Weaver of Iowa as the party's presidential nominee in 1892. His fine showing in a losing cause suggested to the Populists that their movement had a promising future. But hopes were dashed in 1896 when the Democrats chose William Jennings Bryan as their standard-bearer, thereby presenting the Populists with the choice of fusion and absorption or division and collapse. The result was the virtual destruction of the Populist party as a political force. Nor did the Alliance movement fare any better. Many members were opposed to political action at the outset while others became disenchanted as the political battles raged on. In any event the membership evaporated as farmers decamped in droves.

The collapse of populism, of course, did not mean that farm organizations disappeared from the United States. Not only did the Grange live on, but the first years of the twentieth century saw the establishment of several new groups. Most important were the Farmers' Educational and Cooperative Union and the American Society of Equity, founded in 1902 in Texas and Indiana respectively. Other groups, some rather shadowy, included the Farmers' Equity Union and a number of local organizations in southern Illinois.[4]

[3] The foregoing is well summarized in Taylor, *Farmers' Movement*, pp. 193–222. The Farmers' Mutual Benefit Association is discussed in Roy V. Scott, "The Rise of the Farmers' Mutual Benefit Association in Illinois, 1883–1891," *Agricultural History*, XXXII (January, 1958), pp. 44–55. A study of the Agricultural Wheel is Francis C. Elkins, "The Agricultural Wheel in Arkansas, 1882–1890" (Ph.D. Thesis, Syracuse University, 1953). The best figures on the strength of the various organizations are found in Robert L. Tontz, "Memberships of General Farmers' Organizations, United States, 1874–1960," *Agricultural History*, XXXVIII (July, 1964), pp. 145, 147.

[4] Theodore Saloutos and John D. Hicks, *Agricultural Discontent in the Middle West* (Madison, Wis., 1951), pp. 111–148, 219–254; Taylor, *Farmers' Movement*,

All of the farm groups were interested to some degree in the education of their members. The Grange was by far the most active. In fact, a misconception exists concerning the relative significance of the different Grange activities, a misconception that arose because of the attention that early historians of grangerism devoted to the order's role in politics and particularly to the development of public regulation of railroads. It can be argued that in the broad sweep of agricultural history since the Civil War the Grange made its greatest impact on the welfare of the farmer through its educational and social activities.[5]

Later groups—those that formed the Alliance movement—made less of an educational contribution partly because of a subtle difference in their understanding of the causes of the widespread rural distress. More so than the Alliance groups the Grange attributed rural hardship and poverty to the failure of farmers to keep up educationally with other groups. "The loose screw in farming is ignorance," said a Mississippi paper in 1876, observing also that "the monopoly to be feared most by farmers was the monopoly on brains held by other classes."[6] On the other hand alliancemen believed that rural distress and hardship were the result of vicious legislation enacted by legislatures controlled by business elements through bribery and corruption.[7] To men in that frame of mind, education could have only a secondary, not primary, importance.

pp. 391–420. For a discussion of early twentieth-century farm groups in Illinois, see Roy V. Scott, "John Patterson Stelle: Agrarian Crusader from Southern Illinois," Illinois State Historical Society, *Journal*, LV (Autumn, 1962), pp. 246–247.

[5] Such is the contention of a doctoral dissertation in history recently completed at Mississippi State University by Dennis S. Nordin. For a rather idealistic statement by a Grange spokesman, see Jennie Buell, "The Educational Value of the Grange," *Business America*, XIII (January, 1913), pp. 50–54.

[6] Quoted in James S. Ferguson, "The Grange and Farmer Education in Mississippi," *Journal of Southern History*, VIII (November, 1942), pp. 498–499.

[7] Frank M. Drew, "The Present Farmers' Movement," *Political Science Quarterly*, VI (June, 1891), pp. 305–306. This point is well brought out in Homer Clevenger, "The Teaching Techniques of the Farmers' Alliance: An Experiment in Adult Education," *Journal of Southern History*, XI (November, 1945), pp.

In the case of the Grange it seems perfectly clear that its founders formulated the organization with the general idea that it would make its major contribution to agriculture through its educational role. Oliver Hudson Kelley repeatedly stated as much, and in its first printed circular, dated November 1, 1867, the organization announced that it hoped "to encourage and advance education in all branches of agriculture" and as a part of that work to establish "libraries and museums." Later Kelley talked of "mental instruction through the reading of essays, and discussions, lectures, formation of select libraries, circulation of magazines, and other publications touching directly upon . . . the principles governing our operation in the field, orchard, and garden."[8]

Nor did the Grange abandon its educational role during those years when its history was most intimately connected with politics and railroad regulation. In 1870 Grand Master Saunders stated that the ultimate objects of the organization were "to increase the products of the earth by increasing the knowledge of the producer . . .; to learn and apply the revelations of science . . .; and to diffuse the truths and general principles of the science and art of agriculture. . . ." The preamble of the National Grange constitution of January, 1873, maintained that the primary purpose of the society was "mutual instruction and protection." The next year the famous Declaration of Purposes proclaimed that the organization sought to "advance the cause of education among ourselves and for our children. . . ."[9]

Even the farmers' clubs that sprang up and in some areas constituted the political arm of the granger movement had an educational

505–518. See also Alex M. Arnett, *The Populist Movement in Georgia* (New York, 1922), p. 100.

[8] Oliver H. Kelley, *Origin and Progress of the Order of the Patrons of Husbandry in the United States: A History from 1866 to 1873* (Philadelphia, 1875), pp. 38–40; John R. Commons and others, eds., *A Documentary History of American Industrial Society* (10 vols., Cleveland, 1910), X, 74–76.

[9] Charles M. Gardner, *The Grange, Friend of the Farmer* (Washington, 1949), p. 329; Commons, *History of American Industrial Society*, X, 103; Edward W. Martin, *History of the Grange Movement* (Chicago, 1874), p. 462, 470; True, *History of Agricultural Education*, pp. 122–124.

side. The Illinois State Farmers' Association, for instance, stated as one of its objectives the "improvement in the theory and practice of Agriculture and Horticulture. . . ."[10]

A decade or more later there was little change in the Grange's interest in education—it remained paramount. A spokesman in 1890 said that the organization "proposes to educate the American farmer to participate in the discussion of all questions" pertaining to agriculture. More specifically, according to another speaker, the Grange sought to teach farmers to "produce more . . . , to diversify our crops, and to crop no more than we can cultivate. . . ."[11]

The groups that together formed the Alliance movement also professed their interest in education, if of a different type and to a lesser degree. For example, the constitution under which the first locals of the National Farmers' Alliance were organized listed among the objectives of the order "the elevation of agriculture by the mental . . . improvement of its members" and the promotion of "a more rational method of tillage." The Minnesota State Farmers' Alliance proposed "to work for the elevation of agriculture" and "to improve the methods of farming."[12] The Farmers' Mutual Benefit Association expected to "educate the farmer" and to improve the "modes of Agriculture, Horticulture and Stock Raising; [and] to adopt and encourage such rotation of crops as may improve rather than impoverish the soil. . . ."[13]

Southern Alliance groups, too, pledged their interest in education of members. The constitution of the Lincoln Parish Farmers' Club, which was widely used in the Louisiana Farmers' Union in the 1880s, mentioned the elevation of agriculture and the development of better

[10] Jonathan Periam, *The Groundswell: A History of the Origins, Aims, and Progress of the Farmers' Movement* (Cincinnati, 1874), p. 993; *American Agriculturist,* XXXII (September, 1873), p. 355.

[11] *Prairie Farmer,* LXI (February 9, 1890), p. 81; (Springfield) *Illinois State Journal,* December 13, 1887, p. 4; *Hoard's Dairyman,* XLIV (September 20, 1912), pp. 202–203.

[12] *Western Rural,* VIII (October 23, 1880), p. 340; Minnesota State Farmers' Alliance, *Constitution and By-Laws,* 1890 (St. Paul, 1890), p. 5.

[13] *Chicago Tribune,* April 11, 1890, p. 9; Farmers' Mutual Benefit Association, *General Charter, Declaration of Purpose, and Constitution and By-Laws* (Mt. Vernon, Ill., 1890), p. 3.

farming through frequent and free discussions among members. A contemporary historian of the Agricultural Wheel called his organization a "school of education." Speaking in Springfield, Illinois, Ben Terrell of the Southern Alliance claimed that the order might well make its greatest contribution through education.[14] An independent observer, in fact, reported that the Alliance was like a great "national university" in which "college educated as well as self-taught teachers" instructed the mass of the rural population.[15]

Admittedly, southern groups saw "education" as significantly more than the enlightenment of members in better farming methods. To the Farmers' Alliance of South Carolina, education meant "preparation for Service." The Virginia Alliance worked for the "education of the agricultural classes in the science of economical government."[16] In 1890 President Leonidas L. Polk of the National Farmers' Alliance and Industrial Union said that education was "the greatest and most essential need" of his organization, specifically education "in the mutual relations and reciprocal duties between each other . . . , education in the most responsible duties of citizenship . . . , education for higher aspiration, higher thought, and higher manhood. . . ."[17]

The farm groups of the early twentieth century, like those of the preceding decades, were interested in agricultural improvement. For example, the Farmers' Social and Economic Union, an Illinois organization, listed among its purposes the education of its members and the increased productivity of the soil.[18] More influential was the Farmers' Educational and Cooperative Union, which had a substan-

[14] Nelson A. Dunning, ed., *Farmers' Alliance History and Agricultural Digest* (Washington, 1891), p. 222; W. Scott Morgan, *History of the Wheel and Alliance and the Impending Revolution* (New York, 1968), p. 205; *National Economist,* IV (December 6, 1890), pp. 185–186.

[15] Charles S. Walker, "The Farmers' Movement," American Academy of Political and Social Science, *Annals,* IV (March, 1894), pp. 793–794.

[16] Farmers' Alliance of South Carolina, *Proceedings,* 1893 (Spartanburg, 1893), p. 5; Farmers' Alliance of Virginia, *Constitution Adopted August 18–20, 1891* (Richmond, 1891), pp. 3–4.

[17] National Farmers' Alliance and Industrial Union, *Proceedings of the Supreme Council,* 1890 (Washington, 1891), p. 5.

[18] *National Rural,* LVIII (May 17, 1900), p. 615; Scott, "John Patterson Stelle," pp. 246–247.

tial following in the South by 1910. The preamble to its constitution stated that one of its objectives was to "educate the agricultural class in scientific farming."[19]

Doubtlessly farm groups exercised their greatest impact on the mass of rural residents through the ordinary meetings of local bodies. Whether they be granges, alliances, lodges, or unions, all neighborhood groups devoted a part of each meeting to the discussion of some topic of general interest. In numerous instances, admittedly, these topics dealt with politico-economic issues—transportation, money, and monopoly being among the favorites in the nineteenth century—but participants also heard talks on better farming methods, thereby making the gatherings useful mediums for the dissemination of purely agricultural information.

In fact, the topics discussed covered the full spectrum of farming. In the South in the 1870s grangers talked about the overreliance on cotton, the need for diversification, the virtues of manual labor, the use of more and better machinery, and the need for more thorough cultivation.[20] Northern grangers discussed such matters as grain and livestock production, drainage, types of pumps and windmills, and deep versus shallow plowing.[21]

Granger discussion topics in the 1880s were much like those of the preceding decade. In the course of a year in which the Ridott, Illinois, grange met twenty-four times, its members discussed or heard lectures of the structure of plants, the virtues of sulky as opposed to walking plows, plant diseases, and new developments in farm ma-

[19] Charles S. Barrett, *The Mission, History and Times of the Farmers' Union* (Nashville, Tenn., 1909), pp. 87–91, 107; Robert L. Hunt, *A History of Farmers' Movements in the Southwest* (n.p., n.d.), pp. 130–132; George L. Robson, Jr., "The Farmers' Union in Mississippi" (M.A. Thesis, Mississippi State University, 1963), pp. 74–78.

[20] Ralph A. Smith, "The Contribution of the Grangers to Education in Texas," *Southwestern Social Science Quarterly*, XXI (March, 1941), p. 312; William W. Rogers, "The Alabama State Grange," *Alabama Review*, VIII (April, 1955), p. 109; Curtis E. McDaniel, "Educational and Social Interests of the Grange in Texas" (M.A. Thesis, University of Texas, 1938), pp. 40–41.

[21] Illinois State Board of Agriculture, *Transactions*, 1872 (Springfield, Ill., 1873), p. 238; Champaign Grange, Proceedings, March 8, September 23, and November 15, 1875 (Illinois Historical Survey, Urbana).

chinery. In the same state the Edwards County grange talked about dehorning cattle and the "profits in poultry raising."[22]

Nor did topics differ greatly in the monthly or semimonthly meetings of Alliance locals. A group of farmers that met in Centralia, Illinois, in 1887 heard talks on the successful operation of dairies and creameries, the growing of clover, and the practicality of laying tile. A typical program of the Deer Park, Illinois, farmers' alliance included a discussion of sheep and wool production. In 1890—a political year—a local of the Farmers' Mutual Benefit Association devoted one meeting to a discussion of the propagation and cultivation of fruit trees and planned to take up the culture and marketing of small fruit at its next session.[23]

Subjects considered at local gatherings were not left entirely to chance or to the whims of members. Subordinate bodies of both the Grange and the Alliance had an officer known as the lecturer whose duty it was to arrange the educational portion of the program. In Iowa, where the responsibility of the grange lecturer was typical, he was required to be "prepared with some useful information to read, or cause to be read. . . ."[24]

A large percentage of the talks before local granges and alliances were given by members of the group or by other local speakers. Outsiders were occasionally induced to appear. State and national leaders of the various organizations spent considerable time in visiting locals. Ben Terrell of the Southern Alliance and Milton George of the northern group spoke at many gatherings. Hamlin Garland appeared at a grange meeting in Illinois in 1892. Prominent among outside speakers were agricultural college faculty members. In fact, many of them were eager for any opportunity to establish contact with ordinary farmers. Unfortunately, in the 1870s at least, "the an-

[22] *Western Rural,* XXIV (January 27, 1886), p. 133; *Albion* (Ill.) *Journal,* October 25, 1888, p. 4; *ibid.,* July 25, 1889, p. 4; *Marseilles* (Ill.) *Plaindealer,* June 29, 1888, p. 1.

[23] *Centralia* (Ill.) *Daily Sentinel,* January 25, 1887, p. 2; *ibid.,* February 13, 1890, p. 1; *Ottawa* (Ill.) *Free Trader,* January 17, 1891, p. 1; *Western Rural,* XXVII (February 9, 1889), p. 85.

[24] Periam, *Groundswell,* p. 192; John D. Hicks, *The Populist Revolt: A History of the Farmers' Alliance and the People's Party* (Lincoln, Nebr., 1961), p. 129.

nouncement of a professor from the agricultural college had a tendency to empty the hall—rather than fill it with earnest and interested people," a fact that reflected the all too common attitude of farmers toward the land-grant schools.[25]

Some groups went beyond the lecture or informal talk in educating their members. The use of the query box stimulated interest. An Alabama grange in the 1870s appointed a committee to visit the farms of members and to prepare reports concerning each farmer's property, the quality of his stock, fences, and other structures, and the extent to which he used improved methods. The reports formed the basis for discussion at later meetings. The Black's Bend grange in the same state offered prizes for the best garden, promoted a corn contest, and gave awards to women for food preparation. At a multitude of meetings, such as those of the Palmyra, New York, grange in the 1870s, displays of wheat, jellies, fruits, and other produce enhanced the educational value of the sessions.[26]

As grangers and alliancemen participated in local meetings there was a secondary result, perhaps most important of all. Members grasped the need for possessing information, details, and facts, data that could be obtained only by reading. Books and magazines became less alien to them, and they slowly came to understand that knowledge was in itself useful. Farmers went through something of an educational ferment in which apathy and hostility to that which was new gradually dissipated. Without that development, in fact, it is doubtful that they would have turned to agricultural education in the twentieth century with the enthusiasm that they did.[27]

Logically enough, organized farmers interested in education soon devoted attention to other agencies that were involved in the improvement of agriculture. The Grange especially recognized the value of fairs and did much to strengthen existing ones or to launch them

[25] Illinois Farmers' Institute, *Annual Report,* 1904 (Springfield, Ill., 1904), p. 245; Scott, *Agrarian Movement in Illinois,* p. 80.

[26] *Cultivator and Country Gentleman,* XL (August 19, 1875), p. 521; Robert Partin, "Black's Bend Grange, 1873–77: A Case Study of a Subordinate Grange of the Deep South," *Agricultural History,* XXXI (July, 1957), pp. 53–54; *American Agriculturist,* XXXVI (September, 1877), p. 329.

[27] A. E. Paine, *The Granger Movement in Illinois* (Urbana, Ill., 1904), p. 46.

where they had not existed earlier. At a fair held by a Mississippi grange in 1872 members saw exhibits of agricultural products and improved machinery. In neighboring Alabama, grangers had a state fair in 1875, and elsewhere members arranged for a multitude of local and county fairs of varying quality. In 1887 the Texas State Grange specifically urged each of its local and county bodies to hold annual fairs, and in the early 1890s that group was largely responsible for an annual state fair. In other instances the Grange arranged exhibits at fairs held by other groups or by states or counties.

Similarly, Alliance groups were promoters of fairs. As early as 1885 members of the Agricultural Wheel in several Arkansas counties united to hold a district fair, and elsewhere similar exhibitions were common.[28]

In those areas where fairs were well established, grangers and other organized farmers were more than willing to offer advice concerning their management. The farmers were especially concerned with the tendency of county and state fairs to include horse racing and a variety of sideshows, activities which outraged rural sensitivities. According to farm spokesmen fairs should be rescued from the "hossy set" and sideshows should be eliminated so that farmers and their families might attend them "with safety to limb and purse." Horse racing, said a southern farm leader, "breeds a spirit of gambling in the breasts of the young men of the country, and virtually undoes much good that a well conducted fair would otherwise accomplish." In Johnson County, Illinois, the Farmers' Mutual Benefit Association threatened to boycott the local fair if horse racing were allowed. On the other hand, by 1883 the National Grange recognized economic realities and accepted racing at fairs as a means of meeting costs and of increasing attendance.[29]

[28] Buck, *Granger Movement,* pp. 293–294; McDaniel, "Grange in Texas," pp. 61–68; Rogers, "Alabama State Grange," p. 104; *American Agriculturist,* XXXIV (December, 1875), p. 452; W. A. Anderson, "The Granger Movement in the Middle West with Special Reference to Iowa," *Iowa Journal of History and Politics,* XXII (January, 1924), p. 46; Smith, "Contribution of the Grangers to Education," p. 315; Elkins, "Agricultural Wheel," p. 100.

[29] *Prairie Farmer,* LXI (September 7, 1889), p. 569; *Southern Live-Stock Journal,* III (December 14, 1878), p. 2; *Centralia* (Ill.) *Daily Sentinel,* February 1,

Nor did farm groups ignore the state boards and similar agencies established for the benefit of agriculture. In the main, grangers and alliancemen sought to extend the functions of the boards and departments of agriculture or to make them more responsive to what was conceived to be the needs of farmers. In Illinois, for instance, in 1890 the state grange asked that the district vice presidents of the state board of agriculture be made elective. On the other hand, the Farmers' Mutual Benefit Association demanded that the agency be abolished, claiming that it performed no useful function for the mass of dirt farmers. More constructively the West Virginia Grange played a major role in the establishment of a state board of agriculture in the early 1890s, while the Agricultural Wheel called for the creation of that type of agency in Arkansas.[30]

The agrarian crusade also had an impact on farm journalism, both by increasing the number of periodicals available to farmers and by converting thousands of rural residents into avid readers. The appearance of the Grange produced a multitude of short-lived farm papers, many of which were official voices of state or other subdivisions of the order. In the 1870s there were at least three grange papers in Alabama alone. Well-established papers like the *Prairie Farmer* began to carry grange news, partly in the expectation of expanded circulation. The result was that more farmers than ever before were exposed to the farm press, and many of them continued the habit after the initial enthusiasm had evaporated.[31]

The alliance phase of the agrarian movement produced yet another surge in the number and circulation of rural papers. A multitude of organs appeared, headed perhaps by the *National Economist* at Washington. Even the Farmers' Mutual Benefit Association had its *Progressive Farmer*. Milton George's *Western Rural* was for four-

---

1890, p. 2; National Grange of the Patrons of Husbandry, *Proceedings*, 1883 (Philadelphia, 1883), p. 24.

[30] Illinois State Grange, *Proceedings*, 1890 (Peoria, 1891), pp. 72–73; *Alton (Ill.) Sentinel-Democrat*, July 24, 1890, p. 5; Atkeson and Atkeson, *Pioneering in Agriculture*, p. 127; Elkins, "Agricultural Wheel," p. 100.

[31] Buck, *Granger Movement*, pp. 287–290; Anderson, "Granger Movement in the Middle West," p. 45; Smith, "Contribution of the Grangers to Education," p. 313; Rogers, "Alabama State Grange," p. 110.

teen years a powerful voice for agricultural organization. The amount of better farming information that these and other periodicals carried varied widely, but all of them contributed at least to an improvement in the reading habits of farmers.[32]

In their efforts to induce their members to educate themselves some of the farm groups promoted the development of reading circles. In the 1880s the Illinois State Grange recommended that its locals organize reading societies that would function much like the chautauqua. In a like fashion in the mid-1890s the National Farmers' Alliance urged the establishment in each state of a program of progressive reading courses under the direction of the state agricultural colleges. The California State Grange established in 1895 its "Rural Study Club," an arrangement that provided a systematic plan for instruction of members with emphasis on the types of business transactions in which farmers were most often involved.[33]

In other instances the farmers established libraries. In 1876 a Wood County, Texas, grange announced that it had under way the creation of a library that would be limited to volumes on agriculture and literature and that would exclude political tracts. At the same time another library had 100 volumes. By 1883 it was reported that at least a dozen locals in Texas had libraries for their members. In North Carolina, too, locals were encouraged to establish libraries. Most of the libraries, it is evident, not only provided books for farmers but also subscribed to one or more farm papers.[34]

Although the land-grant college movement was well underway when the agrarian crusade appeared on the scene, the different or-

[32] Hicks, *Populist Revolt,* pp. 130–131; Scott, *Agrarian Movement in Illinois,* pp. 82–83; *Jefferson City* (Mo.) *Daily Tribune,* August 7, 1891, p. 2.

[33] *Bloomington* (Ill.) *Daily Pantagraph,* June 14, 1887, p. 3; *Western Rural,* XXXI (February 11, 1893), p. 83; *ibid.,* LII (January 24, 1895), p. 50; (Springfield) *Illinois State Journal,* November 17, 1894, p. 2; *Cultivator and Country Gentleman,* LX (December 26, 1895), p. 938.

[34] McDaniel, "Grange in Texas," pp. 55–60; Sister M. Thomas More Bertels, "The National Grange: Progressives in the Land, 1900–1930" (Ph.D. Thesis, Catholic University, 1962), p. 93; Smith, "Contribution of the Grangers to Education," p. 314; Rogers, "Alabama State Grange," p. 110; Stuart Noblin, *Leonidas LaFayette Polk: Agrarian Crusader* (Chapel Hill, N.C., 1949), p. 101; Periam, *Groundswell,* pp. 147–148.

ganizations played no small role in the early development of the schools. The grangers, especially, were interested in the colleges, considering them as practically their own institutions. The Alliance groups devoted less attention to them, but one historian has contended that the "Populist upheaval of the 1890's did more than any other one thing to convert the agricultural colleges of the Middle West into true institutions of higher learning of a distinctive type."[35]

Probably the primary interest of grangers and their successors was to see that instruction in the land-grant colleges was "practical." In the 1870s the Grange regularly voiced this demand, and rural sentiments did not change in the 1880s and the 1890s. In 1883 the Illinois State Grange was deeply concerned with the lack of agricultural instruction at Urbana, and seven years later that body demanded that the university devote all funds arising from the Morrill Act to the purposes for which they were provided. Milton George periodically denounced "theoretical teachings" in the colleges. Summarizing rural views clearly at its national meeting in 1887 in Minneapolis, the National Farmers' Alliance resolved that "the agricultural colleges, magnificently endowed by the Government and dedicated to the purposes of agricultural and mechanical arts, should be held faithfully to the condition of the grant; and as they have in many cases been diverted, we demand that they be restored."[36]

The demand for "practical" education caused the Grange to be in the forefront of the movement that developed in the 1870s for the establishment of separate agricultural colleges. In numerous instances state authorities had given their states' land-grant funds to existing universities in the belief that such arrangements were more feasible than the establishment of entirely new schools. Although in most cases these decisions were perfectly justifiable, grangers saw

[35] Hunt, *Farmers' Movements in the Southwest,* p. 12; Kenyon L. Butterfield, "The Grange," *Forum,* XXXI (April, 1901), p. 236; Fred A. Shannon, *The Farmer's Last Frontier: Agriculture, 1860–1897* (Vol. V of *The Economic History of the United States,* Henry David and others, eds., New York, 1945), p. 276.

[36] *Decatur* (Ill.) *Daily Republican,* January 19, 1883, p. 3; Illinois State Grange, *Proceedings,* 1890, p. 28; *Western Rural,* XXVIII (January 11, 1890), p. 21; *Chicago Tribune,* October 6, 1887, p. 9.

only the failure to produce immediately a successful program for the teaching of agriculture to farm boys.

In Mississippi, for example, when the state sold its land-grant scrip it gave to the University of Mississippi that portion of the fund that was to be used for the instruction of white boys. The university made an effort to teach agriculture, but results were so poor that in 1876 school authorities discontinued even the pretense of providing education for future farmers. Grangers across the state were outraged and they demanded the establishment of a separate college. At the state grange meeting of 1877 the society stated that it would accept "no further delay nor frittering away" of the agricultural college fund, and it was strongly implied that legislators who ignored the rural demands would do so at their peril. Apparently the message was clearly understood; the next year the legislature chartered the Agricultural and Mechanical College that was later established at Starkville.[37]

A similar situation developed elsewhere. In Rhode Island grangers were successful in wrestling the land-grant fund from the aristocratic Brown University and transferring it to what would become the University of Rhode Island. The University of North Carolina fell under Grange attack as early as 1876, but it was in 1887 that the legislature, moved at least in part by continuing rural agitation, created North Carolina State College. Connecticut grangers waged a struggle with Yale University's Sheffield Scientific School, complaining especially that the institution's admission standards were prohibitive. To remedy the situation, grange officers proposed that a sizeable portion of the land-grant fund be transferred to Storrs Agricultural College, a private school with less restrictive entrance requirements. Several years later, partly in response to granger demands, the legislature made Storrs a state school, thereby creating the foundations for the University of Connecticut.[38]

[37] Ferguson, "Grange and Farmer Education," pp. 502–503. The quotation appears in John K. Bettersworth, *People's College: A History of Mississippi State* (University, Ala., 1953), p. 20.

[38] Herman F. Eschenbacher, *The University of Rhode Island: A History of Land-Grant Education in Rhode Island* (New York, 1967), pp. 18, 28; Armstrong, "Agricultural Education in North Carolina," pp. 24, 32–34, 39, 61; Walter Stem-

Organized farmers played similar roles in other cases where separate agricultural colleges were established. In New Hampshire, Kentucky, Oregon, and South Carolina, grangers and alliancemen raised a clamor that helped to bring about the goal they sought.[39]

In several instances rural opposition to the established universities failed to produce a separate college for agriculturists but led instead to some sort of compromise. It was the Farmers' Alliance that spearheaded an assault on the University of Minnesota. In 1885 the organization discovered that the institution at Minneapolis was ignoring its responsibilities for agricultural instruction and was misusing land-grant funds. Moreover, according to outraged alliancemen who labeled the university an "utter failure," agricultural students were ridiculed by scholars in other fields. The upshot was that in 1886 by an overwhelming vote the state alliance demanded the establishment of a separate agricultural college. The Grange was less militant but it joined in the general hue and cry.

Minnesota authorities rejected the agrarian demands but they recognized that something would have to be done to placate rural taxpayers. Consequently, while instruction in agriculture was left in the hands of the university, a new campus for farm students was established at St. Anthony Park, some two miles across the prairie from the university. In addition, the state launched a farmers' institute program and established a school of agriculture for pupils of less than college grade.[40]

In Wisconsin it was the Grange that carried on a bitter attack on the university in the 1880s, charging discrimination against farmers

mons, *Connecticut Agricultural College: A History* (Storrs, 1931), pp. 56–57, 64–65.

[39] Ross, *Democracy's College*, pp. 80–82.

[40] Andrew Boss, *The Early History and Background of the School of Agriculture at University Farm, St. Paul* (St. Paul, 1941), pp. 32–33; James Gray, *The University of Minnesota, 1851–1951* (Minneapolis, 1951), pp. 96–99; John D. Hicks, "The Origin and Early History of the Farmers' Alliance in Minnesota," *Mississippi Valley Historical Review*, IX (December, 1922), pp. 204–205; *Ariel*, IX (March 15, 1886), pp. 83–84; *St. Paul Pioneer Press,* February 26, 1886, p. 6; *ibid.*, February 5, 1887, p. 5; O. C. Gregg to W. W. Folwell, April 9, 1914 and January 28, 1915, William Watts Folwell Papers (University of Minnesota Archives, Minneapolis).

on the board of trustees and against agricultural students on the campus. The atmosphere at Madison, grangers said, was aristocratic, and they seized upon the fact that since its establishment the agricultural school had produced only one graduate. Upon examination grangers discovered that entrance requirements included preparation in algebra, plane and solid geometry, and natural philosophy, subjects with which few farm boys could be expected to be familiar. Moreover, the grangers contended that college authorities urged boys who went to Madison to study agriculture to go into other fields. In conclusion farm spokesmen believed that their university was simply modeled after the "old priests' schools," and in 1883 the state grange formally demanded that a new college be established for agriculture and the mechanical arts.

Although Professor W. A. Henry, who was in charge of agricultural instruction at the university, personally supported the granger demands, college authorities sought a compromise that would preserve the unity of the university and satisfy the farmers. When a bill providing for a separate agricultural college failed in the legislature, the university moved. It ordered Henry to institute a short course for farmers and threatened him with removal if he continued to talk in terms of a new agricultural college. Finally, the university cooperated with a number of leaders in the state who were in the process of beginning institute work. These innovations in time proved to be popular, and most farm leaders became supporters of the university. Still, as late as 1890 the state grange was calling for the establishment of a separate school.[41]

Something of a similar nature occurred in West Virginia. There the agricultural college of West Virginia was established in 1867, but the next year the institution obtained its modern name, and its leaders proceeded to develop a classical college of the traditional type. When the Grange achieved power in the state it sought the transfer of the land-grant fund to the Jefferson County Agricultural College, an institution chartered in 1875. Failing in that endeavor,

[41] Glover, "Agricultural College Crisis," pp. 17–20; Wisconsin State Agricultural Society, *Transactions*, XXI (Madison, 1884), pp. 183–186, 195–199, 207–210; *Chicago Tribune*, December 11, 1890, p. 2.

in the mid-1880s the Grange demanded a splitting of West Virginia University. Insufficient support in the legislature blocked that proposal, so the organization shifted its attention to reforms within the existing school. The appointment in 1891 of Thomas C. Atkeson, a prominent granger, as professor of agriculture led to the laying out of a course of study. The first graduate received his bachelor of agriculture degree in 1893.[42]

Quite obviously, some of the criticism directed toward the schools was irrational, anti-intellectual, and simply bad tempered. On at least one occasion, for example, the farmers' alliance in New York demanded that the state appropriate no more money for higher education of any kind, contending that the state had no obligation to provide education beyond the common-school level. Professors at institutions supported by the state, the farmers charged, were little better than parasites. A recent historian of the University of Maryland has called the rural criticism of that institution ignorant and vindictive with elements of a class struggle. Without doubt, grangers and alliancemen lacked almost completely any understanding of the value and place of education in the liberal arts. In a number of cases their bitter opposition to changes in the names of institutions displayed a distressing tendency toward irrationality.[43]

Still, in balance rural groups were powerful and useful friends of the agricultural colleges. Often grangers added their voices to demands for larger state appropriations, occasionally going so far as to set out plainly how much money was needed. In Michigan the Grange raised a howl when in 1879 Professor Charles L. Ingersoll decamped for the greener fields of Purdue. His loss, said the grangers, was due to the failure of the legislature to support the school adequately. Later the Grange denounced the legislature for appropriating funds for the wrong purpose; tax dollars should not be spent for

42 William D. Barns, "The Influence of the West Virginia Grange upon Public Agricultural Education of College Grade," *West Virginia History*, IX (January, 1948), pp. 128–138; Atkeson and Atkeson, *Pioneering in Agriculture*, pp. 129–131.

43 *Cultivator and Country Gentleman*, XLII (September 13, 1877), p. 584; Callcott, *University of Maryland*, pp. 181–182; Ross, "The 'Father' of the Land-Grant College," p. 184; University of New Hampshire, *History of the University of New Hampshire* (Durham, N.H., 1941), p. 188.

the construction of a college gymnasium "where the delicate sons of wealth may become athletes."[44]

In a like fashion the Grange and the Alliance expressed their views concerning the Morrill Act of 1890 which provided an additional endowment to each state for resident instruction. The measure was subjected to close scrutiny by organized farm groups. They were determined that the funds would not be used for "ordinary college training in belles-lettres and the dead languages." The curriculum restrictions written into the law reflected very well the aims of the Grange.[45]

Nor were the Grange and other organizations hesitant to advise college authorities concerning the administration of the colleges. The curriculum came in for consideration as grangers sought to make it ever more practical. In the main the organized farmers favored the retention of the manual labor system and they supported an extension of the use of elective courses, believing that each student should be allowed to select those courses which would be most useful to him.

Grangers especially tended to be friends of coeducation. In a number of states—including Minnesota, Michigan, and Connecticut—farmers called for the land-grant schools to admit young ladies, claiming that rural girls badly needed training in "household economy."

In Mississippi, on the other hand, grangers supported the establishment of a state institution for girls. Partly as a result of the organization's agitation in the early 1880s, the legislature in 1884 chartered the Industrial Institute and College, an institution for farm girls that was established in Columbus and that in due course became Mississippi State College for Women.[46]

---

[44] Mississippi Patrons of Husbandry, *The State Grange and the A. and M. College* (n.p., n.d.), pp. 2–3; James S. Ferguson, "Agrarianism in Mississippi, 1871–1900: A Study in Nonconformity" (Ph.D. Thesis, University of North Carolina, 1952), pp. 187–188; Madison Kuhn, *Michigan State: The First Hundred Years, 1855–1955* (East Lansing, 1955), p. 101; Fred Trump, *The Grange in Michigan* (Grand Rapids, 1963), p. 28.

[45] Ross, *Democracy's College,* pp. 178–179; Eddy, *Colleges for Our Land and Time,* pp. 100–102.

[46] Dennis S. Nordin, "The Educational Contributions of the Patrons of Hus-

Organized farmers also used their influence to lower the entrance requirements at the land-grant colleges. Arguing that secondary schools were rarely available to rural children, Grange spokesmen contended that the colleges should admit students directly from the primary grades. There is little doubt that the Grange did make some contributions in that area; the preparatory departments that sprang up in many of the colleges was one result. Elsewhere less rigid admission policies testified to the presence of rural pressures.[47]

The Grange and the Alliance groups made some contribution to the Hatch Act of 1887. In the 1870s the Grange began to talk about the need for experiment stations, and its agitation was instrumental in the establishment of pioneer stations in Ohio, New York, and Wisconsin. When the movement to obtain federal assistance arose in the 1880s, the Grange urged the passage of the Carpenter and Holmes bills. Later, at its 1886 meeting, the National Grange asked Congress to modify the Hatch-George bill to allow those stations that were independent of agricultural colleges to share in the benefits of the measure and to bestow the grant upon boards of agriculture or similar agencies in those states where the colleges had manifestly failed to fulfill the requirements of the Morrill Act. The first request found its way into the final version of the Hatch Act.[48] Among Alliance leaders Milton George was one who consistently added his voice to the clamor for federal assistance in the establishment of a nationwide experiment stations system.[49]

The agrarian crusade also contributed in a major way to the farmers' institute movement, the first popular technique for di-

---

bandry, 1867–1900" (M.A. Thesis, Mississippi State University, 1965), pp. 99–102; Ferguson, "Agrarianism in Mississippi," pp. 189–190.

[47] Nordin, "Educational Contributions of the Patrons of Husbandry," p. 101.

[48] Alfred C. True, *A History of Agricultural Experimentation and Research in the United States, 1607–1925* (Washington, 1937), pp. 94–99; AAACES, *Proceedings*, 1889, pp. 41–42; Gould P. Colman, *Education and Agriculture: A History of the College of Agriculture at Cornell University* (Ithaca, N.Y., 1963), pp. 70–71; Vernon Carstensen, "The Genesis of an Agricultural Experiment Station," *Agricultural History*, XXXIV (January, 1960), p. 18; Nordin, "Educational Contributions of the Patrons of Husbandry," pp. 45–50; Ross, *Democracy's College*, p. 140.

[49] *Western Rural*, XXIII (February 7, 1885), p. 85.

rectly reaching the mass of ordinary farmers. As early as 1875 a rural leader in Illinois asked, "Teachers have institutes, why not farmers?"[50] Later by resolutions and other actions, practically every organized group asked for the creation of an institute system in their state or region. In Colorado grangers demanded that the "professors give . . . plain talks on plain subjects," a demand that pointed directly to the inauguration of institutes. Among the resolutions adopted at the meeting of the Illinois State Grange in 1883 was one that asked for the establishment of institutes in that state. Samuel Sinnett, later a Farmers' Alliance member in Iowa, urged Milton George to make the promotion of institutes one of the purposes of his organization.[51]

The Grange was practically a partner in establishing and maintaining farmers' institutes. The greater interest of the Grange in education, combined with a realization that institute speakers in essence performed the same functions as did grange lecturers, made the organization an avid supporter. According to the Minnesota State Grange in 1893, the "Farmers' Institute is a high school for farmers . . . that is of incalculable benefit. . . ." A few years later the same body hailed the institutes as "the best means of instruction within reach of many thousands of the farmers. . . ." Across the nation grange organizations on all levels voiced similar sentiments.[52]

The Alliance, too, saw a value in institutes. In Virginia, according to the historian of the Alliance in that state, the "institutes were a further example of the educational influence of the Alliance, applied in this instance to technical advance in farming methods. . . ."[53]

[50] *The Golden Era* (McLeansboro, Ill.), March 5, 1875, p. 1.

[51] Michael McGiffert, *Higher Learning in Colorado: An Historical Survey, 1860–1940* (Denver, 1964), p. 98; *American Agriculturist*, XLVI (April, 1887), p. 173; *Decatur* (Ill.) *Daily Republican*, January 17, 1883, p. 4; *Western Rural*, XVII (December 27, 1879), p. 409.

[52] Minnesota State Grange, *Proceedings*, 1892 (Minneapolis, 1893), pp. 21–22; *ibid.*, 1899, p. 47; Lois G. Aldous, "The Grange in Kansas since 1895" (M.A. Thesis, University of Kansas, 1941), p. 58; Paine, *Granger Movement in Illinois*, p. 46; Thomas C. Atkeson, *Semi-Centennial History of the Patrons of Husbandry* (New York, 1916), pp. 226–227.

[53] William D. Sheldon, *Populism in the Old Dominion: Virginia Farm Politics, 1885–1900* (Princeton, N.J., 1935), p. 44.

In numerous cases organized farmers joined with agricultural college personnel and with state boards of agriculture to begin institute work or were influential in inducing the agencies to do so. For instance, the Ithaca, New York, Farmers' Club held a meeting in 1877 which was attended by Cornell professors and which was something of a prototype for later institutes in the state. In Mississippi the Grange played a large role in the beginnings of institute work. In 1884 President Stephen D. Lee of Mississippi Agricultural and Mechanical College announced that upon "invitation from the Grange . . . a delegation of the College Faculty . . . will meet with them for the purpose of presenting papers and taking part in discussions pertaining to agriculture. . . ."[54] In Minnesota while the state farmers' alliance was attacking the university for its failure to provide adequate resident instruction in agriculture, the organization expressed the opinion that "we believe the holding of farmers' institutes in every county . . . to be one of the best methods of educating the people. . . ."[55]

After institute systems were in operation, organized farmers sought in various ways to improve them. Practically all farm groups were willing to urge legislators to increase appropriations for the work and in innumerable instances grangers and alliancemen proposed changes in the operation of institutes to make them more effective as a teaching device or to allow them to reach more farmers. It was the Grange in Minnesota that in the 1890s proposed that the institute management in that state hold sessions in conjunction with the state fair, an innovation that proved to be both effective and popular. Elsewhere in the state, as in every part of the nation, local granges made their halls available for institute meetings.[56] The next decade, after the militant stage of the farmers' organizational move-

---

[54] *Cultivator and Country Gentleman,* XLII (March 15, 1877), p. 167; *Southern Live-Stock Journal,* VIII (May 1, 1884), p. 1; Ferguson, "Grange and Farmer Education," pp. 504, 508; *Prairie Farmer,* LVIV (January 29, 1887), p. 73.

[55] Minnesota State Farmers' Alliance, *Resolutions Adopted at Minneapolis, February 25, 1886* (n.p., 1886), p. 13.

[56] (Springfield) *Illinois State Register,* March 8, 1891, p. 1; *Cultivator and Country Gentleman,* LVIII (October 26, 1893), p. 831; Aldous, "Grange in

ment had disappeared, the National Farmers' Congress added its voice to those who asked that the United States Department of Agriculture enlarge its role in the institute movement.[57]

Conversely, there were a few instances in which farm groups threatened to interfere with the operation of institutes or in which leaders tried to use them for their own benefit. In Minnesota Alliance members elected to the state legislature in 1890 seemed to believe that institute lecture forces should include at least one allianceman who at meetings would discuss economic issues and indeed would seek to use the gatherings to enlist more farmers in the organization, a proposal that met a blunt rebuff. Alabama's Reuben F. Kolb used the institutes to promote the Alliance in that state and to enhance his own political prospects.[58]

Agricultural educators generally recognized and appreciated the role of farm groups in strengthening the emerging institute movement. An institute worker in New Jersey observed in 1901 that in reality the Grange and the institutes were working toward the same end and that cooperation was both natural and desirable. Oren C. Gregg of Minnesota, superintendent of one of the most successful institute systems in the country, reported that the Grange was of considerable assistance in his work. Echoing the Minnesotan's sentiment, C. D. Smith of Michigan found that granges and other organizations furnished institutes with valuable local talent, besides providing members with the parliamentary knowledge needed by institute chairmen. Many institute workers observed that institutes were most successful in those neighborhoods where rural organizations existed. There, farmers were more alert and less inclined to approach the institute in a critical mood. According to F. E. Dawley of New York, a blind institute lecturer would have been able to

Kansas," p. 59; Minnesota State Grange, *Proceedings*, 1892, p. 28; *ibid.*, 1898, p. 5; Illinois State Grange, *Proceedings,* 1890, pp. 66–67.

[57] John Hamilton to A. C. True, October 2, 1905, Office of Experiment Stations Records.

[58] *Farm, Stock, and Home,* VII (January 15, 1891), p. 66; William W. Rogers, "Reuben F. Kolb: Agricultural Leader of the New South," *Agricultural History,* XXXII (April, 1958), p. 116.

ascertain by the reaction of the audience whether he was in such a neighborhood.[59]

Finally, the agrarian crusade coincided with and contributed to a dramatic change in the farmer's view of himself and of his place in society. In the late nineteenth century most farmers were devotees of what Richard Hofstadter has called the agrarian myth. Convinced that agriculture afforded a superior way of life, farmers saw themselves as true democrats, independent of baser elements of society, producing for their own livelihood, and largely immune to the realities of the marketplace. Although such views were counter to the facts, except where geography and the lack of markets made subsistence farming and a pioneer existence inevitable, farmers continued to see themselves as simple and honest tillers of the soil. It was, in fact, a view that enjoyed tremendous vogue among politicians and editorial writers at a time when farmers comprised the majority of the nation's population.[60]

But the failure of the agrarian crusade in its political phase to reverse the trend in the nation toward industrialization and urbanization pointed out very clearly that there was to be no return to the simple existence that the agrarian myth envisioned. Results at the ballot box indicated that rural welfare in the future would rest upon prices, costs of production, and efficiency. Only one course remained for farmers, and that was to forget the past, look to the future, and adopt those ideas that promised to help them fit into the new order of things.

Thus it was that when forward-looking farm periodicals like the *Prairie Farmer* talked in 1868 of agriculture being a business like other businesses, it was a voice crying in the wilderness. But little more than three decades later the *Cornell Countryman* could proclaim that the "object of farming is . . . to make money" and that agriculture "is to be conducted upon the same business basis as any other producing industry." Nor was it an accident that former populist leaders would become advocates of farmer institutes and that

[59] AAFIW, *Proceedings,* 1901, pp. 45–46.

[60] The agrarian myth is set forth in Richard Hofstadter, *The Age of Reform from Bryan to F. D. R.* (New York, 1960), pp. 23–59.

one of those institutes could be described as "a business meeting for business men."[61] In short, by 1900 farmers were ready for a revolution in methods if a successful technique could be found to take agricultural science to them.

[61] U.S. Department of Agriculture, *Yearbook of Agriculture*, 1940 (Washington, 1940), pp. 144–145; Illinois Farmers' Institute, *Annual Report*, 1902, p. 338.

III. ORIGINS
OF THE INSTITUTE MOVEMENT

THE FARMERS' INSTITUTE MOVEMENT stemmed from diverse sources. In some instances, as in New England, institutes represented merely an expansion of the functions of the state agricultural societies or boards of agriculture. From the outset the annual meetings of these agencies had been educational in purpose; sessions held at different points in a state were simply a logical next step. Elsewhere, in such states as Missouri, state boards of agriculture assumed the responsibility for managing separate institute systems when popular demand for such work appeared. In other cases institutes sprang from the newly established agricultural colleges as the administrators of those institutions sought techniques by which they might reach the ordinary farmer. In some of the newer states the agricultural experiment stations, faced with essentially the same farmer distrust and neglect as the colleges, turned to institute work as one means of justifying their existence. In any event institutes could not succeed until at least a few farmers in a given state were interested in them, and the first attempts to establish extension programs through the use of the institute technique more often than not resulted in failure.

The dissemination of agricultural knowledge through itinerant lecturers or open meetings of agricultural societies may be traced to

64

the pre–Civil War period. In 1824 Stephen Van Rensselaer was reported to have employed a speaker to travel throughout rural New York, giving lectures on natural history. More important were the open lectures and discussions held by agricultural and other societies, such as those of the Philadelphia Society for Promoting Agriculture which opened its doors to nonmembers in 1856. Of some significance were the weekly open sessions of the Massachusetts Legislative Agricultural Society that began in 1839. At these gatherings, to which ordinary farmers were invited, members of the state legislature heard talks by such figures as Daniel Webster, Professor Benjamin Silliman, and Henry Colman, then commissioner of the Agricultural Survey of Massachusetts.[1]

As early as 1842 or 1843 the New York State Agricultural Society began a program of itinerant lectures. Secretaries of the group, including such men as Daniel Lee, Joel B. Nutt, and Benjamin P. Johnson, assumed responsibility for visiting and addressing some of the county agricultural societies in the state.[2]

In Ohio in February, 1845, Dr. N. S. Townshend suggested that an appropriate state agency be empowered to employ lecturers competent in the various sciences pertaining to agriculture and to send the men out to speak to farmers throughout the state. The next year the Ohio State Board of Agriculture recommended the establishment of neighborhood clubs and urged knowledgeable farmers to make themselves available as lecturers to those groups that might request their services. Four years later, former governor Allen Trimble, then the president of the state board, called upon the state agricultural chemist and the corresponding secretary of the state board to deliver lectures at different points in Ohio, a task that the officials politely refused, citing lack of time.

Elsewhere in Ohio there were early short-lived lecture programs. In 1847 the Lorain County Agricultural Society offered to send lec-

[1] Stemmons, *Connecticut Agricultural College*, pp. 230–231; Fletcher, *Pennsylvania Agriculture and Country Life*, p. 444; *Experiment Station Record*, VII, p. 636.

[2] Bailey, *Annals of Horticulture*, 1891, pp. 137–138; *Cultivator and Country Gentleman*, LII (November 3, 1887), p. 838.

turers to any town in the county requesting them and willing to provide a hall for their use. Under this arrangement sessions were scheduled at twenty-one points before the program died out from lack of support. Later, in December, 1854, a system of independent agricultural lectures lasting for three months was launched at Oberlin College. James Dascomb, N. S. Townshend, and James H. Fairchild spoke on a variety of topics pertaining to agriculture, including chemistry, feeding and breeding of stock, veterinary medicine and surgery, geology and botany, rural architecture, and farm bookkeeping. When interest was less than had been hoped, promoters in 1855 and 1856 shifted the lectures to Cleveland, only to discontinue them because of lack of support.[3]

These early efforts indicated that education enthusiasts were grasping in the dark. Rural leaders recognized full well that ordinary farmers needed instruction in better methods, but there appeared to be no satisfactory method of reaching into the countryside. Needed was some means by which a sizeable portion of the rural population in any given area could be contacted in a systematic way.

It was in New England that farmers' institutes appeared in the general form that they would take for decades. The first mention of institutes came in Massachusetts, January 21, 1859, when the state board of agriculture appointed a committee "to consider and report upon the propriety of instituting meetings similar to teachers institutes." Although the committee reported favorably, little came immediately of the proposal.[4]

Perhaps John A. Porter's lecture course for farmers at the Sheffield Scientific School helped to stimulate interest; in any event in 1863 the state board of agriculture began holding an annual itinerant meeting in different parts of the state, thereby allowing farmers who could not hope to attend the regular sessions to hear papers and to participate in discussions. These meetings continued for years, and after 1869 they were aided by an annual appropriation to pay the expenses of those who read papers in the sessions.[5]

[3] Ohio State Board of Agriculture, *Brief History*, pp. 25–28.

[4] Bailey, *Annals of Horticulture*, 1891, p. 138.

[5] *Experiment Station Record*, VII, p. 637; Allan Nevins, *The Origins of the Land-*

In 1871 the board asked the twenty-nine agricultural societies in the state to hold annual institutes in their localities. The statewide system was labeled The Farmers' Institutes of Massachusetts. When the local societies failed to act voluntarily, in 1878 the board offered its assistance to those groups willing to undertake the holding of institutes and the next year the board flatly directed those societies receiving support from the state to hold at least three meetings annually. The secretary of the state board was instructed to attend as many of the local gatherings as possible, and in due course his office began to provide some of the lecturers who spoke at those sessions.[6]

Comparable developments took place in Connecticut, Vermont, and elsewhere in New England. The Connecticut State Board of Agriculture, established in 1866, almost immediately began holding its annual meetings at different points in the state. A decade after its formation the agency was cooperating with farmers' clubs and other local groups in conducting numerous programs at smaller towns. Meanwhile, the Vermont State Board of Agriculture, Manufactures and Mining held an open meeting at St. Johnsbury in 1871; by the end of 1874 there had been at least thirty-seven such gatherings. Work of a similar nature began in New Hampshire at Concord in November, 1870, several months before the first session in Vermont. The state boards in Maine and Rhode Island waited until the 1880s, but once under way their programs differed little from those functioning elsewhere in New England.[7]

While state boards of agriculture were establishing the first institutes in New England, the land-grant colleges were the pioneering

Grant Colleges and State Universities (Washington, 1962), p. 15; Bailey, *Annals of Horticulture,* 1891, p. 139; *Cultivator and Country Gentleman,* XXXVIII (November 27, 1873), p. 761; *ibid.,* XL (November 25, 1875), p. 745; *ibid.,* XLII (September 6, 1877), p. 569.

[6] Bailey, *Annals of Horticulture,* 1891, pp. 138–139; *Experiment Station Record,* VII, p. 637; *Cultivator and Country Gentleman,* LI (February 25, 1886), p. 145.

[7] *Experiment Station Record,* VII, p. 637; *Cultivator and Country Gentleman,* XL (January 14, 1875), p. 27; *ibid.* (June 3, 1875), p. 345; *ibid.,* XLIX (May 22, 1884), p. 438; *ibid.,* XLVIII (March 15, 1883), p. 213; *American Agriculturist,* XXXIII (April, 1874), p. 140; Metcalf, *New Hampshire Agriculture,* p. 21; Clarence A. Day, *Farming in Maine, 1860–1940* (Orono, Maine, 1963), p. 205.

agencies in other areas, including the Middle West. These institutions found little favor with farmers, despite early hopes, and at the outset some of the schools created institutes to provide means by which they might come into contact with practicing farmers.

The Illinois Industrial University launched a series of "agricultural lectures and discussions" on the campus, January 12, 1869, only ten months after the institution opened its doors. Proposed by Willard C. Flagg, a prominent farmer from Madison County who served as corresponding secretary of the university's board of trustees, the lectures were modeled after those given at Yale in 1860. Programs consisted of three lectures followed by discussions each day, four days a week, for two weeks. Among the speakers were several professors and Regent John M. Gregory of the university; Norman J. Colman, editor of a well-known Missouri farm paper; and Sanford Howard, secretary of the Michigan State Board of Agriculture. The lectures dealt with broad and general aspects of farming, but their promoters conceived of them as early forms of extension. In opening the first session Regent Gregory said, "We inaugurate today a part of the plan of operations, contemplated from the outset, to extend the benefits of this university . . . out into the fields of adult life and of adult labor." The lectures, according to the regent, represented an important phase of the university's mission, "the diffusion of agricultural science among mankind."[8]

First reactions were favorable; approximately seventy university students attended the lectures, along with a considerable number of Champaign County residents including many farmers. Some came from distant parts of the state. So significant, in fact, did the lecture program appear that the university trustees devoted almost two-thirds of their annual report to recording the lectures and the discussions that followed.

These early hopes were dashed the next year, and innovations

---

[8] Illinois Industrial University, *Second Annual Report of the Board of Trustees, 1868–69* (Springfield, Ill., 1869), pp. 120–123; Fred H. Turner, "The Illinois Industrial University" (Ph.D. Thesis, University of Illinois, 1931), pp. 726, 764–765, 768; Illinois Farmers' Institute, *Annual Report*, 1904, pp. 243–244; *Cultivator and Country Gentleman*, LIV (March 7, 1889), p. 186.

introduced later failed to save the program. When the second series of lectures, held at the university in January, 1870, attracted no more than thirty persons, mostly university students, the college administrators resolved to take the programs to the farmer, rather than to wait for him to come to Urbana. The next year sessions met not only at the university but also at Springfield, Pekin, and Cobden. In 1872 conclaves were held at five points in the state, and in 1873, the final year, lecturers spoke in seven towns. The usual arrangement was for the university to provide the lecturers, while the community furnished a properly lighted and heated hall. Attendance was uniformly less than satisfactory, and by 1873 university administrators were convinced that its personnel could better devote their time to resident instruction. In addition, Willard C. Flagg, the moving spirit in the project, had become involved with the Illinois State Farmers' Association, a politically oriented wing of the granger movement. Some feared that continuation of the program might imperil the university's already difficult relations with the state legislature.[9] The first attempt by the land-grant college in Illinois to take agricultural knowledge to the farmers of the state collapsed in complete failure.

No more lasting were early lecture programs undertaken by the agricultural colleges elsewhere. In Kansas the first step came June 23, 1868, when the board of regents of the state agricultural college, acting in response to a resolution introduced by Elbridge Gale of Manhattan, directed the faculty members to visit different points in the state to give farmers information concerning agricultural principles. That the project was expected to benefit the college as well as the farmers was suggested by a provision of the resolution directing the speakers to impress upon farmers the "aims and character of the State Agricultural College." College speakers appeared before

[9] Illinois Industrial University, *Annual Report of the Board of Trustees,* 1868–69, pp. 120, 122–361; *ibid.,* 1870–71, pp. 139–141; *ibid.,* 1871–72, pp. 163–165; *ibid.,* 1872–73, pp. 176–177; *Prairie Farmer,* XLIV (January 18, 1873), p. 20; Turner, "Illinois Industrial University," pp. 783, 868–869; Illinois Farmers' Institute, *Annual Report,* 1904, p. 245; *Cultivator and Country Gentleman,* LIV (March 7, 1889), p. 186.

a meeting of the Union Agricultural Society in Manhattan, November 14, 1868, in what is considered to have been the first farmers' institute in Kansas. Professors from the college enlightened members on such topics as "Tree Borers" and "Economy on the Farm." A two-day affair was held at Wabaunsee, November 20–21, and thereafter for the next few years a few meetings were held at scattered points in the state.

More important in Kansas was a lecture program, similar to that in Illinois, held in Manhattan annually from 1869 to 1873. These were four-day gatherings in January of each year. They featured out-of-town speakers, including such persons as President A. S. Welch of the Iowa agricultural college. In 1872 the program was reported to have been successful "in some respects," but continuing light attendance showed clearly that Kansas farmers in the early 1870s were not sufficiently interested to leave their homes and travel to Manhattan. When political controversies further complicated matters, the board of regents late in 1873 voted to abandon the whole project.[10]

The Iowa State Agricultural College, under the leadership of President Welch and Professor Isaac P. Roberts, took up institute work in the winter of 1870–71. The first of the three-day affairs opened at Cedar Falls on December 20, 1870; later, similar conclaves were held at Council Bluffs, Washington, and Muscatine. At the latter point farmers attending the gathering took the opportunity to form a rural club, while at Washington 250 persons turned out. These results were sufficiently gratifying to cause President Welch to schedule seven institutes for the 1871–72 winter, but following that season the work seems to have died, and institutes were not placed on a solid footing in Iowa until 1887.[11]

[10] Julius T. Willard, *History of the Kansas State College of Agriculture and Applied Science* (Manhattan, Kans., 1940), pp. 32–33; Kansas State Agricultural College, *Annual Report*, 1868, p. 8; *ibid.*, 1869, p. 13; *ibid.*, 1870, p. 6; *ibid.*, 1872, p. 14.

[11] Iowa State Agricultural College, *Fourth Biennial Report,* 1871 (Des Moines, 1872), p. 49; *Prairie Farmer,* XLIII (January 20, 1872), p. 17; Morgan, *Extension Service of Iowa State College,* pp. 12–14; Earle D. Ross, *A History of the Iowa State College* (Ames, Iowa, 1942), p. 165.

In spite of the rather unpromising results of these first beginnings, other land-grant colleges in the 1870s and early 1880s continued to search for some means by which they might reach the ordinary farmer. In Colorado, for example, the agricultural college held in its chapel a meeting for farmers in November, 1879, less than three months after the institution had opened its doors. Allen R. Benton, the first chancellor of the University of Nebraska, chose the second annual report to the board of regents to suggest that faculty members be sent out to talk to farmers. By November, 1873, professors were in the field, discussing agriculture and imploring Nebraska farmers to send their sons to Lincoln.[12] In 1881, Edward D. Porter, professor of agriculture at the University of Minnesota, hit upon the idea of holding a lecture course on campus for farmers, although a similar project had been a flat failure six years earlier. Porter mapped out a four-week course, with two lectures a day for four days each week, the lectures to be of a "practical" nature, calculated to appeal to ordinary farmers. A total of 225 persons appeared for the first course which opened January 31, 1882. Since the attendance was 400 percent greater than the minimum set by the regents, similar programs were held in 1883 and 1884 when numbers attending rose to 308 and 1,181. Still, in each instance, the majority of those present resided in or near Minneapolis and St. Paul, and Porter concluded somewhat reluctantly that lecture courses conducted on the campus constituted no final answer to the problem of reaching common farmers.[13]

Michigan Agricultural College was the first land-grant school to develop a system of institutes in essentially the form that would continue into the twentieth century. As late as 1875 the college at East

[12] Alvin T. Steinel and D. W. Working, *History of Agriculture in Colorado* (Fort Collins, 1926), pp. 589, 593, 611; McGiffert, *Higher Learning in Colorado,* p. 63; Robert P. Crawford, *These Fifty Years: A History of the College of Agriculture of the University of Nebraska* (Lincoln, 1925), pp. 34–36; *Cultivator and Country Gentleman,* XXXVIII (November 27, 1873), p. 761.

[13] Executive Committee of the University of Minnesota Board of Regents, Minutes, December 17, 1881 (University of Minnesota Archives, Minneapolis); Boss, *School of Agriculture,* pp. 30–31; University of Minnesota Board of Regents, *Fourth Biennial Report, Supplement I* (St. Paul, 1887), pp. 19, 33–34, 36.

Lansing was small, so little known that some neighborhood farmers thought it was a mental institution, and it was generally ignored by the state legislature. Professor Manly Miles, who had served as a lecturer in the programs at the Illinois Industrial University, suggested to President Theophilus C. Abbot that the college undertake something of a similar nature. There was early legislative approval of such a step; in 1861 the institution had been authorized to offer instruction by means of lectures to persons not students at the college.

Faculty members appeared at the towns of Armada and Allegan on January 11, 1876, to launch institute work in Michigan. The first meetings were two-day, five-session affairs, featuring not only college speakers but local talent as well. Early in the work it was made plain that institutes would be held only where a local group, such as a grange or farmers' club, was prepared to act as the sponsoring agency and to provide a hall and a part of the lecture force. The state board of agriculture participated by paying the expenses of the faculty members and by printing in its annual publication the reports of the meetings. In all, six institutes were held in as many places in January, 1876. During the next twenty years, the work continued with little change, except that the number of institutes gradually increased to twenty or more a year and the state board of agriculture increased its annual outlay to $2,000.[14]

By 1880 the concept of the institute as a teaching technique was established in New England and in Michigan, but elsewhere little had been done. The next twenty years saw the flowering of the institute movement and its expansion into every corner of the nation. Moreover, the institute systems created in the various states in the two decades before 1900 in most instances survived until other teach-

[14] W. J. Beal, *History of the Michigan Agricultural College* (East Lansing, 1915), p. 158; Bailey, *Annals of Horticulture*, 1891, pp. 139–194; Kenyon L. Butterfield, "Farmers' Institutes," *The Chautauquan*, XXXIV (March, 1902), p. 638; Michigan State Board of Agriculture, *Fourteenth Annual Report*, 1875 (Lansing, 1876), pp. 72–77; *Cultivator and Country Gentleman*, XL (September 30, 1875), p. 617; American Association of Farmers' Institute Managers, *Report of the Meetings Held at Watertown, Wisconsin, March 13, and at Chicago, October 14, 1896* (Lincoln, Nebr., 1897), pp. 16–17.

ing techniques eliminated the need for institutes. Ohio launched its system in 1880, Maine and Missouri in 1883, and Mississippi in 1884. The board of regents of the University of Wisconsin in 1885 organized that state's institutes and appointed a superintendent to manage them. New York adopted the technique the next year, and Minnesota followed in 1887. Several other states took their first steps in institute work in the 1890s.[15]

In New York, where one of the most successful systems emerged, the institute movement was the result of cooperation between the agricultural college at Cornell and the state agricultural society. When Isaac P. Roberts left Iowa to go to Cornell in February, 1874, he discovered that "farmers of the state had no vital interest in their college of agriculture." Since Roberts had participated in the pioneer institutes in Iowa, he concluded that similar meetings might serve a useful purpose in New York. Under his leadership the Ithaca farmers' club staged an institute in February, 1877, but when response was poor, Roberts abandoned his efforts until the middle of the next decade.

Conditions had changed by the 1880s. New York agriculture was in depression due in part to poor farming practices. Concerned with the deterioration that was evident on every hand, Professor Roberts, President Charles K. Adams of Cornell, and J. S. Woodward of Lockport, a prominent member of the Western New York Horticultural Society, resolved to make a second effort. The institute met February 16–18, 1886, on the Cornell campus.

The affair was well attended, and its sponsors announced that it would become an annual event. Moreover, influenced by Woodward's enthusiastic reports of the Cornell gathering, the state agricultural society agreed to provide $1,050 for similar meetings elsewhere. Finally, in 1887 the state legislature did its part, appropriating $6,000. These funds allowed the state agricultural society, aided by lecturers from Cornell, to hold twenty institutes in 1887–88 and about forty the next winter.

[15] *Experiment Station Record*, VII, p. 638; Liberty H. Bailey, "Farmers' Institutes: History and Status in the United States and Canada," U.S. Office of Experiment Stations, *Bulletin* 79 (Washington, 1900), p. 31.

Expansion and systemization came in the 1890s. Appropriations rose to $20,000 by 1898. The state agricultural society managed the institutes until 1893, when the state department of agriculture was created and the institutes were placed within its jurisdiction. J. S. Woodward supervised the program until 1891, and in 1896 F. E. Dawley, who in time would become one of the nation's best-known institute workers, became director.[16]

The institutes in Pennsylvania were an outgrowth of the state board of agriculture, an agency created in 1876 to administer certain regulatory laws pertaining to agriculture. Almost immediately farmers began to demand that the board undertake a statewide educational program. The first institute met at Harrisburg, May 22, 1877, followed by a few others annually during the next decade. Upon the county representatives who served on the board of agriculture fell the primary responsibility for the management of these early institutes. Those individuals organized and presided over institutes in their counties, aided by speakers from the state agricultural college and by outstanding local farmers. Not until 1885 did the legislature give the board an appropriation of $1,000 for its educational work, but later the financial support increased, reaching $7,000 in 1891 when eighty-one institutes were held. Still, by the 1890s it was apparent that the program needed better organization and more centralized administration.

In 1895 the legislature established the Pennsylvania Department of Agriculture and assigned to it the supervision of institutes. John Hamilton of Centre County became deputy secretary of agriculture and director of institutes, a post he held until 1899.

Under Hamilton's leadership, Pennsylvania developed the institute system that many considered to be the best in the United States. In order to minimize travel expenses Hamilton divided the state into

[16] D. P. Witter, "A History of Farmers' Institutes in New York," New York Department of Farms and Markets, *Bulletin* 109 (Albany, 1918), pp. 211-212, 216–218; New York Bureau of Farmers' Institutes, *Report*, 1908–09 (Albany, 1910), pp. 45–47; *Cultivator and Country Gentleman*, LII (January 27, 1887), p. 66; *ibid.* (October 6, 1887), pp. 766–767; *ibid.*, LIV (October 24, 1889), p. 809; Gould P. Colman, "A History of Agricultural Education at Cornell University" (Ph.D. Thesis, Cornell University, 1962), pp. 59–60, 76–79.

sections and assigned a corps of lecturers to each section, scheduling their appearances so that the speakers might move from one institute to the next without waste of time or funds. Each corps of lecturers consisted of a man from the state department of agriculture, one from the agricultural college, and a group of specialists who attended only those meetings at which their particular knowledge was useful. Local arrangements remained in the hands of local people, but Hamilton created in each county a local arrangements committee which included officers of granges and other farm groups there. Hamilton also exercised some control over lecture topics in an effort to make the sessions more meaningful. Under these plans and with an enlarged appropriation that amounted to $12,500 in 1899, the number of institutes increased rapidly. In 1898–99 there were 308 institutes in Pennsylvania with a total attendance of over 30,000.[17]

In the Middle West, Ohio established the first system of institutes that covered the state in a reasonably complete fashion and that with only minor alterations continued into the twentieth century. At a meeting of the state board of agriculture, September 14, 1880, Dr. William I. Chamberlain, who at that juncture was secretary of the state board, asked permission to address or to send other speakers to meetings of farmers' clubs, granges, or other local rural groups during the coming winter. Such a plan, he pointed out, was in operation in Michigan where it was producing good results. Impressed, the board not only gave its approval but it also provided $1,000 from the earnings of the state fair. President Edward Orton of the state agricultural college sought and received permission for faculty members to serve as lecturers. In launching the undertaking Secretary Chamberlain called upon intelligent farmers to interest themselves and others in the project. He offered to hold an institute in any county where there was sufficient interest to induce local groups to agree to provide a hall, advertising, and entertainment for the

[17] Fletcher, *Pennsylvania Agriculture and Country Life,* p. 444; Pennsylvania Department of Agriculture, *Annual Report,* 1904 (Harrisburg, 1905), p. 42; *American Agriculturist,* LVIII (July 18, 1896), p. 44; *Cultivator and Country Gentleman,* LXII (February 25, 1897), p. 144; Bailey, "Farmers' Institutes," p. 25.

lecturers while they were in the community. The first session under these arrangements opened at Wooster, November 19, 1880. Twenty-six other institutes were held in the course of the winter.

During the first decade of institute work in Ohio, the number of meetings rose steadily but not spectacularly, reaching 62 in the winter of 1889–90. Farmer reaction was by no means always favorable, but it was sufficiently friendly to induce the legislature in 1890 to provide a firmer foundation for the work. The new law called for the establishment in each county of an incorporated society that would assume the responsibility for making all local arrangements. The measure also provided for the allocation from the general funds of each county a per capita allowance of five mills, the total not to exceed $200. Of this amount, 40 percent went to the state board of agriculture, primarily for the payment of per diem and other expenses of lecturers who would be sent out to county institutes. The balance remained in the individual counties to enable the county groups to meet their expenses. Six years later, as the program proved its worth, the legislature raised the institute allowance per capita to six mills and the maximum figure per county to $240, dividing that amount equally between the county group and the state board. The better support and systemization afforded by these measures permitted a rapid increase in the number of institutes; standing at 124 in 1890–91, it doubled to 250 by 1898–99.[18]

Meanwhile, Michigan systematized its institute program. From 1876 through 1890 there were only six institutes a year in the state, and these were conducted in the main by faculty members from the agricultural college with financial aid from the state board of agriculture. In 1891 that aid was expanded, and the number of institutes rose to 22. An important innovation in 1895 was the requirement that twenty or more persons in each county wanting institutes organize themselves into a county institute club to constitute the local group believed to be necessary for proper management. The local agency supplied the hall, advertising, and one-half of the speak-

18 Ohio State Board of Agriculture, *Brief History*, pp. 29–31, 43, 64, 81; Burkett, *Ohio Agriculture*, pp. 208–209; *Breeder's Gazette*, XXVIII (December 11, 1895), p. 423; *American Agriculturist*, LVIII (July 18, 1896), p. 43.

ers. The state board of agriculture, still working through the agricultural college, provided the balance of the lecturers. Kenyon L. Butterfield, soon to be one of the leading agricultural educators in the nation, was appointed superintendent of institutes. In the winter of 1898–99, after the new system had been in operation only four years, there were 174 one and two-day meetings in the state.[19]

Institute work in Ohio and Michigan stimulated developments in Indiana. At a meeting of the state board of agriculture in January, 1882, speakers pointed to the success of institutes to the east and north. A committee instructed to look into the matter proposed that the state board and the agricultural college hold four institutes that year. Two months later, March 8, 1882, the first institute opened at Columbus, followed by one at Crawfordsville. Politics and other issues blocked further efforts until 1888 when members of the state board assumed the responsibility for holding institutes near their homes. At least two were held under this arrangement at Franklin and Anderson.[20]

These pioneer efforts showed, among other things, that an appropriation of sizeable proportions was necessary. When the state board of agriculture hesitated, fearing that the legislature might reduce its financial support of the board's other functions. Purdue University moved forward to take the leadership in the institute movement. An act approved March 9, 1889, appropriated $5,000 for the work and placed the management of the institutes in the hands of the board of trustees of Purdue University. The trustees promptly created a committee consisting of the president of the university, the director of the experiment station, and the professor of agriculture, who happened to be W. C. Latta, and named the latter superintendent of institutes. Under his leadership work was begun

[19] *American Agriculturist*, XXXIX (January, 1880), p. 7; American Association of Farmers' Institute Managers, *Report of the Meetings Held at Watertown, Wisconsin, March 13, and at Chicago, October 14, 1896*, pp. 18–19; Beal, *Michigan Agricultural College*, pp. 150, 158–161; U.S. Commissioner of Education, *Report, 1895–96* (2 vols., Washington, 1897), II, 1249; Bailey, "Farmers' Institutes," p. 17.

[20] William C. Latta, *Indiana Farmers' Institutes from Their Origin in 1882, to 1904* (Lafayette, Ind., 1904), pp. 5–11; *Breeder's Gazette*, XIII (January 11, 1888), p. 36; *ibid.*, XIII (March 21, 1888), p. 282.

on a statewide basis. There was a conscientious effort to work with local groups, and in due course county societies were created which took the responsibility for making all local arrangements. From Purdue University came overall supervision and a portion of the talent that appeared at the various institutes. The first season that the new system was in operation programs were held in only fifty counties, but expansion was steady and beginning in the 1893-94 season, there was at least one institute in each county.[21]

In Missouri, as in Ohio, the state board of agriculture launched institute work. But in Missouri there was no attempt to create and utilize county organizations. As early as 1869 Norman J. Colman, one of the earliest and most eloquent spokesmen for agricultural education in the Middle West, proposed that the Missouri State Board of Agriculture sponsor meetings in which "farmers [would] be called together and lectures delivered to them by men learned in agricultural science, like teachers' institutes."[22] Apathy and lack of funds doomed that proposal, but thirteen years later response was more favorable when Jeremiah W. Sanborn, professor of agriculture at the University of Missouri, suggested that the board hold institutes at those points in the state where local interest warranted the effort. The first institute in Missouri opened at Independence, December 29, 1882. Farmer interest was far from overwhelming, but the board persisted, and by the winter of 1890–91 demands for institutes exceeded the board's financial resources. Partly in response to these pressures, in the spring of 1891 the state legislature reorganized the state board of agriculture and increased its support to the extent that $5,000 became available for the institutes. They increased rapidly in number, so that by the first years of the new century almost 150 were held annually.[23]

[21] William C. Latta, *Outline History of Indiana Agriculture* (Lafayette, Ind., 1938), p. 280; H. E. Stockbridge to W. O. Atwater, August 14, 1889, Office of Experiment Stations Records; Latta, *Indiana Farmers' Institutes*, pp. 19–21; *Cultivator and Country Gentleman*, LIV (September 26, 1889), p. 727; Purdue University, *Twenty-Second Report*, 1896 (Indianapolis, 1897), pp. 19–20.

[22] Missouri State Board of Agriculture, *Annual Report*, 1869 (Jefferson City, 1870), pp. 8–10.

[23] *Ibid.*, 1882, pp. 16–19, 53–84; *ibid.*, 1890–91, pp. 20–21; *ibid.*, 1904, p. 15;

Institutes in Illinois were late in developing, and they never achieved the importance that they had in Indiana, Michigan, Wisconsin, or Minnesota. After the pioneer efforts of the Illinois Industrial University died in 1873, the school sought new methods to take knowledge to farmers in distant parts of the state. In 1875 the board of trustees authorized its corresponding secretary to cooperate with local groups in the holding of institutes. Promoters might call upon the university for lecturers, provided that such service was "without detriment to their work with their classes" and that the local communities assumed the burden of traveling and other expenses connected with institutes. Very few meetings were held under this arrangement. In fact, the university and the farmers drifted far apart in the late 1870s and remained alienated until the end of the 1890s. There was little understanding between them, and suspicion and distrust were common on both sides.

It was in this situation that early in 1882 President J. R. Scott of the state board of agricultural proposed that the state board fill the void. Specifically, he suggested that the board agree to send speakers, including some of its members, to institutes to be held annually in each of the state's congressional districts. The board accepted the plan, and meetings were held at Belleville and Decatur. The number increased slowly, and there were only nine in 1887. By that year the board had agreed to supply speakers at county institutes, and there were thirty-three of that type.[24] The first appropriation came in 1889 when the legislature gave $100 annually for each congressional district institute; two years later the legislature voted to grant county institute organizations $50 per year. To pro-

Levi Chubbuck to J. M. Rusk, January 8, 1889, Office of Experiment Stations Records; *Kansas City Journal*, December 29 and 31, 1882. For a more detailed account of the institutes in Missouri, see Roy V. and Jane B. Scott, "A Forgotten Phase of Agricultural Education: The Institutes in Missouri," *Mississippi Quarterly*, XIV (Fall, 1961), pp. 169–182.

[24] Illinois Industrial University, *Eighth Annual Report,* p. 105; *American Agriculturist,* XXXIX (January, 1880), p. 34; Illinois Farmers' Institute, *Annual Report*, 1904, p. 245; *Breeder's Gazette,* I (May 25, 1882), p. 675; *ibid.*, II (August, 17, 1882), p. 239; *ibid.*, XII (December 29, 1887), p. 1030; *Western Rural,* XXV (December 17, 1887), p. 823.

vide unity to the movement, in 1895 the legislature created a central organization and gave it $7,000 per year to hold an annual round-up institute, issue an annual report, and assist the county groups in their programs. Meanwhile, the state contribution to the county groups rose to $75 a year in 1899. Soon the movement became statewide, with institutes in every county.[25]

The Iowa institute system was unique in that it was completely decentralized. After the failure of the early efforts by the Iowa agricultural college, there were no further attempts to reach the common farmer until 1885. At a meeting of the Improved Stock Breeders' Association that year, President Leigh Hunt of the agricultural college proposed that the various agricultural societies in the state undertake an institute program, a proposal that was seconded later by various agricultural spokesmen, including James Wilson and William I. Chamberlain, Hunt's successor. Beginning in 1887 the College contributed $150 to the work, primarily to pay the traveling expenses of college personnel who participated in the gatherings. The institutes were "deemed a means of advertising the College and advancing its interests."[26] When these first efforts were well received, prominent rural spokesmen, including leading figures in the Farmers' Alliance movement, sought legislative support for the work. In March, 1892, the Iowa institute law was approved. It provided that when forty or more persons in a county established a county farmers' institute organization and held institute meetings of not less than two days duration a year, the state would contribute $50 annually to the local organization. Under this arrangement each county was left absolutely free to do as it pleased, and there was no control from the state level.[27]

In Wisconsin, where in time one of the most successful teaching

[25] Illinois Farmers' Institute, *Annual Report,* 1904, pp. 246–247; *Colman's Rural World,* L (December 2, 1897), p. 376; *Wallaces' Farmer,* XXVI (March 22, 1901), p. 404.

[26] *Ibid.,* XXXI (April 6, 1906), p. 487; Ross, *Iowa State College,* p. 165; Iowa State Agricultural College, *Twelfth Biennial Report,* 1886–87, p. 181.

[27] W. A. Coleman, *Farmers' Institute and Reference Book for Page County, Iowa* (Des Moines, 1900), p. 6; Morgan, *A History of the Extension Service of Iowa State College,* p. 15; *Wallaces' Farmer,* XXVI (December 20, 1901), p. 1481.

systems developed, friends of the agricultural college established institutes as a means to protect the institution from agrarian attacks. In the early 1870s when W. W. Daniels, the first professor of agriculture at the University of Wisconsin, found that there was little demand for his services, the university regents suggested that he attend rural meetings of various types to establish favorable relations with farmers and that he hold institutes in those communities where sufficient interest existed. Nothing came of this early interest in institutes; farmers attending the 1878 and 1879 meetings of the state agricultural society were loud in their denunciation of the university, claiming that it had failed completely to meet the needs of rural residents. Attending these gatherings was Hiram Smith, prominent dairy farmer and member of the university board of regents. He worked out a plan for the university and three statewide agricultural societies to hold six two-day conventions, really institutes, in 1880. Farmers came out in considerable numbers, but the experiment was not repeated. University faculty members believed that the gatherings were of questionable value, since they felt that there was insufficient material to present at them.[28]

Professor William A. Henry, who joined the university faculty in 1880, spent a good part of his time participating in the meetings of various rural societies, primarily to stress the importance of agricultural education. His efforts did little to appease discontented farmers, and in 1885 a bill was presented before the legislature providing for the separation of the agricultural college from the university. It was in this situation that Charles E. Estabrook, a legislator who had heard one of Henry's talks, pushed through a bill directing the regents of the university to establish a system of farmers' institutes in the state and providing $5,000 for the work. The regents promptly named William H. Morrison, a farmer from Walworth County who was past president of the State Dairymen's Association, as superintendent and gave him almost complete

[28] Carstensen, "Genesis of an Agricultural Experiment Station," pp. 14–16; Wilbur H. Glover, *Farm and College: The College of Agriculture of the University of Wisconsin* (Madison, 1952), p. 94; "Fifty Years of Cooperative Extension in Wisconsin," pp. 18–19.

freedom in managing the institutes. Response was more than satisfactory, and in 1887 the state legislature raised the annual appropriation to $12,000, giving Wisconsin one of the best supported institute systems in the United States. By 1899, 127 institutes drew 55,000 people.

Morrison died in 1893 and he was replaced by George McKerrow, a farmer from Waukesha County who served as superintendent for over twenty years. Both men were outstanding leaders and innovators in institute work, and the system in Wisconsin became a model for other states.[29]

In Minnesota the Northwestern Dairymen's Association held gatherings called institutes as early as 1884, but the launching of the formal institute system in the state came in response to a militant attack on the university by the Farmers' Alliance. In 1885 and 1886 that organization harshly criticized the university, claiming that agricultural education was being subordinated, that land-grant funds were being misused, and that agricultural students were ridiculed by other scholars on the university's Minneapolis campus. The Alliance called for the separation of the agricultural college from the university, a demand that was seconded by the state grange.[30] To fight the agrarian move John S. Pillsbury, a former governor and then president of the university's board of regents, sought the advice of Oren C. Gregg, a Lyon County dairyman known for his advanced farming techniques. When Gregg suggested that a system of farmers' institutes would be a step toward a solution of the problem, the regents quickly instructed Edward D. Porter to hold a series of meetings in the state. They also appropriated $1,000 for the work and authorized Porter to secure Gregg's assistance.[31]

[29] Ernest L. Luther, "Farmers' Institutes in Wisconsin, 1885–1933," *Wisconsin Magazine of History,* XXX (September, 1946), pp. 60–63; W. H. Morrison to W. O. Atwater, December 17, 1888, and A. C. True to R. W. Silvester, January 28, 1895, Office of Experiment Stations Records.

[30] Hicks, "Origin and Early History of the Farmers' Alliance," pp. 204–205; James Gray, *The University of Minnesota, 1851–1951* (Minneapolis, 1951), pp. 96–99; *Ariel,* IX (March 15, 1886), pp. 83–84; Boss, *School of Agriculture,* pp. 32–33; *St. Paul Pioneer Press,* February 26, 1886.

[31] O. C. Gregg to W. W. Folwell, April 19, 1914, William Watts Folwell Papers;

Reaction to the 31 institutes held in 1886 justified an expansion of the work the following year. On March 2, 1887, the governor signed a measure establishing a board of administration to manage an institute system and providing an annual appropriation of $7,500. The new agency was empowered to appoint for two-year terms a superintendent to direct the work. In April the board named Gregg to the post, thus bringing into the work in a supervisory capacity the man who for twenty years headed the institute movement in Minnesota. Gregg was given almost complete freedom in carrying on the work, and he promptly developed patterns that proved to be highly successful and that continued in use until 1907 when he retired. Under his direction the program expanded steadily and legislative support rose, amounting to $18,000 a year as early as 1903. The number of institutes increased from 58 in 1887–88 to 158 in 1903–04.[32]

From Minnesota, institute work spread into the Dakotas. In fact Minnesota lecturers appeared at a meeting called an institute in Bismarck as early as 1889, and it appears that scattered meetings were held by farm groups of various types during the next few years. A further step came in May and June, 1894, when the Society of United Farmers of Cass County, North Dakota, sponsored meetings at Casselton and Fargo. Members of the agricultural college served as lecturers. During the next four years a few institutes were held; the total probably did not exceed fifteen. Late in 1896 President John

---

Northwestern Agriculturist, XXII (May 11, 1907), pp. 463–464; Executive Committee of the University of Minnesota Board of Regents, Minutes, February 10, 1886; The Development of Agricultural Extension Work in Minnesota, manuscript dated December 14, 1937 (University of Minnesota Institute of Agriculture Archives, St. Paul).

[32] Minnesota Farmers' Institutes, *Annual* 1 (St. Paul, 1888), p. 8; *St. Paul Pioneer Press,* February 4, 10, and 17, 1887; O. C. Gregg to W. W. Folwell, January 28, 1915, William Watts Folwell Papers; W. P. Kirkwood, "The Man Who Roused the Farmer," *Northwestern Miller,* LXXXVIII (December 13, 1911), p. 650; Roy V. Scott, "Pioneering in Agricultural Education: Oren C. Gregg and Farmers' Institutes," *Minnesota History,* XXXVII (March, 1960), p. 24. The movement in Minnesota is discussed in greater detail in Roy V. Scott, "Early Agricultural Education in Minnesota: The Institute Phase," *Agricultural History,* XXXVII (January, 1963), pp. 21–34.

H. Worst of North Dakota Agricultural College began a campaign to obtain an appropriation for the work. Three years later the legislature gave $1,000 and placed the responsibility for institutes in the hands of the state dairy commissioner. These innovations permitted the holding of twenty meetings during the winter of 1899–1900.[33] In neighboring South Dakota the first step came in 1891 when the board of trustees of the state agricultural college authorized that institution to hold institutes. There were five meetings in the winter of 1891–92, a number that had risen to twenty-three nine years later when the state provided $2,000 a year to finance the activity.[34]

In Kansas, after the pioneer efforts failed in the early 1870s, there were no further attempts to establish an institute system until 1881 when President George T. Fairchild of Kansas State Agricultural College took the lead. Fairchild had participated in the early institutes in Michigan, and in his suggestions he drew heavily upon the Michigan experience. He proposed that the college send lecturers to meetings called by any farmers' club, grange, or other local group that agreed to provide facilities and make the necessary local arrangements. Response was prompt and, following approval of the project by the trustees of the college, institutes were held in six counties during the first months of 1882. President Fairchild reported "the series to be on the whole a successful inauguration of an effort for closer relations between the College and the farmers of the State." During the first few years following the launching of institutes, an average of six a year was held, and the college contributed between $300 and $500 a year as well as the bulk of the speakers for the work. The number of meetings gradually increased, until there were thirty in 1897–98, but college officials and others were fully aware of the fact that Kansas lagged behind such neighboring states as Missouri in the work. In 1899 the legislature granted a

---

[33] *Farm, Stock, and Home,* V (March 1, 1889), p. 133; North Dakota Farmers' Institute, *Annual,* 1900 (Fargo, 1900), p. 18; William C. Hunter, *Beacon across the Prairie: North Dakota's Land-Grant College* (Fargo, 1961), pp. 29, 54.

[34] Bailey, "Farmers' Institutes," p. 27; William H. Powers, ed., *A History of South Dakota State College* (Brookings, S. Dak., 1931), pp. 91–92.

biennial appropriation of $2,000 to allow the college to undertake a dramatic expansion of its institute activity.[35]

Nebraska's institutes originated with the farmers themselves. On February 7, 1882, farmers organized the Nemaha County Farmers' Institute Association in order that they might have a forum for discussing common problems. Before the year was out Johnson and Kearney counties had similar groups. Faculty members from the University of Nebraska participated as lecturers from the outset, but the university provided nothing in the way of central supervision. The movement simply grew as farm groups of various types called meetings which they labeled as institutes. About fifty such gatherings were scheduled for the winter of 1891–92. Some systemization was introduced following a meeting of farm groups at Lincoln, November 5, 1896, where it was decided that the various state agricultural societies would provide a number of lecturers for a central pool from which speakers would be assigned by a new state superintendent of institutes to be stationed at the university. F. W. Taylor, a faculty member, was chosen for the post. His academic associates continued to serve as lecturers, supplementing the efforts of speakers provided by the state agricultural societies. The state legislature in 1897 played its part by appropriating $3,000 biennially to finance the work. Under these arrangements, fifty-one institutes were held in thirty-three counties in 1899–1900.[36]

Logically enough, institute work lagged in the mountain states and in the Far West, despite the pioneer efforts in Colorado. The sparse population, the lack of transportation and communications, and the primitive condition of agriculture made the holding of institutes impossible or impractical in many areas.

[35] Willard, *Kansas State College*, pp. 65–66; Kansas State Agricultural College, *Biennial Report*, 1881–82, p. 16; *ibid.*, 1887–88, p. 78; *ibid.*, 1897–98, p. 44; *American Agriculturist*, LVIII (July 18, 1896), p. 44; *National Rural*, XXXVII (August 10, 1899), p. 1043; Aldous, "Grange in Kansas," pp. 58–59.

[36] Crawford, *University of Nebraska*, pp. 64, 113–114; Nebraska Farmers' Institute, *First Report* (Lincoln, 1906), pp. 3, 20–21; *American Agriculturist*, LVIII (July 18, 1896), p. 45; *Cultivator and Country Gentleman*, XLVIII (February 22, 1883), p. 151.

Still, in several states there were earnest efforts to establish institute systems. The Washington State Agricultural College had little more than opened its doors when the first of a series of farmers' institutes was held at Colton on January 30, 1892. Faculty members read papers to the sixty farmers and their wives who appeared. At a second meeting, held at Garfield in Whitman County, President Lilly told the assembled farmers that he and his colleagues had come to Garfield "to discuss . . . various subjects of interest to the farmer, and also to explain the objects and workings of the College." Thereafter, the college and experiment station personnel held from ten to twenty meetings a year until 1902, when the legislature appropriated $2,500 a year for the work and directed that one institute be held in each county in the state.[37]

In California, agricultural extension activity may be traced to a resolution of the university board of regents of June 21, 1870, directing the professor of agriculture to visit different parts of the state and to deliver lectures "in order to extend the advantages of the Agricultural College of the University of the largest number of citizens possible. . . ." Accordingly, various faculty members spent no little time attending meetings of agricultural societies and other rural groups. The primary purpose, however, seems to have been to bring "the University and its purposes directly to the attention of that large part of the population of the state for whose benefit it was created."[38] In fact, it was not until 1891 that institute work in the normal sense began in California. During the first years the work was concentrated in the southern part of the state. Until 1903 the university alone carried the financial burden. By the late 1890s approximately eighty institutes a year were held under the direct management of a superintendent who was also a faculty member of the university.[39]

[37] Enoch A. Bryan, *Historical Sketch of the State College of Washington* (Spokane, 1928), pp. 531–532; Washington Agricultural Experiment Station, *Bulletin* 2 (Pullman, Wash., 1892), pp. 21, 35–36; *ibid.*, 3, p. 41; John Hamilton, "History of the Farmers' Institutes in the United States," United States Office of Experiment Stations, *Bulletin* 174 (Washington, 1906), p. 89.

[38] Quoted in William W. Ferrier, *Origin and Development of the University of California* (Berkeley, 1930), pp. 615–616.

[39] AAFIW, *Proceedings*, 1902, p. 16; "The Work of the College of Agriculture

The Oregon agricultural college managed its first institute at Salem in December, 1888. Later, the college and the experiment station sponsored about twenty meetings a year. They enjoyed no direct financial assistance from the state.

Elsewhere in the mountain west little was done before 1900. Wyoming, Nevada, and New Mexico, for example, made little or no pretense before that date. The first affair in Idaho that could be called an institute was held at Moreland in December, 1898. Two years later, the legislature gave $1,000 a year, and the professor of agriculture at the state college assumed the job as superintendent. Institutes appeared in Montana in 1893, the year that the state's agricultural college was organized, and meetings were held irregularly until 1901 when the state legislature systematized the work and provided funds for expenses.[40] Arizona agricultural experiment station personnel met farmers and ranchers in meetings in Phoenix in 1895 and 1898 and the next year they held sessions at seven points. Not until 1903 did the legislature provide a specific appropriation for the work. Meanwhile, in a measure approved March 28, 1896, the Utah legislature directed the state agricultural college to hold at least one annual institute in each county and appropriated $1,500 for the purpose. Under this act thirty-six programs were held in 1899.[41]

The institute movement in the South was also significantly less successful than in the East and the North. The reasons were not hard to see; the unique characteristics of southern agriculture, combined

and Experiment Stations," California Agricultural Experiment Station, *Bulletin* 111 (Berkeley, 1896), p. 9; *Pacific Rural Press*, XLV (March 4, 1893), p. 188; *ibid.*, LII (September 5, 1896), pp. 146, 148.

[40] Hamilton, "History of Farmers' Institutes," pp. 60–61, 64, 76–77, 96; H. T. French to A. C. True, February 12, 1898, Office of Experiment Stations Records; Bailey, "Farmers' Institutes," p. 18; Montana Farmers' Institutes, *First Annual Report*, 1902 (Helena, 1902), pp. 4–5; Idaho State Farmers' Institutes, *Year Book*, 1901–02 (Moscow, n.d.), pp. 4–5.

[41] Hamilton, "History of Farmers' Institutes," pp. 19–20; Bailey, "Farmers' Institutes," pp. 9, 28; U.S. Commissioner of Education, *Report*, 1899–1900, II, 2048; Utah Farmers' Institutes, *First Annual Report*, 1897 (n.p., n.d.), n.p.; Joel E. Ricks, *The Utah State Agricultural College: A History of Fifty Years, 1888–1938* (Salt Lake City, 1938), p. 51.

with the general poverty of the area, simply made impossible institute systems like those in Minnesota, Wisconsin, Pennsylvania, and New York. Even President J. C. Hardy of Mississippi Agricultural and Mechanical College, an outstanding institute leader in the South, said in 1902 that his part of the country had done little to educate the common farmer prior to 1900.[42]

In those cases—in South Carolina and Mississippi—where an early start was made, proponents were unable to continue the momentum, and the work was interrupted for several years. The state agricultural society in South Carolina inaugurated summer meetings in various parts of the state as early as the mid-1870s, but nothing of a continuous nature was done until 1890 when the state legislature saddled the trustees of Clemson Agricultural College with the task of conducting institutes. By 1899 fewer than twenty institutes were being conducted annually with funds and lecturers provided by the agricultural college and experiment station.[43] Mississippi's institute work originated in that state's agricultural college. In March, 1884, the faculty recommended that institutes be held, and two weeks later the board of trustees gave its approval, directing the faculty to hold at least six institutes annually in those communities requesting them. The college provided the bulk of the speakers, and a $500 annual appropriation by the board of trustees was expected to pay their traveling expenses, an arrangement that permitted the holding of forty-six institutes by the end of 1888. But during the next decade financial and other problems severely restricted the number that the college could sponsor; in some years there were none at all. A revival came in 1897 when $500 of experiment station funds were used, and in 1900 the state legislature made its first direct appropriation.[44]

[42] AAACES, *Proceedings,* 1902, p. 70.
[43] Bailey, "Farmers' Institutes," p. 21; AAFIW, *Proceedings,* 1902, p. 29.
[44] S. D. Lee to A. W. Harris, December 28, 1888, and S. D. Lee to J. M. Rusk, May 27, 1889, Office of Experiment Stations Records; Mississippi Agricultural and Mechanical College, *Annual Catalogue,* 1883–84 (Jackson, 1884), p. 46; *ibid.,* 1897–98, p. 48; Bettersworth, *Mississippi State,* pp. 114–115, 118, 215. The institutes in Mississippi are discussed in greater detail in George J. Pope, "Agricultural

Elsewhere in the South, North Carolina, Alabama, and Texas began institute programs in the late 1880s, but developments in those states could hardly be called flourishing or continuous. In North Carolina a state law in 1887 directed the state board of agriculture to hold institutes in every county at least every two years, but it failed to provide funds for the work. Little was actually done, and there was nothing in the way of a systematic program. In 1890 the Alabama legislature authorized the state department of agriculture, then headed by Reuben F. Kolb, to hold institutes. Kolb was energetic, but he was deeply involved in politics which handicapped his efforts. Meanwhile, in 1892 faculty members at Alabama Polytechnic Institute began holding institutes at scattered points in the state. Finally, in 1897 the board of trustees of the agricultural college allocated $400 for the work, an amount that permitted school personnel to hold twenty-one institutes during the 1898–99 year. Institutes in Texas began in the late 1880s under the auspices of *Farm and Ranch*, a Dallas farm paper that offered to provide lecturers to any community willing to furnish a hall and other necessities. A few meetings were held under this arrangement, but when the legislature failed to provide funds to continue the work it disappeared, and nothing further was done until 1902.[45]

Virginia, Kentucky, and West Virginia began their programs in the early 1890s. The pioneer institute in Virginia was held near Richmond in September, 1890. The affair was so well attended that the state board of agriculture the next year appropriated $250 to support an institute in each of the state's congressional districts. Only four were held in 1891 and five in 1892. Subsequently, interest waned, and by the late 1890s the institute movement was practically dead in Virginia. A revival commenced in 1900, when the state board set aside $3,000 for the project. The first state appropriation

Extension in Mississippi Prior to 1914" (M.A. Thesis, Mississippi State University, 1963), pp. 33–58.

[45] Hamilton, "History of Farmers' Institutes," pp. 18–19, 69, 83; Rogers, "Reuben F. Kolb," pp. 115–116; Samuel L. Evans, "Texas Agriculture, 1880–1930" (Ph.D. Thesis, University of Texas, 1960), p. 352.

for institutes in Kentucky came in May, 1890. The task of conducting them was handed to the commissioner of agriculture, an overworked individual who was handicapped by a lack of money and rural support and by political troubles. A total of twenty-one institutes were held in the winter of 1898–99, but throughout the decade there had been no systematic planning for the work. Slightly more productive was the program in West Virginia, which stemmed from a meeting of private individuals in February, 1889. The conclave worked out an elaborate organization with associations at the local, county, and state level, called the Farmers' Institute Society of West Virginia. The first regular institute sponsored by the group was held at Parkersburg in July. Subsequently, the state board of agriculture and the agricultural college joined in the work, with members of the board chiefly responsible for promoting institutes in their congressional districts. Finally, in 1897 the board named a director of institutes and instructed him to hold at least one institute in each county, a step that for the first time gave the state the beginnings of an effective system.[46]

In Georgia, Florida, Arkansas, and Oklahoma, there was practically nothing in the way of institutes before 1900. In Tennessee a few meetings were held early as 1891 but it was not until 1898 that the legislature provided an appropriation amounting to $2,500 a year. Louisiana institute work began in June, 1897, after the newly established board of agriculture and immigration joined with the agricultural college and the experiment station to support and manage the work. A total of twenty-one institutes were held the first year.[47]

[46] Thomas Whitehead to J. M. Rusk, September 30, 1890, and J. H. Stewart to A. C. True, September 20, 1901, Office of Experiment Stations Records; Bailey, "Farmers' Institutes," pp. 14, 29; *Southern Planter,* LIII (November, 1892), p. 640; *ibid.,* LXI (August, 1900), p. 458; Hamilton, "History of Farmers' Institutes," pp. 41, 89–94; Atkeson and Atkeson, *Pioneering in Agriculture,* pp. 135–136.

[47] Wheeler, *Agricultural Education in Georgia,* pp. 73–75; *Hoard's Dairyman,* XXXIV (January 29, 1904), pp. 1174–1175; "Farmers' Institutes in Georgia," University of Georgia, *Bulletin* IX (Athens, 1909), p. 4; Samuel Proctor, "The Early Years of the Florida Experiment Station," *Agricultural History,* XXXVI

As the institute movement became national and even international in scope, prominent leaders in the United States and Canada concluded that the creation of a central organization was the next logical step. R. E. A. Leach, superintendent of institutes in Manitoba, first proposed the establishment of such an organization, and his proposal was warmly supported by Oren C. Gregg of Minnesota, Kenyon L. Butterfield of Michigan, George McKerrow of Wisconsin, and other leaders in the United States. McKerrow called the organizational meeting at Watertown, Wisconsin, March 13, 1896. There delegates from six midwestern states established the American Association of Farmers' Institute Managers. The first regular session convened in Chicago in October, 1896; thereafter, the organization met annually until its disappearance in the World War era. In 1899 members changed the name to the American Association of Farmers' Institute Workers in order to enlarge the scope of the organization and to expand the number of persons interested in it. The association was expected to function as a clearinghouse, gathering information and statistics and helping to produce some uniformity in the movement.[48]

By the end of the 1890s the institute movement had spread to practically all parts of the continental and contiguous United States. Of the forty-five states and three territories, only Arkansas, Nevada, New Mexico, Oklahoma, Texas, Virginia, and Wyoming had made what amounted to no progress toward establishing an institute sys-

(October, 1962), p. 220; Bailey, "Farmers' Institutes," pp. 9, 11; Harrison Hale, *University of Arkansas, 1871–1948* (Fayetteville, 1948), p. 195; U.S. Commissioner of Education, *Report*, 1899–1900, II, 2050; Hamilton, "History of Farmers' Institutes," p. 82; Watts, "Agricultural Extension in Tennessee," pp. 99–102; Roy V. Scott, "Farmers' Institutes in Louisiana, 1897–1906," *Journal of Southern History*, XXV (February, 1959), pp. 77–80.

[48] *Farm, Stock, and Home*, XII (February 15, 1896), p. 97; *Farm Students' Review*, I (March, 1896), p. 35; *Cultivator and Country Gentleman*, LX (November 21, 1895), p. 841; *American Agriculturist*, LVIII (July 18, 1896), p. 41; American Association of Farmers' Institute Managers, *Report of the Meetings Held at Watertown, Wisconsin, March 13, and at Chicago, October 14, 1896*, pp. 5, 9; A. C. True, Report of the Bibliographer to the Association of American Agricultural Colleges and Experiment Stations, 1915, Office of Experiment Stations Records.

tem. Elsewhere, there existed programs enjoying varying degrees of success. The most successful in terms of attendance were found in Michigan, Ohio, New York, Wisconsin, and Pennsylvania, where the numbers of farmers instructed in institutes ranged from 100,000 downward to 50,000. Those states, too, provided the best support for their systems. New York expended the most money—$20,000 a year—closely followed by Ohio, Minnesota, Pennsylvania, and Wisconsin. In Alabama, Arizona, Florida, Georgia, Idaho, Montana, and North Carolina the institutes by 1899 were in no more than their infancy or they were so inadequately supported as to reduce materially their effectiveness.

By 1900 the organizational structures of many state institute systems had achieved the shape that they would retain. They fell into two major categories with a number of special arrangements. In seventeen states the boards or departments of agriculture managed the institutes, selected all or part of the speakers, determined sites, and established general policies. More than half of these states were found in the East, reflecting the better organized government structures in that area. The institutes were managed by the agricultural colleges or experiment stations in nineteen southern, midwestern, and western states. In many of those states, no other agricultural educational agencies existed, so the burden fell upon the colleges and experiment stations by default. Delaware and Iowa left the management of their institutes entirely to the counties, but the programs were subsidized in part by the states. In Illinois a state body supervised the work after the late 1890s, but actual management of local meetings rested in the counties. Minnesota's institute system was perhaps as unique as any in the nation. There the work was headed by an independent board of administration, an agency that at the outset included two regents of the state university and officers of four rural organizations. The board appointed the superintendent in whose hands rested the actual direction of institutes in the state.[49]

[49] Bailey, "Farmers' Institutes," pp. 8–31; The Development of Agricultural Extension Work in Minnesota, manuscript dated December 14, 1937; Minnesota, *General Laws,* 1887 (Minneapolis, 1887), pp. 380–381; *St. Paul Pioneer Press,* February 27, 1887.

Despite differences in organization the institute movement had an essential unity in terms of purposes and methods. Everywhere, the institutes could be described most simply as "the adult farmers' school" where practicing agriculturists gathered to learn of newer techniques and methods from trained scientists and from progressive and successful farmers. According to the 1887 law establishing Minnesota's institutes on a firm basis, "the sole object and purpose of these institutes [is] to disseminate practical knowledge pertaining to agriculture, horticulture, and stock and dairy farming, with the least expense or inconvenience to the people of the state." A decade later Jordan G. Lee said that the institutes in Louisiana "are . . . a place of meeting or exchange where facts, experiences, and ideas are the commodities transferred; an association that makes men and women more useful and more progressive members of the communities in which they live; a place where all meet on a common level, each to learn from the other."[50] Similar purposes and descriptions were stated or implied wherever institutes developed.

In form, most institutes followed a standard pattern. They generally were two or three-day affairs, although one-day meetings later became common, and each day's work consisted of two or three sessions, depending upon whether there was an evening program. The first morning session usually opened with an address of welcome by a local dignitary recruited for the occasion and with a talk by an institute official on the general purposes of the meeting. Then would begin the real work, with five or six lecturers speaking on as many topics and with a discussion or question and answer period following each talk. In the course of an institute, lecturers generally spoke on more than one subject. Most lectures required no more than thirty minutes, since the primary objective was to present general principles that the ordinary farmer could use in his own operations. At the conclusion of the morning program, a dinner break provided an opportunity for the ladies to display their culinary talents and permitted considerable socializing among neighbors and members

[50] U.S. Department of Agriculture, *Yearbook,* 1897 (Washington, 1898), p. 280; Minnesota, *General Laws,* 1887, p. 381; Louisiana Farmers' Institute, *Bulletin* 1 (Baton Rouge, 1898), pp. 7–8.

of the lecture force. Afternoon sessions were similar to those in the morning. Evening programs on the other hand generally offered lighter fare. Many were educational, but others sought only to entertain those farmers and their wives who returned to the meeting after completing their chores. Some institutes also provided amusement for children who accompanied their parents. Baseball games and various types of races were common; they presumably kept boys and girls occupied while the adults were learning of better methods.

Institute managers attempted to hold meetings in those places and at those times in the year that would allow the largest number of farmers to attend. Most often institutes met in centrally located town halls or courthouses, but others were held in grange halls, churches, or even shady groves. There were many exceptions, but at the outset institutes in the northern states usually met in the winter months when farmers would have more free time. In the South, on the other hand, institutes commonly met in the late summer after the corn had been laid by and before cotton picking began. As the programs became better established, however, many northern states began a summer institute season, usually in June or July, and winter institutes became common in the South.[51]

In those states where some agency exercised general control over the institutes, that office or individual generally selected sites for meetings and assigned speakers to them. But local people were expected to play important roles. They were called upon to provide suitable accommodations, to advertise the meeting, and to arrange a part of the program. Quite often they were solely responsible for the evening sessions, those that stressed entertainment instead of education.[52]

[51] *Farm, Stock, and Home,* IV (November 15, 1887), p. 3; *Northwestern Farmer,* XI (June 1, 1893), p. 202; Louisiana State Agricultural Society, *Proceedings,* 1897 (Baton Rouge, 1898), p. 50; *ibid.,* 1898, pp. 60–61; Williamson, *Agricultural Extension in Louisiana,* pp. 34–35; N. J. Colman to J. S. Lawson, February 8, 1889, Office of Experiment Stations Records; Butterfield, *Chapters in Rural Progress,* pp. 92–93.

[52] Missouri State Board of Agriculture, *Monthly Bulletin,* I (September, 1901), p. 12; *ibid.,* II (June, 1902), p. 9; *Colman's Rural World,* XLVII (November 15,

Topics discussed from lecture platforms covered the entire spectrum of American agriculture and occasionally wandered into subjects only vaguely related to farming. Massachusetts farmers in 1875 were hearing talks on breeding and management of horses, cultivation of fruits, and restoration of forests. At Milford, Massachusetts, the same year, a speaker expounded upon the attractiveness of eastern farm life as compared with that in the West. Everywhere agrarian fundamentalism was strongly expressed. In the early 1880s the principal topics discussed in Missouri were rotation of crops, drainage, the uses of clover, methods of raising stock, fruit growing, animal diseases, and beekeeping. The advantages or diversification and improved farm practices received primary attention during the early years in Minnesota and Louisiana, but there as elsewhere a farmer could expect lectures on any phase of agriculture and rural life. Among the speakers at the first institute at Shenandoah, Iowa, was Henry Wallace; his topic, "Raising Hogs." At an institute in Greene County, Illinois, in 1890, a meeting typical of many others, the topics included cattle feeding, dairying, wheat and corn growing, the outlook for small farmers, the evils of trusts, and the need for better rural roads. One speaker contented himself with a paper on "Old Greene County."[53]

Evening sessions usually included entertainment provided by local performers. In other instances they featured lectures on matters of general interest. In Minnesota William Dickson, who had served with the British government in India, discussed his experiences there. Professor John A. Myers of Mississippi Agricultural and Mechanical College used lantern slides to illustrate talks that he gave on the

1894), p. 359; (Opelousas, La.) *Courier*, June 5, 1897; Louisiana State Agricultural Society, *Proceedings*, 1898, pp. 60–61; *ibid.*, 1902, pp. 20–27; *Pacific Rural Press*, LII (September 12, 1896), p. 162; N. J. Colman to J. S. Lawson, February 8, 1889, and H. E. Stockbridge to W. O. Atwater, August 14, 1889, Office of Experiment Stations Records.

[53] *Cultivator and Country Gentleman*, XL (November 25, 1875), p. 745; *ibid.*, XL (February 11, 1875), pp. 83–84; Missouri State Board of Agriculture, *Annual Report*, 1892, p. 8; *Farm, Stock, and Home*, V (January 1, 1889), p. 55; *Lafayette Advertiser*, XXXIV (July 22, 1899), p. 1; Coleman, *Farmers' Institute and Reference Book*, p. 8; *White Hall Register*, XXI (February 14, 1890), p. 4.

Near East. At Lafayette, Louisiana, the town orchestra and local singers entertained evening audiences. Less common was the event at an institute in Forest County, Pennsylvania, where a young couple was married.[54]

Institute lecturers generally favored discussion following the formal talks; through informal comments audiences often gained valuable insights into the methods of successful neighborhood farmers. To overcome the ordinary farmer's hesitancy to speak out, many institutes used the question box, a technique that often brought out practical information and encouraged spirited discussion.[55]

Many lecturers enhanced their talks with various kinds of demonstration materials. In Wisconsin and Minnesota, for instance, livestock judging and churning exhibitions were common. Speakers there habitually carried blackboards with them as well as dairy tools, models of buildings, and other materials useful in demonstrating points in their presentations. As early as 1877 the institute management in Massachusetts held field meetings in which farmers were taken into fields, orchards, and barnyards where they observed as well as heard of improved methods.[56]

To supplement the information given in lectures, many state institute systems provided various types of printed matter which were distributed among farmers. By the late 1890s some twenty states published all or part of the papers read at institute sessions, while most of the remainder gave more or less detailed reports of meetings to local papers and the rural press. In Maine as early as 1884 the state board of agriculture printed several addresses in its annual report, while in New York the state agricultural society performed a similar service. Wisconsin's and Minnesota's institute manage-

---

[54] *Farm, Stock, and Home,* VII (February 1, 1891), p. 85; Bettersworth, *Mississippi State,* p. 117; *Lafayette Advertiser,* XXXIV (July 22, 1899), p. 1; Fletcher, *Pennsylvania Agriculture and Country Life,* II, p. 445.

[55] U.S. Department of Agriculture, *Yearbook,* 1897, p. 281; *Farm, Stock, and Home,* X (January 1, 1894), p. 51; (Abbeville) *Meridional,* XLI (July 3, 1897), p. 3.

[56] *Cultivator and Country Gentleman,* LII (December 8, 1887), pp. 924–925; *ibid.,* XLII (September 6, 1877), p. 569; *ibid.,* LII (September 1, 1887), p. 673; Scott, "Early Agricultural Education in Minnesota," p. 26.

ments each produced annual volumes. In the former state the publication contained the verbatim record of the entire proceedings of a single institute considered typical of others held during the season. Minnesota's annual, on the other hand, was a "compilation of practical experience and personal observation" consisting of articles written by institute lecturers, agricultural college men, and others. In both cases the volumes were issued in editions of several thousand copies and distributed free of charge. Publication costs were met by selling advertising space to railroads, implement concerns, and other businesses.[57]

While other factors entered into the results, the most important single ingredient in a successful institute program was the quality of the speakers who appeared on lecture platforms. Institute lecturers fell into two or three types. In those states where the college of agriculture or the experiment station managed the institutes or in those cases where little money was available for institute support, agricultural scientists—college and station staff members—constituted the bulk of the lecture corps. Elsewhere, lecturers tended to be successful farmers who by their achievements in field and barnyard had acquired enviable reputations. Within this group, however, there was a rather clear division. In some instances, where institutes were in their infancy or where support was lacking, there was an inclination to rely upon purely local farmers whose services were limited to institutes in their immediate neighborhood. On the other hand, such states as Minnesota used widely known farmers whose abilities made them acceptable everywhere. Some men of this type, in fact, were so well regarded that they were in demand in other states.

There was no agreement among institute administrators concerning the best type of lecturers. Some authorities were convinced that only trained scientists could provide the type of information that farmers needed, although they recognized that success in the labora-

[57] *Experiment Station Record,* VII, p. 640; N. J. Colman to J. S. Lawson, February 8, 1889, Office of Experiment Stations Records; Minnesota Farmers' Institute, *Annual* 1, p. 7; *Farm, Stock, and Home,* IV (November 1, 1888), p. 379; *Cultivator and Country Gentleman,* L (October 8, 1885), p. 822; *ibid.,* LII (December 8, 1887), p. 924; *ibid.,* LVIII (December 21, 1893), pp. 998–999.

tory or classroom did not insure success on the lecture platform. What was wanted, these men felt, were speakers who combined training, scientific knowledge, and eloquence. Moreover, the use of college men was "one of the levers that [experiment] stations can use . . . to elevate the agriculture of the country" and to justify their existence in the eyes of farmers.[58]

Other institute administrators maintained that most farmers had a deep suspicion of scientists and that few trained men could speak in a manner intelligible to farmers. After the first few years in Minnesota, for instance, Oren C. Gregg insisted upon the use of purely "practical" lecturers, men who had actually performed on the farm the tasks and methods they described. Too often, he thought, the academic man was a "wet blanket," and after he spoke "there would not be enough left of the audience to pronounce a benediction on."[59]

All would have agreed upon the characteristics of the ideal institute speaker. A fifteen-minute man, he spoke authoritatively and to the point in language that the most uninformed could understand. He terminated his talk in the allotted time, leaving a few minutes for discussion and questions. Finally, he never hesitated to admit his inability to answer a query, since few things would drive farmers from an institute quicker than a speaker who tried to bluff his audience.

Among the nation's successful institute lecturers were men of all types. Typical of the academic lecturer were William C. Stubbs, William R. Dodson, and William H. Dalrymple of Louisiana; Willet M. Hays, Thomas Shaw, and Juniata L. Shepperd of Minnesota; J. C. Whitten and Frederick B. Mumford of Missouri; N. S. Townshend of Ohio and Isaac P. Roberts of Cornell; and E. R. Lloyd of Mississippi.[60] Two of the better known farmer lecturers were T. B.

[58] *Experiment Station Record,* VII, p. 641; AAACES, *Proceedings,* 1889, p. 90; AAFIW, *Proceedings,* 1901, pp. 34–36.

[59] *Cultivator and Country Gentleman,* LXII (December 9, 1897), p. 965; Minnesota State Horticultural Society, *Annual Report,* 1892, pp. 70–72; *ibid.,* 1895, pp. 289–292; *Farm, Stock, and Home,* XVII (February 15, 1901), p. 123; *Northwestern Farmer,* XI (November 15, 1893), p. 391; Minnesota Farmers' Institutes, *Annual 5,* p. 17.

[60] *Colman's Rural World,* L (November 18, 1897), p. 362; Louisiana State

Terry and John Gould, both of Ohio. After several years experience in their native states, they served in Minnesota, Wisconsin, Louisiana, and elsewhere. Representative of many others was Theodore Louis, a self-made Wisconsin hog farmer who served in Minnesota. Louis had no formal training but he knew hog raising as few men did, and with his evangelical desire to instruct others in the art he was in Oren C. Gregg's opinion the ideal institute lecturer. Outstanding also was Missouri's Samuel M. Jordan, the "corn man" whose career in agricultural education began with institutes and extended into more sophisticated methods.[61]

The life of an institute lecturer was by no means an easy one. The early days were especially discouraging. Years later T. B. Terry recalled that in the 1880s the reception and treatment of speakers in communities they visited were less than generous. After being taken to the cheapest restaurant in town, where the food was inedible, lecturers were housed with neighborhood farmers whose homes were reached only after a long, cold drive in the dark. Following one of the pioneer institutes at Council Bluffs, Isaac P. Roberts and President Welch of the Iowa agricultural college found themselves sharing the same freezing bed in a farmer's home. Seemingly, institutes in the North were always held when the temperature was at its lowest. In 1886 when John Gould arrived in Wisconsin for institute service there, the thermometer registered sixteen degrees below zero.

The work was hard and tiring, and it required a great deal of traveling, sometimes under primitive conditions. In 1885–86 W. I.

Agricultural Society, *Proceedings,* 1898, p. 60; *Farm, Stock, and Home,* VI (December 1, 1889), p. 18; *Northwestern Farmer,* XI (November 22, 1893), p. 391; *ibid.,* XV (December 1, 1897), p. 379; Minnesota Farmers' Institutes, *Annual* 15, pp. 18–19; *Farmington Times and Herald,* December 25, 1902, p. 3; Ohio State Board of Agriculture, *Brief History,* p. 41; *Breeder's Gazette,* XI (January 13, 1887), p. 49; S. D. Lee to W. C. Stubbs, February 22, 1899, W. C. Stubbs Papers (Louisiana State University Library, Baton Rouge).

[61] *Farm, Stock, and Home,* VI (March 1, 1890), p. 128; *ibid.,* XVI (January 15, 1900), p. 67; *Cultivator and Country Gentleman,* LIX (February 8, 1894), p. 111; Scott, "Early Agricultural Education in Minnesota," p. 28; Roy V. Scott, "The Career of Samuel M. Jordan: A Study in the Evolution of Agricultural Education," Missouri Historical Society, *Bulletin,* XVIII (April, 1962), pp. 239–259.

Chamberlain attended thirty-six of the forty-six institutes held in Ohio, lecturing twice each day. During one week in particular, he spoke twelve times at six different points, traveled over 600 miles, and got only one full night's sleep in the course of the period.

Occasionally, too, local leaders failed to advertise adequately forthcoming institutes or to make other necessary arrangements. In such situations, lecturers arrived in town to find that no hall had been set aside for the institute and that few people had been informed that an institute was scheduled. Speakers had to make their own arrangements, locate a room, build a fire to heat it, and try in one way or the other to attract a crowd.

That in itself was at times a task. In one Ohio town T. B. Terry sent a boy into the street with a bell in an attempt to get an audience; in another instance the sheriff merely called out a window, succeeding in attracting only the local loafers who gathered on the courthouse steps. In Minnesota Oren C. Gregg and his lecturers resorted to showmanship to attract a crowd; the breaking of a vicious horse in the street was a favorite technique in the early days.[62]

Across the nation, in fact, rural apathy, suspicion, and prejudice handicapped the pioneer institutes. In the 1880s especially, many farmers tended to believe that institute speakers had some kind of livestock or machinery to sell or some political doctrines to teach. As late as 1896 farmers avoided institutes in southern Indiana because they believed the meetings to be connected with the Farmers' Mutual Benefit Association, a group that had become little more than a wing of the Populist party.[63]

In a few instances, admittedly, there was some basis for the rural belief that politics was involved in the institutes. Reuben F. Kolb tried to use his position as an institute speaker in Alabama to build

[62] *National Rural,* XXXVII (September 21, 1899), pp. 1234–1235; Roberts, *Autobiography of a Farm Boy,* p. 100; *Cultivator and Country Gentleman,* LI (January 28, 1886), p. 65; *ibid.,* (March 25, 1886), p. 225; *Farm Students' Review,* I (March, 1896), pp. 34–35.

[63] N. J. Colman to J. S. Lamson, February 8, 1889, Office of Experiment Stations Records; State of Indiana, *Report on Farmers' Institutes for the Year 1896* (Lafayette, 1896), p. 5; *Breeder's Gazette,* XVII (January 29, 1890), p. 99; *Pacific Rural Press,* XLVIII (September 15, 1894), pp. 162–163.

his political fortunes. Occasionally, such groups as the farmers' alliances openly opposed the institutes, apparently considering them rivals for the loyalty of farmers. In Illinois a Populist periodical denounced the institutes as "humbugs" dominated by college professors and millionaire landowners with scant knowledge of the harsh realities of farming.[64]

But the most serious problem before 1900 was simple rural inertia and apathy. In 1889 Missouri's Levi Chubbuck wrote, "The chief obstacle to the complete success of Farmers' Institute work is the lack of appreciation on the part of farmers of their need for more information which will enable them to produce greater yields . . . at less cost . . . ." Nor were Missouri farmers unique. Even in such states as Ohio and Minnesota, where institutes enjoyed solid success, early lecturers found that they "couldn't get the farmers out." In the 1890s a Virginia lecturer reported that farmers there had "in almost every case shown a most disheartening appreciation for the efforts made to help them."[65] Even when farmers did appear at the first institutes in an area, according to one observer, "two-thirds of the . . . audiences were present to be amused or entertained. . . ."[66]

Still, the spread of the institute movement across the nation, the steadily growing attendance, and the increasing financial support for the work showed that it was serving a useful purpose in the countryside. Many rural leaders became convinced that institutes satisfied a need that could be met in no other way. Speaking in 1888 a president of the Minnesota State Horticultural Society said, "The Farmers' Institute I consider one of the best mediums . . . for the dissemination of practical horticultural information among our farming population." A few years later William H. Morrison of Wisconsin observed that he was "more and more convinced that the most prac-

[64] Rogers, "Reuben F. Kolb," p. 116; *The Farmer,* XXV (December 1, 1907), p. 772; *Farmers' Voice,* VII (January 21, 1893), p. 4.

[65] Levi Chubbuck to J. M. Rusk, January 8, 1889, Office of Experiment Stations Records; AAACES, *Proceedings,* 1889, p. 43; *Farm, Stock, and Home,* IV (March 1, 1888), pp. 114–115; *Southern Planter,* LIII (October, 1892), p. 569.

[66] Quoted in Crawford, *University of Nebraska,* p. 115.

tical and successful method of reaching the farmer is through the farmers' institute. . . ."[67]

Even before 1900 institute enthusiasts could cite a long list of accomplishments attributed to the meetings. According to Norman J. Colman they played a major role in destroying the one-crop agriculture that prevailed in much of the western Middle West. In areas where institutes had been held regularly he found more and better livestock, numerous silos, and other indications of diversification. In addition he claimed that institutes helped to break down rural isolation and had led to the formation of farmers' clubs. Other spokesmen, of whom J. C. Hardy of Mississippi was typical, attributed to institutes the rising demand for experiment station publications and the increased enrollment in the agricultural colleges. William H. Morrison contended that institutes stimulated pride in and respect for agriculture, made farm life more attractive, and helped to keep the best of rural youth on the farm, besides accomplishing in admirable fashion their primary task of educating farmers.[68]

Most important to professional agricultural educators was the widespread indication that institutes had driven a wedge into the wall of suspicion and distrust that had characterized rural attitudes toward science. "At least," according to a University of Georgia spokesman, "we have come to realize that instruction in agriculture is not a joke." Moreover, the institutes represented a step to the future. "No one expects that attendance upon a few of these institute meetings will graduate a finished scholar in all of the arts of successful husbandry, but they go far in giving intelligent direction to the thoughts of those whose minds are not hopelessly clogged," editorialized a midwestern farm paper in 1892.[69]

Perhaps a New York farmer best summarized the ordinary rural

[67] Minnesota State Horticultural Society, *Annual Report,* 1888, pp. 67–68; *Breeder's Gazette,* XVII (January 8, 1890), p. 34.

[68] N. J. Colman to J. S. Lamson, February 8, 1889, Office of Experiment Stations Records; AAACES, *Proceedings,* 1902, p. 71; Bailey, *Annals of Horticulture,* 1891, p. 143.

[69] Quoted in Wheeler, *Agricultural Education in Georgia,* p. 77; *Northwestern Farmer,* X (February 1, 1892), p. 36.

reaction to the early institutes. According to Cornell's Liberty Hyde Bailey, as he was leaving an institute in the western part of New York, he heard one farmer ask another, "Well, Sam, how did you like it?" Replied his companion, "Oh, I don't know. It hain't hurt me none."[70] If by 1900 a majority of the nation's farmers shared that New Yorker's sentiments, the pioneer institutes had achieved a major goal. Rural residents in reality were ready for the next step in the evolution of agricultural extension.

[70] Quoted in W. O. Hedrick, "The Tutored Farmer," *Scientific Monthly*, VII (August, 1918), p. 159.

IV. FARMERS' INSTITUTES
COME OF AGE

By 1900 FARMERS' INSTITUTES had secured a significant place in the education of rural adults in the United States. Functioning more or less effectively in most areas, they were tremendously popular in New York, Pennsylvania, Minnesota, Wisconsin, and several other states. With few exceptions farmers everywhere at least refrained from rejecting them out of hand. The pioneer institutes, in short, had broken through the wall of indifference and apathy that for so long had prevented farmers from adopting new ideas.

If the years before 1900 had been successful ones for farmers' institutes, the next dozen were even more so. The number of institutes, appropriations for their support, and attendance all increased. A larger percentage of the nation's farmers were brought into contact with science through them. Institute administrators sought to make institute work more effective and meaningful. Innovations came in profusion as managers and lecturers developed a variety of new teaching methods. Finally, institute officials expanded both quantitatively and qualitatively their programs for women and for the first time they tried in a systematic way to reach and instruct rural youth.

The number of institutes, attendance, and appropriations for their support rose dramatically after 1900. Of these, only appropriations failed to climb steadily to 1914, instead reaching a peak in 1911–12. In the 1901–02 institute year the United States Department of Agriculture reported that 2,772 institutes were held; in 1913–14 the figure was 8,861, a gain of 230 percent. Attendance rose even more rapidly, increasing 265 percent between those years; reported figures were 819,995 and 3,050,151 respectively. Appropriations climbed from $163,125 to $449,883, or 176 percent, showing the growing public acceptance of institutes and suggesting that farmers were getting more for their money as time progressed.

GROWTH OF FARMERS' INSTITUTE MOVEMENT IN THE UNITED STATES, 1901–14

YEAR	NUMBER OF INSTITUTES	ATTENDANCE	APPROPRIATIONS
1901–02	2,772	819,995	$163,125
1902–03	3,179	904,654	187,226
1903–04	3,306	841,698	210,211
1904–05	3,271	995,192	225,739
1905–06	3,521	1,299,172	269,672
1906–07	3,927	1,596,877	284,451
1907–08	4,643	2,098,268	325,570
1908–09	5,014	2,240,925	345,666
1909–10	5,651	2,395,908	432,374
1910–11	5,889	2,291,857	432,693
1911–12	7,598	2,549,199	533,972
1912–13	7,926	2,897,391	510,785
1913–14	8,861	3,050,151	449,883

The expansion of the institute movement between 1900 and 1914 was especially pronounced in the South and the West. Quite obviously, before 1900 institutes were most popular in the Middle West and the Northeast; consequently, after the turn of the century they grew most rapidly in the South and the West as those states sought to provide better instruction for their farmers. In the South the appearance of Seaman A. Knapp's demonstration work, instead of eliminating the need for institutes, seemed to stimulate their development. In the West the sparseness of the population forced states to hold a large number of institutes to reach a satisfactory percentage of their farmers. The following tables, giving figures for selected

105

years, illustrate the general trends in the different sections of the country.[1]

NUMBER OF INSTITUTES

	1901–02	1903–04	1909–10	1913–14
Nine northeastern states	743	738	866	935
Sixteen southern and border states	364	747	1,843	3,179
Twelve midwestern states	1,419	1,849	2,427	3,252
Eleven western states	242	328	506	1,492

INSTITUTE ATTENDANCE

	1901–02	1903–04	1909–10	1913–14
Nine northeastern states	271,245	172,899	371,741	326,501
Sixteen southern and border states	69,621	111,282	337,753	802,111
Twelve midwestern states	435,564	470,742	1,534,295	1,672,579
Eleven western states	43,385	86,575	151,699	248,960

NUMBER OF INSTITUTES PER MILLION RURAL POPULATION

	1901–02	1903–04	1909–10	1913–14
Nine northeastern states	114	113	128	140
Sixteen southern and border states	18	37	81	140
Twelve midwestern states	87	92	148	198
Eleven western states	101	137	145	426

As important as the growth in number of institutes and of people who attended them was the increasing sophistication of the instruction offered after 1900. The pioneer institutes had aroused an interest in scientific agriculture, they had suggested to farmers that agricultural colleges had something to offer, and they had shown

[1] Figures taken or compiled from U.S. Office of Experiment Stations, *Annual Report*, 1902, p. 408; *ibid.*, 1905, p. 411; *ibid.*, 1912, pp. 382–383; John Hamilton, "Farmers' Institutes and Agricultural Extension Work in the United States in 1913," U.S. Department of Agriculture, *Bulletin* 83 (Washington, 1914), pp. 32–33; J. M. Stedman, "Farmers' Institute Work in the United States in 1914, and Notes on Agricultural Extension Work in Foreign Countries," U.S. Department of Agriculture, *Bulletin* 269 (Washington, 1915), p. 18; U.S. *Thirteenth Census: Population*, I, 56.

that farmers with no more than common-school educations were capable of understanding the basic principles of better farming.

Having accomplished these goals, institute managers needed to improve their medium if it were to continue to serve a useful purpose. The quality of the instruction had to be strengthened. The early institutes had presented little more than elementary information with large doses of sensational approaches and entertainment added. After the farmers' interest was aroused it was both possible and necessary to present a more thorough type of instruction. Farmers wanted to know not only the principles but also the details, and they wanted them discussed in terms of their own particular circumstances. The new type of instruction demanded more careful selection and preparation of lecturers. Increasingly, farmers refused to attend institutes where the speakers in reality knew little more than the members of the audience. Instead, farmers wanted to hear specialists. Finally, institute managers needed to develop methods by which a larger percentage of the rural population might be contacted and by which instruction and advice would be more regularly available than through the old annual institute.

In some instances even before 1900 institute managers had held meetings that differed from the standard institute. In Minnesota, for instance, Oren C. Gregg inaugurated what he called his "fair institute" in 1890. These were ordinary institutes held on the state fair grounds in conjunction with that annual event. The subjects covered were essentially those of the ordinary meetings, but lecturers were able to use many fair exhibits to improve the quality of their talks. In 1891 Gregg encouraged farmers to bring their families to the fair and camp on the grounds. The idea proved a popular one and for several years the "tented city" was a prominent feature of the fair.[2]

Before 1900, too, Gregg began holding special institutes, meetings devoted to the production of a single crop or type of livestock. In the spring of 1890 lecturers held four meetings in the Red River

[2] *Farm, Stock, and Home,* VI (October 1, 1890), p. 369; *ibid.,* XV (August 15, 1899), p. 305; Darwin S. Hall and R. I. Holcombe, *History of the Minnesota State Agricultural Society* (St. Paul, 1910), pp. 233, 240, 245–246, 265, 269, 275.

Valley at which the single topic under consideration was the growing of spring wheat. In 1897 a number of institutes were devoted
exclusively to sheep husbandry. At those sessions speakers pointed
out that sugar beet raising and sheep production made a feasible
combination. In the 1890s, also, special dairy institutes were held
upon the request of a sufficient number of farmers in a community.
Speakers at these affairs took with them cream separators, Babcock
milk testers, churns, and other articles useful in the operation of a
successful dairy farm.[3]

When audiences in Missouri became sufficiently sophisticated to
demand more thorough instruction, the special institute appeared
in that state. In 1898–99 one speaker held fifteen meetings devoted
exclusively to the improvement and maintenance of rural roads.
Another lecturer conducted a series of institutes on dairying and
creamery operation at various points in the state where farmers
displayed an interest in milk production.[4]

In time special institutes developed in practically every state,
their number and subject material limited only by funds available
and the ingenuity of institute officials. Fruit and vegetable institutes, for instance, were held for growers in the Florida parishes of
Louisiana in 1905; the same year farmers in the Shreveport area
attended a meeting devoted to the use of fertilizers and the growing
of feed stuffs. Potato and clover institutes were popular in Minnesota after 1900. In Pennsylvania and New York dairying was one
of the aspects of farming that received detailed treatment in special
institutes.[5]

Reflecting the desire to reach a larger number of farmers, some
institute officials began holding special one-day meetings, or school-
house institutes as they were called in Minnesota. To these gather-

[3] *Farm, Stock, and Home,* VI (March 1, 1890), p. 129; *ibid.,* VI (April 15,
1890), p. 187; *ibid.,* XIII (April 1, 1897), p. 146; Minnesota Farmers' Institutes,
Annual 8, pp. 43–48.

[4] Missouri State Board of Agriculture, *Annual Report,* 1899, p. 10.

[5] Louisiana Farmers' Institute, *Bulletin* 9, pp. 10–11; *Farm, Stock, and Home,*
XXV (February 15, 1909), p. 107; Witter, "Farmers' Institutes in New York," p.
218; *Hoard's Dairyman,* XXXVI (December 8, 1905), p. 1117.

ings the institute management sent only one man. Local talent was used extensively. After arriving in a locality a single state lecturer could preside over meetings at several different points, many of them inaccessible by railroad, thereby reaching in perhaps a superficial way many more farmers than was possible with ordinary institutes.[6]

A logical outgrowth of the special institute and the growing demand among farmers for more thorough instruction was the movable short course. Ranging in length from one week to a month or more, these affairs were held in at least ten states by 1909–10, when they attracted more than 65,000 people.[7]

Pennsylvania was one of the leaders in using movable short courses, but others were not far behind. In 1904 the institute management in Pennsylvania began sending out lecturers who held three or four-day programs dealing in depth with such topics as dairying, poultry raising, and fruit growing. In New York in the spring of 1909 the institute management arranged three "institute schools" in Delaware, Monroe, and Allegany counties. As many as sixteen speakers appeared at each point, and two or three lectures were presented simultaneously to classes of enrollees. So successful were the affairs that ten were scheduled for the winter of 1909–10. Minnesota held its first movable short course in 1906, and the next year the Missouri State Board of Agriculture organized four of the schools. At each of the six-day affairs, morning lectures and afternoon demonstrations dealt with a diversity of subjects in greater detail than was possible in ordinary institutes.[8]

Management of movable short courses required the guaranteed support of local people and often involved effective cooperation between institute authorities and other agencies working toward agri-

[6] *Farm, Stock, and Home*, XXII (January 1, 1906), p. 5; Minnesota Farmers' Institutes, *Annual* 22, pp. 8–10; *Experiment Station Record*, XXI, p. 298.

[7] U.S. Office of Experiment Stations, *Annual Report*, 1908, p. 296; *ibid.*, 1910, p. 388.

[8] Fletcher, *Pennsylvania Agriculture and Country Life*, p. 445; Witter, "Farmers' Institutes in New York," p. 224; *Hoard's Dairyman*, XL (March 19, 1909), p. 214; AAFIW, *Proceedings*, 1906, p. 26; *ibid.*, 1908, p. 27; Missouri State Board of Agriculture, *Annual Report*, 1907, p. 23; *ibid.*, 1908, pp. 23–25.

cultural improvement. In Colorado, for instance, where in 1907–08 the institute management held twelve of the schools for men and five for women, communities were required to provide facilities and local advertising and minimums of 100 men and 50 women had to pledge their attendance. Fees were charged, amounting to two dollars for men and one dollar for women. In Mississippi by 1910 the institute management was cooperating with the newly established agricultural high schools in offering movable short courses to farmers.[9]

In a number of states institute officials tried to make their work more effective and the results more continuous by establishing local farmers' clubs. Such groups, they reasoned, could meet regularly, disseminating information and sustaining interest in the intervals between formal institutes. Moreover, they could guarantee the local support always necessary for a successful program. In 1902 President J. C. Hardy of Mississippi Agricultural and Mechanical College believed that these clubs might become study groups using Department of Agriculture and college bulletins as textbooks. He also proposed that clubs select from among their membership farmers who would test the findings reported in agricultural experiment station publications. Ten years later several counties in the state had established clubs.[10]

Other states, too, adopted the local club technique. At Abbeville in Louisiana's Vermilion Parish a club was organized in 1898, and in 1904 state authorities counted five in existence, including one made up of Negro farmers in West Baton Rouge Parish. A club formed in Meeker County, Minnesota, in 1908 was interested in more than education. It announced that its members had banded together "for the purpose of advancing our material welfare through the medium of cooperation in county and district organization, to improve market facilities, and to disseminate useful information on agricultural subjects. . . ." In DeKalb County, Illinois, the secretary of the county institute in 1911 reported the establishment of several

[9] U.S. Office of Experiment Stations, *Annual Report*, 1908, p. 306; *ibid.*, 1910, p. 408.

[10] AAFIW, *Proceedings*, 1902, p. 73; *ibid.*, 1903, p. 20; U.S. Office of Experiment Stations, *Annual Report*, 1907, p. 335.

township clubs. Each expected to hold monthly meetings during the winter, making a total of from twenty to thirty educational gatherings in the county.[11]

A number of other innovations reflected the general effort to improve institutes as a teaching system. The Illinois Farmers' Institute, for instance, inaugurated a free library plan. In that arrangement the state organization acquired sets of agricultural books consisting of about forty-five volumes and loaned them to any community requesting them. Some institute systems developed correspondence courses in agriculture, much like those being offered by some of the land-grant colleges. In 1906 the Maryland institute management employed one lecturer who devoted his time to visiting the rural schools where he spoke to pupils on country life and improved agriculture. Later, Arizona and a number of other states used members of their regular institute forces to aid in the introduction of nature study into rural schools. Perhaps most important in terms of the number of farmers contacted was the extensive cooperation beginning in 1904 between institute men and the railroads in the operation of educational trains.[12]

The need for a new type of lecturer presented institute managers with perhaps their most serious problem. Especially in the Middle West and the Northeast, where institutes were most advanced, the successful, practical farmer was no longer wholly satisfactory as a lecturer. Audiences demanded specialists, men whose knowledge of a subject was far more thorough than that of even the most outstanding farmer. Essentially, farmers wanted college-trained men with sufficient speaking ability to make their talks clear and to the point.

This demand seemed to point toward increased reliance upon agricultural college faculty members and experiment station per-

---

[11] (Abbeville) *Meridional*, XLII (July 16, 1898), p. 3; Louisiana State Board of Agriculture and Immigration, *Annual Report*, 1903 (Baton Rouge, 1904), pp. 8–9; *Farm, Stock, and Home*, XXIV (March 15, 1908), p. 173; *Breeder's Gazette*, LIX (February 15, 1911), p. 421.

[12] Hamilton, "History of Farmers' Institutes," p. 30; U.S. Office of Experiment Stations, *Annual Report*, 1904, pp. 598, 623; *Ibid.*, 1906, p. 308; *Ibid.*, 1909, p. 331.

sonnel. In a number of states, especially those in the South and the West, such men had been used extensively from the outset, in some cases constituting the entire lecture force. It was not until 1903, for instance, that Mississippi used its first "practical lecturer," and as late as 1906 professors did about 75 percent of the work.[13]

Unfortunately, after 1900 a number of factors tended to reduce the availability of college personnel and to cause college administrators to be less enthusiastic about sending their staff members out into the hinterlands. The increasing enrollment at the agricultural colleges, the expansion of experiment station work made possible by the Adams Act, and the establishment of formal extension programs by the colleges meant that professors and station workers found it more and more difficult to leave campus to participate in institutes.

In a few instances the colleges sought to restrict the institute work of their personnel without antagonizing farmers and other rural educational agencies. In New York, Dean Liberty Hyde Bailey concluded at a relatively early date that institutes were of limited value and that the manpower at Cornell could be used to better advantage in other ways. He made institute assignments a personal decision for individual staff members and he directed that under no circumstances should they accept a salary from the Bureau of Farmers' Institutes for any appearance. As Cornell's extension program grew in popularity and scope, Bailey objected even more vigorously to the sending of a highly trained speaker far out into the state to give a forty-five minute talk when his talents could be put to better advantage on campus.[14]

Still, taking the nation as a whole, the demand for trained speakers caused the role of college personnel to grow after 1900. In 1903, for instance, faculty and experiment station men constituted 24 per-

[13] *Prairie Farmer*, LXXVII (October 19, 1905), p. 6; AAFIW, *Proceedings*, 1903, p. 21; "Farmers' Institute Bulletin, 1906," Mississippi Agricultural Experiment Station, *Bulletin* 100 (Agricultural College, 1906), p. 3.

[14] F. E. Dawley to G. W. Cavanaugh, November 7, 1903, L. H. Bailey to A. C. True, December 5, 1910, C. H. Tuck to Bailey, March 21, 1911, and Bailey to R. A. Pearson, August 23, 1911, Liberty Hyde Bailey Papers (Cornell University Library, Ithaca, N.Y.).

cent of the salaried institute lecture force in the United States; by 1914 the percentage was 41.[15]

In some states the pressure for the new type of lecturer was sufficient to produce a significant shift in leadership. Such was the case in Minnesota where in 1907 after twenty years of successful work Oren C. Gregg was sacked as superintendent of institutes. Other factors were involved in his dismissal, but his opponents were especially critical of his use of practical farmers as lecturers. His successor, Archie D. Wilson, was an agricultural college staff member, and under his direction the institutes in Minnesota turned increasingly to the college for their manpower.[16]

Faced with the demand for specialists but at the same time finding many colleges reluctant to allow their staff members to spend their time in institutes, some states sought to create a body of trained speakers, using a technique known as the normal institute. At these gatherings, usually held on the campuses of the agricultural colleges, members of the lecture force who were of the practical-farmer type heard addresses by specialists, including scientists from other states and from the United States Department of Agriculture. The first normal institute was held at Cornell in 1899, Pennsylvania had one the same year, Illinois and West Virginia used the technique by 1905, and other states followed.

The normal institutes in Pennsylvania and New York set the patterns for others. The affair in 1907 in Pennsylvania included representatives from county institute organizations, the entire corps of institute lecturers, and delegates from the various farm organizations in the state. The meeting featured an address by Alfred C. True, head of the United States Office of Experiment Stations. The 1903 normal institute in New York was a two-week conclave; participants divided their time between the Cornell campus and the experiment station at Geneva.[17]

[15] Alfred C. True, *A History of Agricultural Extension Work in the United States, 1785–1923* (Washington, 1928), pp. 32–33.

[16] Minnesota Agricultural Experiment Station, *Annual Report*, 1907–08 (n.p., 1908), p. ci; *Northwestern Agriculturist*, XXII (August 31, 1907), p. 724.

[17] U.S. Office of Experiment Stations, *Annual Report*, 1905, p. 363; *ibid.*, 1907,

Other states tried to improve the quality of their lecturers through the use of roundup institutes. These gatherings were usually held at the end of an institute season at the agricultural colleges or at some central location. Members of the lecture corps participated, as in the case of normal institutes, but the roundups invited ordinary farmers as well, giving the meetings many of the characteristics of an ordinary institute and of farmers' week, a type of program soon to be held by many of the colleges. Wisconsin was the pioneer in roundup institutes, holding one in 1887, and by 1903 fourteen states had them. Alabama held its first roundup that year, attracting 130 farmers to Auburn. The conclave became an annual event, with an attendance that reached 835 in 1910. The 1911 convention, known formally as the Farmers' Summer School and Round-Up Institute, was an eight-day affair, featuring exhibits, judging of livestock, lectures, and demonstrations by agricultural college faculty members.[18] Mississippi originated its roundup in 1902. In 1906 the farmers who attended heard addresses by Seaman A. Knapp; W. M. Bamberg, one of Knapp's subordinates; Assistant Secretary of Agriculture Willet M. Hays; Farmers' Institute specialist John Hamilton; and M. V. Richards and J. C. Clair, officials of the Southern and Illinois Central railroads whose duties fell into the general category of traffic generation.[19]

As the institute movement gained strength in the South after 1900, it came face to face with the race question. Increasingly, institute leaders recognized that any general improvement of farming meth-

---

p. 309; Hamilton, "History of Farmers' Institutes," p. 68; Fletcher, *Pennsylvania Agriculture and Country Life*, p. 445; A. L. Martin to A. C. True, March 19, 1907, and True to Martin, March 26, 1907, Office of Experiment Stations Records; *Experiment Station Record*, XV, p. 4.

[18] True, *Agricultural Extension Work*, pp. 16–17; U.S. Office of Experiment Stations, *Annual Report*, 1903, p. 637; George McKerrow to W. M. Hays, February 6 and March 6, 1906, and J. J. Esch to Hays, February 7, 1905, Records of the Office of Secretary of Agriculture (National Archives, Washington); *Breeder's Gazette*, LVII (March 11, 1910), p. 1121; *Progressive Farmer and Southern Farm Gazette*, XVII (July 29, 1911), p. 668.

[19] *Southern Farm Gazette*, VII (August 15, 1902), p. 5; *ibid.*, XI (September 1, 1906), p. 2; *ibid*, XII (September 15, 1907), p. 9; J. C. Hardy to W. M. Hays, February 10 and June 16, 1905, Office of Experiment Stations Records.

ods in the region would entail the education of the Negro in at least the basic aspects of crop production. This enlightened view met considerable opposition in some areas of the South, especially in those districts where the sharecropper system was most firmly established. In Mississippi J. C. Hardy of the agricultural college complained about the negative attitude of Delta planters who thought of agriculture only in terms of the "negro and mule," a combination to which they assumed scientific farming had little relevance.[20]

Nevertheless, the institute movement did reach increasing numbers of Negroes. From the outset a few Negro farmers had appeared at ordinary institutes, and their number grew after 1900. In North Carolina, in fact, it was reported that "some Negro farmers are present at all of the institutes, and they are always welcome." Mississippi was another state where officially at least Negroes were invited to attend ordinary institutes.

Probably more important near the turn of the century and later was the appearance of institutes held exclusively for Negroes. In some instances these programs were arranged by the institute management in particular states; in other cases institutes of various types were held by the Negro land-grant colleges. The institute officials in North Carolina reported the holding of four well-attended institutes for Negroes in 1903; five years later there were several comparable meetings in Mississippi.[21] Meanwhile, the Negro agricultural colleges were pushing ahead as rapidly as funds and manpower permitted. As early as 1899–1900 there were gatherings called institutes on the campuses of the Negro agricultural colleges in Georgia and Kentucky, while in South Carolina two faculty members attended local meetings in twenty-two counties. The Florida Agricultural and Mechanical College for Negroes at Tallahassee in January 1910, held its first winter institute for Negroes. The Negro college

---

[20] *Experiment Station Record,* XXIV, p. 97; AAFIW, *Proceedings,* 1904, p. 26; "Farmers' Institute Bulletin, 1902," Mississippi Agricultural Experiment Station, *Bulletin 80,* p. 57.

[21] Hamilton, "History of Farmers' Institutes," p. 70; "Farmers' Institute Bulletin, 1907 and 1908," Mississippi Agricultural Experiment Station, *Bulletin 120,* p. 3.

in North Carolina was another of the schools of its type that arranged an annual conference on campus for Negro farmers.[22]

Tuskegee Institute had what appeared to be the most successful institute system among the Negro colleges. The objectives of the institutes, after all, squared very well with the philosophy of Booker T. Washington. Tuskegee's program consisted of two types of gatherings. First was a series of institutes held throughout the area by Tuskegee faculty members. In form these were similar to those held across the United States. The state provided no money for the work, expenses being met instead by the different localities. Meetings were advertised by the usual circulars and personal letters and by preachers who played a role unknown among white farmers. As early as 1903 Tuskegee officials reported an aggregate attendance of 10,000; the next year there were 139 separate meetings.

Tuskegee also held an annual winter meeting on the campus known in 1907 as the Tuskegee Normal and Industrial Institute. Attended by farmers from all parts of the state, the gathering featured speeches not only by Tuskegee faculty members but also by authorities from the Department of Agriculture and elsewhere. Some 2,000 Negro farmers and their families attended the affair in 1910.[23]

In a few instances other minorities received some attention. New York's director of institutes reported in 1907 that his organization was attempting to aid Indians residing on reservations in the state. One of the major difficulties he found was unusual apathy and a general refusal to assume responsibility for making necessary local arrangements.[24]

While institute directors were expanding their programs for the nation's farmers, it was only reasonable that they would ultimately try to reach farm wives. Certainly there was as great a need for instruction in the kitchen as there was in the barn. In fact, accord-

[22] U.S. Commissioner of Education, *Report,* 1899–1900, II, 2048–2049, 2051; *Experiment Station Record,* XXII, p. 500; J. H. Bluford to D. J. Crosby, January 3, 1913, Extension Service Records (National Archives, Washington).

[23] U.S. Office of Experiment Stations, *Annual Report,* 1903, pp. 650–651; *ibid.,* 1907, p. 316; *ibid.,* 1910, p. 359.

[24] New York Farmers' Institute and Normal Institutes, *Report,* 1906 (Albany, 1907), p. v.

ing to Maria L. Sanford, a University of Minnesota staff member who appeared on the state's institute platform in the 1880s, "one great cause of trouble was the lack of good, thrifty wives in the farmers' homes. Some men were dissipated and worthless because they never had anything fit to eat."[25] Probably Miss Sanford's view of the matter was an oversimplification; but throughout the country, authorities recognized that farm wives had much to learn, not only about food preparation and sanitation but also about the caring for children, the making and repairing of clothes, and the providing of desirable home conditions.

Almost from the beginning of institute work, women had appeared on institute platforms, speaking on subjects pertaining to the home or discussing other topics with which they were more familiar. At the pioneer institutes in Iowa, in fact, one of the speakers was Mrs. Ellan Tupper, the "Bee Woman," who accompanied President Welch and Professor Isaac P. Roberts of Iowa agricultural college to Council Bluffs and other points in the state. A decade later, in the early 1880s the first instructor of home economics at the Iowa college gave a series of lectures on her specialty at Des Moines. Even before Miss Sanford took her post in Minnesota, she had served as an institute lecturer in the East.[26]

By the 1890s it was common for institute directors to include in one or more sessions at institutes women who spoke on a variety of topics related to the farm home. At Opelousas, Louisiana, in 1898, a woman entitled her talk "The Farmer's Wife" and called upon rural women to make the farm home more healthy and pleasant. Ten years later in Mississippi women still appeared on the same platform with men and discussed food preparation and diets.

Other women addressed institute audiences on purely agricultural subjects. As early as 1887 two women lectured in Wisconsin on dairying, and for more than a decade after 1891 Mrs. Ida E. Til-

[25] *Farm, Stock, and Home,* III (February 15, 1887), p. 99.

[26] John Hamilton, "Farmers' Institutes for Women," U.S. Office of Experiment Stations, *Circular* 85 (Washington, 1909), pp. 3–4; Roberts, *Autobiography of a Farm Boy,* p. 100; Morgan, *Extension Service of Iowa State College,* p. 16; D. Williams to M. L. Sanford, July 13, 1880, and F. F. Graham to Sanford, July 28, 1880, M. L. Sanford Papers (University of Minnesota Archives, Minneapolis).

son, a Wisconsin schoolteacher who abandoned the classroom for the poultry yard, was a regular member of the lecture force in Minnesota.

Interspacing men and women as lecturers at institute sessions was considered by some authorities to have a special value. By presenting programs of interest to both sexes, attendance was increased. Moreover, the presence of women on the platforms and in the audiences uplifted the whole program, giving it a tone usually lacking in other institutes.[27]

Still, institute managers sought more effective means of instructing farm wives. One innovation was the cooking school, held in conjunction with ordinary institutes but with wives meeting separately from their husbands. As early as 1890 Oren C. Gregg began holding cooking schools in Minnesota, using as instructors university faculty members and wives of staff members who had been trained in "domestic science." The schools introduced farm wives to the mysteries of balanced and varied diets, different cuts of meats, new recipes, and food analysis. Programs consisted of practical demonstrations as well as talks by the instructors, and often the products of the demonstrations were consumed in a social hour later in the day. Instruction of a similar nature appeared in other states as well. The first cooking school in Wisconsin met at Portage in March, 1892, and until 1907 the institute management in that state scheduled eleven each year. Meanwhile, in Michigan Mrs. Mary A. Mayo held twenty cooking schools with an attendance of 5,300 during the period 1896–97.[28]

All of these types of efforts to reach rural women continued after 1900 and numerous innovations were introduced. In Minnesota, for

[27] Louisiana Farmers' Institute, *Bulletin* 2, pp. 84–87; (Opelousas) *Courier*, XLV (July 16, 1898), p. 1; *Southern Farm Gazette*, IX (October 1, 1904), n.p.; "Fifty Years of Cooperative Extension in Wisconsin," p. 27; Scott, "Early Agricultural Education in Minnesota," p. 29; Oren C. Gregg to Juniata L. Shepperd, nd., Juniata L. Shepperd Papers (University of Minnesota Archives, Minneapolis).

[28] Minnesota Farmers' Institutes, *Annual* 10, pp. 48–49; *Northwestern Farmer*, XIII (November 15, 1895), p. 357; *Farm, Stock, and Home*, VI (February 1, 1890), p. 96; "Fifty Years of Cooperative Extension in Wisconsin," p. 28; Beal, *Michigan Agricultural College*, pp. 159–160; G. M. McKerrow to J. S. Morton, November 7, 1895, Office of Experiment Stations Records.

example, Oren C. Gregg found that interest in cooking schools declined after 1900, so in 1903 he undertook to organize women's clubs in rural communities. Special instructors went into the countryside, called neighborhood women together, and helped to establish the organizations. Once clubs were established, Gregg hoped that they would function without direct supervision from his office and would serve as a medium through which information might be funneled to rural women.[29]

Illinois began organizing women's clubs even earlier. In 1898 institute authorities established an organization for women similar in form to that for men. A central body, known as the Illinois Association of Domestic Science, united a growing number of county organizations that had as their primary function the managing of annual institutes for women. But in addition each county group was encouraged to establish any number of local clubs that would meet more frequently and constitute a medium for continuing education and social intercourse. A decade later seventy-nine counties had organizations, and some of them had as many as eight subordinate clubs. The state contributed financially to the work, authorizing the use of a portion of its appropriation for men's institutes.

The institute management in Missouri held no meetings specifically for women until 1908 when four were scheduled. By 1911 a number of speakers were assigned to the work, and that year they held ninety-two meetings and spoke to 6,897 women on such subjects as "Cooling Meat" and "Balanced Diet for Man." The next year the state board of agriculture directed one lecturer to devote her full time to the work with rural women. Increasingly, the emphasis came to be upon the organization of homemakers' clubs among the women contacted.[30]

[29] AAFIW, *Proceedings*, 1906, p. 39; Minnesota Farmers' Institutes, *Annual* 17, p. 5; Hamilton, "Farmers' Institutes for Women," pp. 13–14; *Wallaces' Farmer*, XXX (October 20, 1905), p. 1230.

[30] Illinois Farmers' Institute, *Annual Report*, 1902, p. 247; Butterfield, "Farmers' Institutes," p. 640; *Breeder's Gazette*, LVII (March 9, 1910), p. 648; Hamilton, "Farmers' Institutes for Women," pp. 7, 9–10, 13, 14; Missouri State Board of Agriculture, *Monthly Bulletin*, IX, p. 93; Missouri State Board of Agriculture, *Annual Report*, 1912, p. 27.

In the South, North Carolina had the best system of separate institutes for women. Under the energetic leadership of Tait Butler, the first meetings exclusively for women were held in 1906. Two years later there were sixty-eight of them scattered across the state, and the number more than tripled by 1912. In the main these affairs consisted of one or two sessions held concurrently with farmers' institutes but in different halls. The state institute management provided two speakers, and local talent also served on the programs. Beginning in 1908 instruction was improved at many points through the use of a domestic science demonstration car provided by the Southern Railway. A converted passenger coach, the car served as an auditorium and carried a well-stocked kitchen as well as labor-saving utensils and devices of use in the rural home.[31]

New York had separate institutes for women from 1906 to 1908, but later the state returned to the older practice of assigning women to regular institutes, allowing them to speak on topics of the home to mixed audiences. The work was considered to be uncommonly successful in New York, partly because it was well supported financially by the state. Among the speakers who lectured before the state's women was Martha Van Rensselaer, one of the nation's pioneers in home economics.[32]

Still, a study made in 1909 by the United States Department of Agriculture showed that the women's institute movement had failed to become nationwide in scope. Of the forty-eight states and territories, twenty made no provision for the instruction of women. Thirteen others used female lecturers at ordinary farmers' institutes, giving instruction on farm home topics to mixed audiences. Only nine states held separate institutes exclusively for women, while women's auxiliary organizations, similar to that in Illinois, func-

[31] "Women's Institutes in North Carolina," *Journal of Home Economics,* I (April, 1909), pp. 161–163; *Southern Planter,* LXIX (September, 1908), pp. 770–771; *Progressive Farmer and Southern Farm Gazette,* XXIX (August 15, 1914), p. 864.

[32] F. E. Dawley to L. H. Bailey, January 23, 1904, Bailey Papers; *Farm, Stock, and Home,* XXIII (June 15, 1907), p. 373; Witter, "Farmers' Institutes in New York," p. 221; Colman, "History of Agricultural Education," p. 258.

tioned in four states. Finally, by 1909 two states were attempting to reach a few women through the use of movable schools of domestic science, an extension technique that was growing in popularity with the land-grant colleges.

Institutes for women failed completely to achieve the popularity of those for men. Women consistently displayed a greater reluctance to attend sessions, and apparently fewer of them continued their interest after the first meetings. Between 1910 and 1912, for example, the number of states holding separate institutes for women declined, and in 1913–14 total attendance was only 78,237.[33] Most promising of the various approaches to the problem of educating rural women was the neighborhood club, meeting more or less regularly and featuring demonstrations by trained workers.

Soon after 1900 institute managers began to devote some attention to instructing the youth of their states. In several instances the new departure took the form of simply providing special lectures at ordinary institutes for farm boys and girls. Such was the case in New York when in 1903 the director of institutes reported that in a few instances as many as 300 children had attended sessions designed for them. The closing of schools on the appointed days no doubt contributed to the attendance. In Missouri lecturers were directed to handle certain topics in such a way as to make them understandable and appealing to children. Elsewhere, as for example in Louisiana, institute officials made concerted efforts to attract schoolteachers to institutes, reasoning that their attendance might contribute to the teaching of agriculture in the rural schools.[34]

More common than ordinary institutes were the various types of

[33] Hamilton, "Farmers' Institutes for Women," pp. 12–15; U.S. Office of Experiment Stations, *Annual Report*, 1910, pp. 421–422; *ibid.*, 1911, pp. 382–383; *ibid.*, 1912, pp. 376–377; A. C. True to P. S. Spence, December 1, 1914, Office of Experiment Stations Records.

[34] F. E. Dawley to James Wilson, November 27, 1903, Records of the Office of the Secretary of Agriculture; John Hamilton and J. M. Stedman, "Farmers' Institutes for Young People," U.S. Office of Experiment Stations, *Circular* 99 (Washington, 1910), pp. 16, 19; Missouri State Board of Agriculture, *Annual Report*, 1906, p. 19; *Lafayette Advertiser*, XXXVIII (June 27, 1903), p. 1.

clubs established for farm boys and girls. The club approach allowed promoters to arrange contests and introduce a competitive element, thereby arousing greater interest than was possible any other way. The organization of club work involved cooperation with other agencies to a far greater degree than did institutes for men or women. School officials and the land-grant colleges played significant roles. In various states these agencies joined with the institute managements in establishing and supervising clubs; in other instances they originated club work, while the institute officials were little more than pleased observers.

In any event club work generally had two clear-cut objectives. Promoters hoped on the one hand to teach young people methods of improved agriculture, both through the actual producing of a commodity and through the changing of school programs to make them more meaningful to rural students. On the other hand, officials who established youth clubs hoped to use them to influence adult farmers. Presumably, few farmers could fail to be impressed by techniques that proved successful for their sons.

Legend has it that corn clubs, the direct ancestor of the 4-H movement, originated in the South. That view is counter to the facts although southerners made the technique dramatically more effective as a teaching device. The idea of organizing farm boys and girls into clubs for the growing of corn or other crops first appeared in relatively recognizable form in the Middle West, where a few schoolteachers, school superintendents, and institute workers took up the work.

The origins of boys' club work in the North is almost as obscure as those of farmers' institutes. As early as 1828 a private-school teacher near Cincinnati induced his pupils to grow corn, garden truck, and flowers, and at various points near mid-century farm boys won awards at local and state fairs. In 1882 Professor J. A. Reinhart of Delaware College tried to organize in a systematic way the farm boys of that state. Any boy under eighteen was eligible to enter a statewide contest and to compete for prizes. Participants were required to raise one-fourth of an acre of corn, harvest it, and measure accurately the results. Winners received one of thirty-five prizes, ranging downward from twenty dollars.[35]

Professor Reinhart's experiment was apparently of short duration, but a number of others developed soon after 1900 that constituted important steps toward the establishment of continuing instruction for farm youth. Foremost among these was the work of Will B. Otwell of Macoupin County, Illinois.

When Otwell found it impossible to induce Macoupin County farmers to attend institutes, he turned to their children. In 1900 he distributed first-quality seed corn to 500 boys, arranged for a number of prizes, and announced that there would be a corn show at the fall institute. The innovation achieved its primary purpose; several hundred farmers appeared at the courthouse on the appointed day, along with scores of boys who brought samples of their corn to be judged. More important in the long run was the discovery of a technique for reaching future farmers. The experiment was repeated in 1901 and 1902 with better results each year. Otwell's growing reputation caused the governor of Illinois to place him in charge of the state's agricultural exhibit at the St. Louis fair, scheduled for 1904. Otwell could think of nothing better than a statewide corn growing contest. Some 8,000 packages of seed went out to boys, and 1,250 provided samples of their crops in the fall. Artistically arranged, these constituted the Illinois exhibit at the St. Louis exposition.[36]

After his success at St. Louis, Otwell expanded his work with farm boys, enlisting participants beyond the borders of Illinois, providing them with seed, and holding annual roundups or exhibitions at Carlinville in Macoupin County. To keep in touch with the boys, he established in 1905 a periodical, *Otwell's Farmer Boy*, which in 1911 was reported to have been read by 40,000 boys. That year Otwell claimed to have 25,000 names in thirty-two states on his

---

[35] *American Agriculturist,* XLI (May, 1882), p. 196; Franklin M. Reck, *The 4-H Story: A History of 4-H Club Work* (Ames, Iowa, 1951), pp. 4–5.

[36] Dick J. Crosby, "Boys' Agricultural Clubs," U.S. Department of Agriculture, *Yearbook,* 1904 (Washington, 1905), pp. 489–491.

contestants' list. Otwell formed no clubs and there were no regular meetings of the boys who planted the seeds he sent them. But his work played a role in stimulating interest in better agriculture and it suggested to others the possibility of more systematic programs for farm youth.[37]

Among other early pioneers in the work was O. J. Kern, superintendent of schools in Winnebago County, Illinois. Like many of his contemporaries, Kern was concerned with the failure of the rural schools to meet the needs of farm children and with the ordinary farmer's lack of interest in the schools his children attended.

As Kern was considering his problems in Winnebago County, the University of Illinois, which had underway significant work in corn breeding, was facing the usual difficulty in placing its new knowledge in the hands of farmers. To meet its needs, in fact, the institution had created an extension department, one of the first in the nation, and had appointed Fred H. Rankin as its first head. Consequently, the University of Illinois and O. J. Kern struck a partnership when each became convinced that the other could help in the solution of existing problems.

The result was the establishment on February 22, 1902, in Rockford of a boys' experiment club with thirty-seven charter members. Each boy received samples of seed corn and sugar beets, with suggestions from the university concerning the growing of the crops. In the course of the growing season the boys made simple tests, went on an excursion to the university, and calculated the costs of producing their crops.

Kern carried on his work in later years, adding to it from time to time and urging other county superintendents to undertake similar programs. In September, 1903, for instance, he organized 216 girls in Winnebago County into a Girls' Home Culture Club. School officials elsewhere accepted his suggestions, and as early as 1904 there were clubs in a dozen counties in the state with about 2,000

[37] *Corn,* I (May, 1912), p. 92; *N. W. Ayer and Son's American Newspaper Annual,* 1912 (Philadelphia, 1912), p. 165; Benjamin M. Davis, "Agricultural Education: Boys' Agricultural Clubs," *Elementary School Teacher,* XI (March, 1911), pp. 378–379.

boys enrolled. By that time the work to some degree had been systematized: the Illinois Farmers' Institute, a sugar beet company, and the university furnished seed; the university provided informative bulletins; and local school teachers and superintendents guided the programs at the local level. The schoolmen and local institute officials arranged fall corn shows where outstanding boys received awards as recognition for their work.[38]

Similar in many ways to Kern's work was that of Albert B. Graham in Ohio. A country schoolteacher, Graham had become aware of the inadequacies in rural education as early as 1896. The introduction of nature study offered some possibilities, he thought, but it was not until 1902, when he was superintendent in Clark County, that he found a method for putting his ideas into practice. Calling a meeting at the county building in Springfield on January 15, 1902, he organized a group of farm boys into a club, showed them some litmus paper, and sent them out to test the soil on their fathers' farms. Later he urged them to plant experimental corn plots, using seed selected at their homes. Occasional meetings allowed the boys to discuss better methods and to report on their progress.

The next year Graham wrote to Dean Thomas L. Hunt of the Ohio College of Agriculture, asking for advice and assistance in improving his program. Hunt was more than willing. Like other agricultural college administrators, he was eager to adopt any technique that promised to take science into the countryside. It was decided that the Agricultural Students' Union, an organization of former college students established earlier to test new ideas developed at the experiment station, would cooperate with Graham in his project. Soon eighty-five boys were growing improved varieties of corn; in addition, they continued soil testing exercises and gathered collections of various types of seeds, weeds, and insects. Graham spent a good part of his summer visiting his boys and their projects, and in June he took them to Columbus where they visited the college

[38] Crosby, "Boys' Agricultural Clubs," pp. 491–492; O. J. Kern, " 'Learning by Doing' for the Farmer Boy," *Review of Reviews*, XXVIII (October, 1903), pp. 456–459; Reck, *4-H Story*, pp. 16–19. Kern's work is discussed in detail in his *Among Country Schools* (Boston, 1906), pp. 129–150, 158–174.

of agriculture. Although Graham refused to emphasize the awarding of prizes, the year ended with an exhibit of the boys' work. Meanwhile, in the course of the year twenty girls were organized to grow vegetables and flowers.

From Clark County the idea of organizing Ohio farm boys spread. By 1904 some 1,000 were working in thirteen clubs in as many counties. Two years later there were 3,000 enrollees. Increasingly, the college of agriculture assumed overall direction of the work; in July, 1905, Graham was appointed the first superintendent of agricultural extension at Ohio State. At the outset his duties consisted primarily of extending his club work throughout the state. He worked through local school officials, providing them with information and general supervision, while they took direct charge of the "school agricultural clubs" in their jurisdiction.[39]

The activities of Otwell, Kern, and Graham were only the first steps in a broad and complex rural club movement that arose in the North in the decade before 1914. In most midwestern states some efforts were made during those years to reach farm youth. The work developed spontaneously, with institute officials, school superintendents, and agricultural college personnel participating in different ways. There was no uniformity to these programs until the United States Department of Agriculture entered the work. Circumstances alone dictated the nature of the clubs.

In Missouri it was the Missouri Corn Growers' Association that in 1906 joined with school superintendents and teachers, commercial clubs, bankers, and local newspapers to organize a statewide contest for boys and young adults. The basic unit was the local or county club, made up of any number of boys supervised in the main by rural teachers. Enrollees who won prizes in contests at those levels were permitted to enter contests at the state fair and at an annual corn show held on the University of Missouri campus. Awards offered by businessmen and given for the best displays of five or ten

[39] Reck, *4-H Story*, pp. 12–15; Thomas C. Mendenhall, ed., *History of the Ohio State University* (3 vols., Columbus, Ohio, 1920–26), II, 148–149; A. B. Graham, "Boys' and Girls' Agricultural Clubs," *Agricultural History*, XV (April, 1941), pp. 65–66; Crosby, "Boys' Agricultural Clubs," p. 493.

ears of corn stimulated the boys to do their best. The agricultural college and the farmers' institute management provided enrollees with literature and sent men to act as judges at the shows. Some 500 boys participated in 1906; over 800 the next year.[40]

Boys' clubs in Minnesota originated as they did in Ohio, with rural school officials. As early as 1904 Superintendents T. A. Erickson in Douglas County, George F. Howard in Olmstead County, and L. P. Harrington in McLeod County had the work underway. The first year Erickson spent twenty dollars of his own money to acquire seed for distribution to boys in Douglas County. According to the rules that he established, each participant was to show ten ears of corn at a fair to be held in his school; winners there entered a similar exhibit at a countywide affair held at Alexandria. Later some boys grew potatoes and girls who enrolled produced poultry and tomatoes. Businessmen provided the prizes awarded to winners; in 1906 James J. Hill of the Great Northern Railroad donated twenty-five dollars for the purpose.[41]

Similar in form but broader in scope was the Minnesota Industrial Contest. Resting upon the early efforts of school officials, it became organized on a statewide basis in 1907. The program was launched by the college of agriculture. The farmers' institute management detailed to one man the task of supervising the work, visiting many of the local clubs, and consulting with county superintendents and local teachers. Enrollees were not limited to corn growing; boys producing a variety of grains, fruit, and vegetables were allowed to compete, while girls could enter the results of their cooking and sewing in contests. As in Missouri, prizes were awarded on the county level and winners went on to compete in a state contest.[42]

In Iowa schoolteachers, farmers' institute officials, and agricultural

[40] *Breeder's Gazette*, LI (March 13, 1907), p. 603; *ibid.*, LIX (March 15, 1911), pp. 725–726; *Wallaces' Farmer*, XXX (May 22, 1908), p. 719; AAFIW, *Proceedings*, 1911, pp. 18–19.

[41] T. A. Erickson, *My Sixty Years with Rural Youth* (Minneapolis, 1956), pp. 56–59; C. L. McNelly, *The County Agent Story: The Impact of Extension Work on Farming and Country Life* (Anoka, Minn., 1960), p. 21.

[42] *Breeder's Gazette*, LV (January 13, 1909), p. 77; Hamilton and Stedman, "Farmers' Institutes for Young People," pp. 19–20.

college men combined to introduce club work into the state. One of the leaders was Cap E. Miller, superintendent of schools in Keokuk County, who formed his first club at Signourney in March, 1904. In the fall of that year the 147 school districts in the county held school fairs at which boy club members displayed the grain, fruit, and vegetables that they had grown. Girls' projects included sewing, baking, and basketmaking. First and second-prize winners in each category later entered their exhibits in a county show. All participants kept notes on the progress of their projects, using them to produce compositions that constituted one means of directing ordinary schoolwork toward the needs of farm life. Other features included an annual excursion to Ames, meetings at which club members heard talks by leading figures in Iowa agriculture, and occasional visits to prosperous and well-run farms in the neighborhood. At the end of 1905 Miller reported that 600 boys were enrolled in his Boys' Agricultural Club and there were 500 members in the Girls' Home Culture Club.[43]

Similar developments were underway elsewhere in Iowa. In Wright County Superintendent O. H. Benson inaugurated club work in 1904, introducing at the same time a three-leaf clover as a symbol for his organization. Seven years later, when Benson went to Washington to take a position with the United States Department of Agriculture, his symbol became a four-leaf clover which would soon be the emblem of the nationwide 4-H movement. In Sioux County in 1905 the farmers' institute officers encouraged the superintendent and the teachers to interest young people in better agriculture. That year sixty-one boys and girls grew one-fourth acre plots of corn under the supervision of adults. Miss Jessie Field, superintendent in Page County, launched the work there in March, 1907, after being encouraged to do so by Perry G. Holden of the Iowa agricultural college.[44]

[43] Crosby, "Boys' Agricultural Clubs," pp. 494–495; *Wallaces' Farmer*, XXXIX (June 17, 1904), p. 802; *ibid.*, XXX (December 22, 1905), p. 1530; *ibid.*, XXXIII (October 16, 1908), p. 1261.

[44] Morgan, *Extension Service of Iowa State College*, pp. 44–46; *Wallaces' Farmer*, XXXIV (December 31, 1909), p. 1706; *ibid.*, XXXII (November 29, 1907), p.

These independent but similar programs suggested the practicality of establishing a statewide movement, administered by some central office. As early as 1902 *Wallaces' Farmer*, the leading agricultural paper in Iowa, had begun a seed corn contest for boys, furnishing all who applied with first-quality seed and offering prizes to those who submitted samples in the fall. More importantly the newly established agricultural extension office at Iowa State College was deeply interested in the club work, aiding it whenever possible. When that office was ready to assume direct responsibility for programs with farm youth, it found that the early clubs had prepared the ground.[45]

E. C. Bishop, assistant superintendent of public instruction, took the lead in club work in Nebraska. Working with the college of agriculture, he began the organization of boys' and girls' clubs in 1905. The Department of Public Instruction and the farmers' institute management furnished literature, suggestions for club organization, and general supervision of the work. School superintendents and teachers provided local management. The boys grew corn and vegetables; the girls practiced baking with directions prepared by the domestic science department of the university. There were fall contests at the district, township, county, and state levels. Between 2,000 and 3,000 boys and girls were attending the state shows at Lincoln by 1909. By that year practically every county in Nebraska had at least one youth club, and some observers believed that Nebraska had the best club system in the nation.[46]

Elsewhere in the Middle West the club movement followed in a general way the developments in Iowa, Illinois, and Minnesota. In Wisconsin R. A. Moore and the Wisconsin Agricultural Experiment Association were the pioneers, establishing clubs that by 1912

1401; *ibid.,* (December 20, 1907), p. 1500; *Hoard's Dairyman,* XL (April 2, 1909), p. 295.

[45] Morgan, *Extension Service of Iowa State College,* pp. 46–47.

[46] Davis, "Boys' Agricultural Clubs," p. 380; Crawford, *University of Nebraska,* p. 116; Hamilton and Stedman, "Farmers' Institutes for Young People," p. 21; F. W. Howe, "Boys' and Girls' Agricultural Clubs," U.S. Department of Agriculture, *Farmers' Bulletin* 385 (Washington, 1910), pp. 8–9; *Wallaces' Farmer,* XXXIV (August 27, 1909), p. 1071.

existed in fifty counties and had about 23,000 members. Michigan's first clubs developed in Muskegon and Mason counties in 1908. In Indiana a number of county institute associations soon after 1900 held separate sessions for farm youths, and others cooperated with local school officials to establish clubs. A broader club movement began in 1905 under the general direction of the state institute organization and Purdue University. By 1910 there were clubs in forty-five counties with an enrollment of 12,000. Contests for farm girls in Indiana involved the baking of bread. Professor G. W. Randlett of North Dakota Agricultural College took the lead in his state, establishing the first rural club in 1905. By 1910 boys and girls who had won at county contests were treated to a week at the agricultural college when they attended a short course held for them.[47]

Youth club work also appeared in the East and the West, but there was considerably less there than in the Middle West. Massachusetts may well have been the leader in the East when in 1908 a potato club was organized in New Hampshire County. By 1913 there were boys' and girls' corn and potato clubs in 208 towns with an aggregate membership of over 15,000. The program was administered by the agricultural college, with the assistance of the state board of agriculture. Boys' work in Connecticut began in 1911 when the state board of agriculture organized a corn club. It had only 70 members in 1912. The first clubs in Maine were established the next year. E. M. Rapp, superintendent of schools in Berks County, Pennsylvania, had some 1,500 boys and girls in clubs in 1910.[48] Colorado

[47] "Fifty Years of Cooperative Extension in Wisconsin," p. 53; "History of Cooperative Extension Work in Michigan," pp. 6, 20; Hamilton and Stedman, "Farmers' Institutes for Young People," pp. 17–18; *Wallaces' Farmer*, XXXVI (June 23, 1911), p. 976; Hunter, *North Dakota's Land-Grant College*, p. 57; *Breeder's Gazette*, LVII (April 27, 1910), p. 1036; *ibid.*, LIX (January 11, 1911), p. 87; *ibid.*, LXIII (January 8, 1913), p. 86.

[48] "Boys' and Girls' Clubs in Agriculture and Home Economics in Massachusetts," *The School Review*, XXIV (December, 1916), pp. 765–766; Harold W. Cary, *The University of Massachusetts: A History of One Hundred Years* (Amherst, Mass., 1962), p. 118; Caswell, *Massachusetts Agricultural College*, p. 70; Stemmons, *Connecticut Agricultural College*, p. 247; Merritt C. Fernald, *History of Maine State College and the University of Maine* (Orono, Maine, 1916), p. 265; *Breeder's Gazette*, LIX (January 4, 1911), p. 22.

began the work the same year. The agricultural college assumed direction of the program, but as elsewhere immediate responsibility fell upon school superintendents and teachers. The first year boys grew potatoes and sugar beets in irrigated areas, milo in dry land sections. Girls baked bread and sewed. The first county to be effectively organized was Sedgwick; there Miss Elma O. Law, an energetic superintendent, pushed the work.[49]

The youth club movement outside of the South suffered from a lack of uniformity. It developed spontaneously, and a diversity of organizations, agencies, and individuals played supervisory roles in it. Nor were there any uniform standards applied to the contests in which club members competed. But out of the pioneer efforts in the North, coupled with developments in the cotton states, would come in time modern 4-H work. That development could occur only after Congress and the Department of Agriculture provided basic guidelines.

Meanwhile, some rural leaders found another method to reach rural boys. The originator of the farm encampment was A. P. Grout of Scott County, Illinois. A prominent farmer and institute worker, Grout became convinced that the institutes were not doing enough for farm boys. To fill the void he resolved to hold on his farm near Winchester a week-long encampment for the winners of the county corn judging contests in his congressional district. Grout offered to feed the boys and provide tents and other necessities. The program, which began July 30, 1906, included visits to fields and feed lots and lectures by University of Illinois staff members and by institute workers. In all, fifty-four boys participated, and Grout concluded that the affair was a success. So enthusiastic was he, in fact, that he suggested that similar encampments be held in each congressional district in the state.

Grout repeated his experiment in 1907. Again the Scott County farmer met most of the costs, reducing his outlay somewhat by charging each boy twenty cents a meal. Among other lessons, the boys learned how to care for and adjust various farm machines, in-

[49] *Ibid.*, LVII (April 20, 1910), p. 985; *ibid.*, LXII (May 4, 1910), p. 1087.

cluding plows and grain binders. Among the speakers were Fred R. Crane of the University of Illinois, Superintendent H. G. Russell of the Greenfield public schools, and W. S. Corsa, a widely known horse breeder from White Hall. Only thirty-one boys appeared for the encampment, and Grout discontinued the gatherings after the 1907 affair.[50]

Elsewhere other leaders experimented with Grout's technique. In August, 1907, Samuel M. Jordan, then an employee of the Missouri State Board of Agriculture, held an encampment on his farm in Gentry County. Boys between the ages of ten and twenty were invited. Mornings were devoted to instruction featuring lectures by Jordan and men from the college of agriculture. Demonstrations and various kinds of exhibits enhanced the instruction. Afternoons were devoted to organized sports. At the end of the five-day affair each boy was required to write a report summarizing the week's activities. A total of 132 boys participated, and the effort was deemed a success by its promoters. The next year the state board sponsored five encampments in various parts of the state, instructing in all almost 1,000 boys. Two encampments in Missouri were held in 1909, but thereafter emphasis shifted to clubs organized on the local level, a method by which a larger number of boys might be contacted.[51]

Boys' encampments were held in a few other states. In Page County, Iowa, Superintendent of Education Jessie Page organized one in conjunction with chautauqua week at Clarinda, August 10–19, 1910. There, as in Illinois and Missouri, agricultural college personnel and others instructed boys between the ages of nine and eighteen in grain production, livestock raising, and other aspects of farming. The same year the Illinois State Board of Agriculture invited each county to

[50] *Wallaces' Farmer*, XXXI (December 21, 1906), 1515; *ibid.*, XXXII (September 13, 1907), p. 1004; *Hoard's Dairyman*, XXXIX (August 28, 1908), p. 814; Arthur J. Bell, *The Grout Farm Encampment* (Urbana, Ill., 1906), pp. 5–6; *Breeder's Gazette*, LII (September 11, 1907), p. 477.

[51] Missouri State Board of Agriculture, *Annual Report*, 1907, pp. 24–25; *ibid.*, 1908, p. 22; *ibid.*, 1909, p. 24; *Stanberry Owl*, July 23 and September 3 and 17, 1907; interview by the author with M. F. Miller, June 27, 1960; *University Missourian*, March 8, 1909, p. 3.

send two boys to a school and encampment held in conjunction with the state fair. Each boy was expected to keep a record of his experiences and to make a report at the next farmers' institute in his county. Similar was an encampment held at Valley City, North Dakota, in 1913. The affair was arranged by the North Dakota Better Farming Association, an agency that in a diversity of ways sought to improve farm life in that state.[52]

As various agencies and individuals tried to improve the institute systems in the different states, the United States Department of Agriculture assumed a role in the movement. As early as 1889 some local institute officials had pointed out that a national institute office could constitute an important link between the Department of Agriculture and ordinary farmers.[53] In February of that year John C. Spooner of Wisconsin introduced in the Senate a bill calling for the appointment of a national superintendent of institutes and the appropriation of up to $500,000 to enable him to hold in conjunction with state and local agencies as many as 400 institutes annually. The bill had no hope of passage, but its introduction no doubt played some role in turning the attention of the Department of Agriculture to the institutes. In any event, in 1889 the Office of Experiment Stations began to gather information concerning institutes, a function that it continued for almost thirty years.[54]

But institute leaders wanted the federal government to do more. In 1897 the American Association of Farmers' Institute Managers asked the Secretary of Agriculture to establish in his department a division of farmers' institutes. Alfred C. True, head of the Office of

[52] *Wallaces' Farmer*, XXXV (July 15, 1910), p. 988; ibid., (September 2, 1910), pp. 1148–1150; *Breeder's Gazette*, LVIII (August 24, 1910), p. 311; *ibid.*, LXIV (July 30, 1913), p. 186.

[53] W. H. Morrison to A. W. Harris, January 14, 1889, Office of Experiment Stations Records.

[54] *American Agriculturist*, XLVIII (April, 1889), p. 209; *Breeder's Gazette*, XV (February 27, 1889), p. 226; *Cultivator and Country Gentleman*, LIV (February 21, 1889), p. 150; *ibid.*, LIV (May 16, 1889), 385; *Congressional Record*, February 15, 1889, p. 1919; Levi Chubbuck to J. M. Rusk, January 8, 1889 and N. J. Colman to J. S. Lawson, February 8, 1889, Office of Experiment Stations Records; U.S. Office of Experiment Stations, *Circular 9* (Washington, n.d.), n.p.

Experiment Stations, was converted to the cause in 1900, and the next year Secretary James Wilson asked for $5,000 for the work. Congress responded with $2,000, an amount that permitted the Office of Experiment Stations to publish the annual proceedings of the institute workers' association but to do little else.[55]

Two years later Congress was more generous. It provided the desired appropriation and directed the Secretary of Agriculture to "investigate and report upon the organization and progress of farmers' institutes . . . with special suggestions of plans and methods for making such organizations more effective for the dissemination of the results of the Department of Agriculture and the agricultural experiment stations and of improved methods of agricultural practice." Accordingly, Secretary Wilson created the Office of Farmers' Institute Specialist and assigned it to the Office of Experiment Stations for supervision.[56]

To fill the new post Secretary Wilson first turned to Kenyon L. Butterfield of Michigan, but when he declined Wilson named John Hamilton of Pennsylvania. Born in 1843 on a farm in Juniata County, Hamilton received general and secondary education at a local school near his home and at academies in the state. The Civil War interrupted his education. Enlisting in April, 1861, he served three and a half years as a cavalryman in the Army of the Potomac. After discharge from service Hamilton entered Pennsylvania Agricultural College where he was almost immediately placed in charge of the college farm. The work postponed his graduation until 1871 when he began a teaching and administrative career at the agricultural college that lasted for a quarter of a century. In 1895 Hamilton became deputy secretary of the Pennsylvania State Department of Agriculture. His role in making the Pennsylvania institutes among

[55] Milton Conover, *The Office of Experiment Stations: Its History, Activities and Organization* (Baltimore, 1924), p. 73; K. L. Butterfield to J. H. Bingham, August 17, 1900, A. C. True to Butterfield, October 29, 1900, and True to James Wilson, May 3, 1902, Office of Experiment Stations Records; U.S. Department of Agriculture, *Yearbook*, 1901, p. 81.

[56] *U.S. Statutes at Large,* XXXII, part 1, p. 1164; A. C. True to Wesley Webb, December 1, 1902, Office of Experiment Stations Records; Conover, *Office of Experiment Stations,* p. 73.

the best in the nation made him a logical choice for the new position of Farmers' Institute Specialist.[57]

Hamilton and his associates in the Office of Experiment Stations had no desire to establish a national system of institutes nor did they intend to interfere with the states' management of the work. The language of the 1903 authorization and Hamilton's inclinations meant that his office would simply cooperate with state officials, aiding them in any possible way. He continued the gathering of information pertaining to institutes, and he participated in as many institutes in various parts of the country as time would allow. Beyond these functions Hamilton's office attempted to improve the quality of institute programs by preparing and distributing lists of competent lecturers and by producing or editing a series of bulletins, illustrated lectures, and syllabuses for use by institute lecturers. Finally, Hamilton was a regular and active participant in the annual meetings of the institute workers' association, serving as secretary for almost a decade, and his office published its proceedings.[58]

Hamilton continued as Farmers' Institute Specialist until January 1, 1914, when he retired to private life. John M. Stedman of Missouri, Hamilton's deputy since 1909, was his replacement. Hamilton's retirement in a sense marked the end of an era. In the fall of 1913 the United States Department of Agriculture announced its decision to cease publishing the annual proceedings of the institute workers' association and not to allow in the future any of its personnel to serve as officers of the association. Alfred C. True of the

[57] A. C. True to James Wilson, December 20, 1902, and February 18, 1903, Office of Experiment Stations Records; *Experiment Station Record*, XLV, pp. 98–99; Thomas I. Mairs, *Some Pennsylvania Pioneers in Agricultural Science* (State College, Pa., 1928), pp. 100–101, 113–116; *Cultivator and Country Gentleman,* LXII (November 11, 1897), pp. 890–891; *Breeder's Gazette,* XLIII (March 4, 1903), p. 430.

[58] U.S. Office of Experiment Stations, *Annual Report,* 1903, pp. 36–37; AAA-CES, *Proceedings,* 1903, pp. 28–29; C. P. Norgard to A. C. True, May 23, 1908, and John Hamilton to A. C. True, March 10, 1913, Office of Experiment Stations Records; U.S. Department of Agriculture, *Yearbook,* 1906, p. 105; Conover, *Office of Experiment Stations,* p. 73; Alfred C. True and Dick J. Crosby, "The American System of Agricultural Education," U.S. Office of Experiment Stations, *Circular* 106 (Washington, 1911), pp. 8–9.

Office of Experiment Stations maintained that these decisions were in accordance with precedents established earlier in the relationship between the Department of Agriculture and the Association of American Agricultural Colleges and Experiment Stations. But perhaps more to the point was the recognition by astute observers that institutes were of declining importance, destined to be rendered completely obsolete by new teaching methods.[59]

By 1914, in fact, rural leaders across the nation were questioning the value of institutes, asking whether they had outlived their usefulness or whether they could be further improved to meet the growing rural demand for education. Critics of institutes were able to find many faults in them. Too often, it was said, speakers talked in "glittering generalities" but could not give answers to specific questions. Evening sessions especially tended to be so general in nature that they provided little more than entertainment.[60] Other observers claimed that institutes failed to attract young, energetic farmers. Instead, far too large a percentage of the audiences consisted of elderly men. "Talking to them is a good deal like preaching to seasoned saints," wrote one lecturer who had served on many platforms. Older farmers seemed to be willing to attend institutes but they "gathered around the stove and talked and swapped stories, waiting for someone to come and pour some information into them, which would run out, or rather off, as fast as it was poured on."[61]

More important, institutes could not provide specific information to meet an individual farmer's problems and they could not insure that material presented on the lecture platform would be put into practice by members of their audiences. Institute lecturers presented the principles of better agriculture but they could go little further. Even trained scientists who had only recently left the classroom and the laboratory for the lecture platform could not prescribe

[59] John Hamilton to A. C. True, September 24, 1913, and Hamilton to State Directors of Farmers' Institutes, December 24, 1913, Office of Experiment Stations Records; AAFIW, *Proceedings,* 1913, pp. 38, 61.

[60] *Farm, Stock, and Home,* XXVII (August 1, 1911), p. 589; *Southern Planter,* LXIX (August, 1908), pp. 712–713.

[61] *Wallaces' Farmer,* XXXI (November 30, 1906), p. 1429; *ibid.,* XXXIII (November 27, 1908), p. 1454; *ibid.,* XXXV (October 28, 1910), p. 1427.

cures for a farmer's sick cow without seeing the cow. Practical lecturers could do even less. Moreover, the habitual hesitancy of farmers to accept oral teaching meant that much that was presented on institute platforms was ignored. Farmers found it easy to convince themselves that practices proven elsewhere would be total failures in their fields and barnyards. The infrequency of institute meetings further contributed to the loss that occurred in the transmission of information from lecturer to farmer.

Essentially, institutes were a transitional stage in the evolution of a teaching method for the countryside. By suggesting better methods of farming, they opened the eyes of many farmers to the opportunities that science offered. They provided the first effective contact between the agricultural college and experiment station on the one hand and the farmer on the other. But experience showed that institutes constituted no final answer to the problem of giving farmers the guidance they needed.

V. THE COLLEGES TURN
TO AGRICULTURAL EXTENSION

THE LEADERS of land-grant colleges were fully cognizant of both the value and limitations of farmers' institutes. At the same time most agricultural educators were painfully aware of the general failure of their schools to influence effectively any large number of ordinary farmers. The widespread disregard and even contempt for the colleges, so common among actual farmers, was but the foremost indication that the institutions had failed to reach the goals that their founders had set for them.

By 1900, at least, it had become perfectly clear that the colleges needed to develop new approaches if they were to become a powerful force in agricultural change and improvement. Classroom instruction offered to the relatively few farm boys who came to the campuses would not by itself give the colleges the role that they wanted. The schools needed to establish a close, functional relationship with adult farmers. The institute provided one means—an ineffective one—of doing so; formal agricultural extension constituted another approach.

Almost from the outset agricultural colleges and the experiment stations had attempted to reach farmers through the distribution of bulletins. The Hatch Act specifically directed that the results of experiment station research be made available to farmers. Soon a flood of publications poured out from the campuses. In 1900 Alfred C. True reported that the fifty-six experiment stations then in existence produced annually an average of 400 bulletins and reports that went to over a half a million addresses. "Nowhere else in the world," he claimed, "is there any university extension work which can at all compare with that which is carried on through the publications of the . . . experiment stations. In variety of subjects treated, in the wideness and magnitude of its distribution of information, and in the substantial backing of scientific investigation and general accuracy of statement, it exceeds by far any university extension system yet devised."[1]

The use of experiment station bulletins as a means of instruction presented educators with a number of difficult questions. Included was the mode of distribution. Some stations compiled mailing lists by any and all means. In Missouri, station men obtained names of farmers from state legislators, county officials, and other office-holders. Maine's W. T. Jordan reported that he sought the assistance of members of the state board of agriculture. On the other hand, Isaac P. Roberts of Cornell and H. E. Stockbridge of Indiana maintained that it was a waste to send bulletins to any farmer who was not sufficiently interested to ask for them, a view that was widely shared.

Most station directors saw to it that bulletins went to the farm papers in their states. Editors of agricultural periodicals were eager to receive the publications, which they distilled for their subscribers. More research results reached farmers in that fashion than in any other.[2]

The inability of farmers to understand the material presented in ordinary station bulletins, raised other problems. Some station personnel concluded that two sets of bulletins were needed. One should be highly technical, describing the processes and methods of given

[1] *U.S. Statutes at Large*, XXIV, p. 441; *Breeder's Gazette*, XVII (January 22, 1890), p. 74; Alfred C. True, "University Extension in Agriculture," *Forum*, XXVIII (February, 1900), p. 702.

[2] AAACES, *Proceedings*, 1889, pp. 39, 41, 105; *Breeder's Gazette*, XV (May 15, 1889), p. 523.

experiments and useful primarily to other researchers. A second set, issued especially for farmers, should summarize the results of experiments in language that the ordinary farmer could understand and should suggest how those new ideas might be utilized in actual practice. Such bulletins should be designed to arouse interest and they should be brief. "The farmer is scared by large volumes," said one critic of typical station bulletins.[3]

Despite all innovations it continued to be apparent that bulletins of any type were less than effective as a teaching device. Relatively few farmers asked for bulletins; fewer still gave any indication that they absorbed the information or followed the suggestions contained in them. Many years later a prominent agricultural leader who grew to manhood in Tennessee in the 1890s reported that in his youth he never saw an experiment station publication and, in fact, knew no one who had.[4]

From the outset and increasingly as time went on, agricultural colleges instructed farmers through direct correspondence. Only three years after Mississippi Agricultural and Mechanical College opened its doors, President Stephen D. Lee reported that queries were so numerous that the institution had become a "bureau of information." Some of his faculty members, he claimed, spent almost as much time answering questions posed by farmers as they did in classroom instruction. Three decades later President Kenyon L. Butterfield of the Massachusetts college stated that the "general practice . . . has been to help people whenever we could, in their individual problems . . . by correspondence." In 1911, according to Eugene Davenport of Illinois, a good agricultural school had a correspondence of up to 75,000 letters a year, mostly in response to requests for information.[5]

[3] AAACES, *Proceedings*, 1889, p. 42; *ibid.*, 1894, pp. 69–71; *Cultivator and Country Gentleman*, LVIII (November 16, 1893), p. 894.

[4] AAACES, *Proceedings*, 1910, pp. 154–155; interview by the author with C. A. Cobb, October 6, 1966.

[5] Mississippi Agricultural and Mechanical College, *Report*, 1883 (Jackson, 1883), p. 13; K. L. Butterfield to H. J. Webber, May 26, 1910, Bailey Papers;

Useful though such work was, agricultural educators found it time consuming or worse. "It is often a most laborious task to unravel the unknown and, in most cases, poorly described facts . . . upon which the answer must be based," complained a California professor in 1896. "Learning by driblets, as they now do by our letters, is as laborious for them as it is for us, [and] we could just as easily teach a score as the one correspondent to whom we write."[6]

As the land-grant colleges' administrators searched for satisfactory methods for reaching adult farmers, they began to experiment with reading and correspondence courses in agriculture. These teaching techniques were by no means original with the agricultural colleges; they borrowed them from the older colleges and from other educational systems.

As early as 1856 two scholars in Berlin attempted to teach languages by correspondence, and those innovators were soon followed by others. Twenty years later in the United States the Society to Encourage Studies at Home failed in an effort to instruct enrollees through assigned readings and correspondence. More successful was the work of William Rainey Harper. In the 1880s he established correspondence courses to continue and supplement instruction given through the chautauqua. Later, when Harper became president of the University of Chicago, he created a correspondence office in the university's department of extension. There techniques were hammered out that became standard procedure for schools throughout the nation.[7]

Correspondence instruction in agriculture originated at Canada's Ontario Agricultural College in 1882. Under the leadership of President James Mills, the college outlined courses of readings and awarded certificates to persons completing them. Prizes encouraged enrollees to greater efforts. The number of persons instructed in the

Breeder's Gazette, LIX (March 1, 1911), p. 556; _Wallaces' Farmer_, XXXVIII (February 14, 1913), p. 300.

[6] "Work of the College of Agriculture and Experiment Stations," p. 10.

[7] John S. Noffsinger, _Correspondence Schools, Lyceums, Chautauquas_ (New York, 1926), pp. 3–11.

program was never great, and most of the participants were graduates of the college who desired to continue their education. After a few years the project was abandoned.[8]

In the United States, the Pennsylvania Agricultural College was the leader. In 1891 the director of the experiment station urged the establishment of a correspondence program, believing that it would help to popularize the work of the station. The next year the college launched its Chautauqua Course of Home Readings in Agriculture.

At the outset the Pennsylvania program consisted of prescribing readings in volumes that served as textbooks, giving examinations when students were ready for them, and bestowing certificates or diplomas on those who completed the work. At some places reading circles gave opportunity for discussion. The texts were provided at cost, but many farmers had difficulty in obtaining them, either because the books were in short supply or because of their price. Moreover, the books tended to be too detailed. In 1897 the college replaced textbooks with several series of printed lessons. These were geared more to the ability of farmers; they emphasized fundamentals and suggested simple experiments that farmers could carry out on their own land. With these alterations came a change in name. The project became simply the Pennsylvania Correspondence Courses in Agriculture.

The modified program proved to be reasonably popular. The number of participants grew rapidly: in 1893 some 340 students were enrolled, but by 1899 officials counted 3,416. The number of subject areas in which instruction was offered grew from five in 1898 to thirty-nine in 1907.[9]

Even more popular was a correspondence program at Cornell. After first attempting to instruct farmers in an informal way by

[8] Michigan State Farmers' Institutes, *Institute Bulletin* 12 (Agricultural College, 1906), p. 70; Liberty Hyde Bailey, "Farmers' Reading Courses," U.S. Office of Experiment Stations, *Bulletin* 72 (Washington, 1899), pp. 5–6.

[9] *Ibid.*, p. 8; AAACES, *Proceedings*, 1896, p. 50; *Breeder's Gazette*, XXII (October 12, 1892), p. 250; *Cultivator and Country Gentleman*, LIV (November 15, 1894), p. 825; Fletcher, *Pennsylvania Agriculture and Country Life*, p. 469; *Country Gentleman*, LXXVII (June 1, 1912), p. 11.

providing them with sets of books and station bulletins pertaining to specific topics, the college inaugurated a more systematic plan in 1897. It involved the sending of detailed lesson plans to farmers, the quizzing of enrollees on the subjects covered, and the organizing of reading clubs where possible. Students received materials discussing some general principle of a given type of farming. When they had studied the lesson, they completed a quiz sheet and returned it to the college for grading. Speakers from the college occasionally visited the reading clubs, giving general direction to the work and encouraging students to devote the necessary attention to it. In 1901 the college began to award certificates to those enrollees who satisfactorily completed a prescribed course. In the same year, Cornell began a course for farm women. By that time some 27,000 rural residents had received instruction through correspondence, and the number continued to grow.[10]

Other schools in the northeastern quadrant of the nation soon followed the example of Pennsylvania and New York. In 1896 Connecticut Agricultural College earmarked $500 for support of reading courses for both men and women. The Connecticut system tended to be more formal than in the neighboring states; the courses were two years in length and involved the reading of up to eight books and the passing of an examination on each book. Organization of reading circles, visits by faculty members, and the use of traveling libraries strengthened the program. The goal was to make the level of instruction as close to that of resident instruction as possible. Enrollees who completed a course participated in the annual graduation ceremonies on the college campus.

The New Hampshire College of Agriculture began a program in 1894. Lack of support lead to the abandonment of the work in 1900,

[10] Cornell University Agricultural Experiment Station, *Bulletin* 159 (Ithaca, 1899), p. 265; Colman, "History of Agricultural Education," p. 163; Bailey, "Farmers' Reading Courses," pp. 16–17; L. H. Bailey to A. C. True, March 2, 1899, Office of Experiment Stations Records; John Craig, "Teaching Farmers at Home," *World's Work,* II (June, 1901), p. 811; AAFIW, *Proceedings, 1901,* pp. 50–52.

but it was revived in 1911 and subsequently enjoyed some success. A program that began in Maine in 1893 had to be reorganized ten years later before it was reasonably well received.[11]

Elsewhere the record of extension work in agriculture through correspondence instruction was mixed. Michigan had perhaps the best program in the Middle West, beginning in 1892 with a plan called the Farm Home Reading Circle. It was modeled after the arrangement in Pennsylvania. South Dakota launched a reading course in 1899. Two years earlier Wyoming reported that it had one in operation, and California started a program in 1904. Among other states that at one time or another had such projects were Indiana, Texas, Missouri, Tennessee, Virginia, West Virginia, and Mississippi. With the exception of a few areas rural apathy limited the work, and in the South especially the lack of funds and manpower and the generally low educational level were additional handicaps.[12]

Some agricultural college faculty members contributed to extension by working with commercial correspondence schools. In 1904 the Office of Experiment Stations listed three such enterprises considered to be reliable. Included were the Columbian Correspondence College of Washington, D.C., which had fourteen courses in agriculture; the Correspondence College of Agriculture in Sioux City, Iowa; and the Home Correspondence School of Springfield, Massachusetts. The latter institution was established in 1878, enrolling students primarily in business courses. It first offered instruction in agriculture in September, 1901; officials claimed that almost 1,000 people enrolled during the first six months. Cost per student

[11] Stemmons, *Connecticut Agricultural College*, pp. 233–234; University of New Hampshire, *History*, pp. 128–129; Bailey, "Farmers' Reading Courses," pp. 11–14; Fernald, *University of Maine*, p. 262; AAACES, *Proceedings, 1907*, p. 28.

[12] Bailey, "Farmers' Reading Courses," pp. 10–11, 18–19; *Breeder's Gazette*, XXIV (November 1, 1893), p. 290; Powers, *South Dakota State College*, p. 92; Hamilton, "History of Farmers' Institutes," p. 21; AAACES, *Proceedings*, 1897, p. 57; *ibid.*, 1902, p. 71; *ibid.*, 1907, p. 32; Michigan State Farmers' Institutes, *Institute Bulletin* 12, p. 70; A. C. True to L. J. Eastwood, December 12, 1904, Office of Experiment Stations Records; U.S. Office of Experiment Stations, *Annual Report*, 1904, pp. 597–598; Mississippi Agricultural and Mechanical College, *Annual Catalogue*, 1913–14, pp. 54–57.

was fifteen dollars a course. Like the others the Home Correspondence School employed prominent agricultural educators to lay out the courses and required written papers from the students. These were graded by the professor or by assistants provided for him.[13]

Essentially, correspondence instruction in agriculture fell into two types. One plan, typified by that in Connecticut, presented technical instruction of a relatively high level. It appealed to those rural residents who already had some knowledge of and appreciation for scientific agriculture. On the other hand, the Cornell plan and those modeled after it were less technical and were aimed at the mass of farmers. The purpose was more to arouse the common farmer from his apathy than it was to instruct the few who were already convinced of the value of agricultural knowledge.

Teaching of either type served a useful purpose, but most agricultural educators recognized that correspondence instruction had severe limitations. With few exceptions teaching by correspondence presented no opportunity for the student to come into direct contact with the teacher. It was difficult to maintain interest and enthusiasm. Moreover, all correspondence instruction fell into the category of book farming and carried a stigma sufficient to prevent the mass of farmers, those who most needed instruction, from availing themselves of the opportunities offered.[14]

Quite often, new extension techniques developed out of obvious necessity; such was the case in New York where some of the ablest pioneers in agricultural education formulated new methods after 1894. A year earlier grape growers in Chautauqua County encountered serious problems, and they turned to the experiment station for assistance. Informed that money for additional work was not available, the farmers went to the legislature. There Assemblyman S. F. Nixon fought successfully for a bill that provided $8,000 for "the purpose of horticultural experiments, investigations, and instruction in western New York."

[13] U.S. Office of Experiment Stations, *Annual Report,* 1904, p. 598; S. L. Morse to D. J. Crosby, May 9, 1902, Office of Experiment Stations Records.
[14] Bailey, "Farmers' Reading Courses," p. 21; AAACES, *Proceedings,* 1905, pp. 128–129.

Although the new responsibility had not been solicited by New York's college of agriculture, its horticulturists, headed by Liberty Hyde Bailey, took up the task eagerly. From the outset it was apparent that, while some new knowledge pertaining to local conditions was necessary, experimentation alone would not meet the needs; even more important was the teaching of farmers in matters basic to good agriculture. As a result during its first year the Cornell program included cooperative experiments in plant diseases, entomology, and other aspects of horticulture; instruction through lectures and itinerant schools; and the publication of popularly written bulletins.[15]

First reactions were more than favorable, so in the later years of the decade the program was broadened. Annual appropriations increased steadily and stood at $35,000 in 1900. In 1897 the legislature altered the law to read "for the promotion of agricultural knowledge in the state," indicating that the lawmakers were aware that all farmers, not just horticulturists in the western part of the state, were in need of instruction. To the original features of the work—cooperative experiments, itinerant schools, and bulletins—were added two others, correspondence courses in agriculture and nature study in the rural schools.[16]

The latter was an especially significant innovation; indeed, it marked the earliest effort by a land-grant college to reach country school children in an effective way. The program stemmed from the realization that all efforts at rural adult instruction suffered from the farmers' lack of the most elementary knowledge of science. In fact, the ability to reason logically was beyond many. Men like Isaac P. Roberts and Liberty Hyde Bailey were convinced that all adult teach-

[15] Cornell University Agricultural Experiment Station, *Bulletin* 110 (Ithaca, 1896), pp. 125–127; *Cultivator and Country Gentleman*, LIX (March 22, 1894), p. 230; Alfred C. True, "Popular Education for the Farmer in the United States," U.S. Department of Agriculture, *Yearbook*, 1897 (Washington, 1898), p. 283; Cornell University Agricultural Experiment Station, *Bulletin* 159, pp. 241–242; *Cornell Countryman*, XII (December, 1914), pp. 217–218.

[16] AAACES, *Proceedings*, 1900, p. 137; True, "Popular Education for the Farmer," p. 283.

ing techniques were doomed to no more than limited success until the rural population could be introduced to the basic concepts of science. Quite obviously the place to begin was at the elementary school level.

Much of the responsibility for implementing the nature study program fell upon John W. Spencer, a New York farmer who was appointed to the Cornell staff late in 1896. Born in 1843 at Cherry Valley, New York, Spencer had been an active farmer for thirty years before he became interested in agricultural science. Though self-taught, Spencer became widely known among the state's farmers and was considered to be an authority on progressive agriculture. Perhaps of even greater importance was his personality. A jovial, friendly man, he became the beloved "Uncle John" to thousands of New York's school children. Moreover, he was astute; he recognized that not only country school children but also their teachers needed instruction.[17]

During 1897, the first full year of the nature study program, the work consisted of the issuance of leaflets and bulletins especially designed for the country child and his teacher. By 1900 these publications were being issued quarterly in editions of 30,000 copies. Representative titles included "What is Nature Study?," "The Leaves and Acorns of Our Common Oaks," "The Life History of the Toad," and "How a Candle Burns."

These efforts led to the organization of Junior Naturalist Clubs at country schools, not only in New York State but elsewhere as well. Primary responsibility for organizing and leading the clubs fell upon the local teachers. Clubs received charters from Cornell's Bureau of Nature Study, and members received copies of the *Junior Naturalist*, a monthly issued for their use. Generally clubs functioned only during the school year and were disbanded at the end of the term. By 1901 the bureau claimed 30,000 members.

Spencer often attended ordinary farmers' institutes where he held special sessions for children and discussed elementary aspects of farming. Teachers' institutes and similar meetings provided other

[17] AAACES, *Proceedings*, 1897, p. 56; *Cornell Countryman*, V (June, 1905), pp. 285–287; *ibid.*, XII (December, 1914), p. 218.

147

opportunities to explain the work to those who had the closest contact with pupils.[18]

Although Liberty Hyde Bailey and his associates believed the nature study program to be the most important of Cornell's early extension work, the other activities were pushed vigorously. In its cooperative experiments in 1898 Cornell conducted tests with sugar beets in fifteen counties, provided seed and fertilizer, and sent experts out to visit the 438 farmers engaged in the work. The college managed spraying demonstrations and made soil tests in various parts of the state. The itinerant schools were designed to give more detailed instruction than was possible in ordinary farmers' institutes. During the summer of 1898, for instance, dairy schools were held at thirteen points. In connection with this program an instructor visited cheese factories and creameries. Important, too, were the thirty-five special bulletins issued between 1894 and 1899. These publications, in contrast to the usual station bulletin, were readable, well-illustrated, summary statements of findings. Most of them pertained to the fruit industry.[19]

Men at Cornell were among the more important innovators in agricultural extension techniques, but as a popularizer of agricultural knowledge among farmers the most successful was Perry G. Holden. Many years later M. L. Wilson, then director of extension work in the United States Department of Agriculture, wrote, "When the history of agricultural education is written in this country, it will pay great tribute to you. You pioneered more than anyone I know with the idea of really taking to farmers the scientific knowledge of agricultural research and the successful experience of operating farmers in a manner that was acceptable and of practical value. . . ."[20]

[18] AAACES, *Proceedings*, 1897, p. 56; *ibid.*, 1900, p. 138; Cornell University Agricultural Experiment Station, *Bulletin* 159, pp. 256; AAFIW, *Proceedings*, 1901, p. 43; Howe, "Boys' and Girls' Agricultural Clubs," p. 7.

[19] Cornell University Agricultural Experiment Station, *Bulletin* 159, pp. 242–253; *Cultivator and Country Gentleman*, LX (January 17, 1895), p. 50; *ibid.*, LX (June 27, 1895), p. 489; Liberty Hyde Bailey, "The Revolution in Farming," *World's Work,* II (July, 1901), p. 947.

[20] M. L. Wilson to P. G. Holden, October 12, 1943, P. G. Holden Papers (Michigan State University Library, East Lansing).

Born in 1865 in Dodge County, Minnesota, Holden's family soon took him to a farm in Benzie County, Michigan, where he grew to manhood. Life was primitive there—Holden later recalled that he had not worn a white shirt until he went to college—but his first contact with agricultural education came as early as 1876 when he attended a farmers' institute at Traverse City. Holden entered Michigan Agricultural College in 1885, graduating in 1889. During his undergraduate days he supported himself by teaching the winter terms in country schools. In 1889 he became an assistant in agriculture at Michigan Agricultural College; between 1893 and 1896 he was back in public school education but he took his M.S. degree in 1895. A year later he became assistant professor of agronomy at the University of Illinois, earning the rank of professor in 1899. While at Illinois he engaged in significant but long-ignored work in corn breeding. In addition, he helped organize a state corn growers' association, established the nation's first corn judging school, and assisted in building up the institutes in the state.[21]

In 1900 Holden resigned from the University of Illinois to become agriculturist with the Illinois Sugar Refining Company of Pekin. His new position launched him on a career in extension. Field work brought him into contact with hundreds of farmers, and he acquired a reputation as a popular speaker at short courses. It was in that capacity, in fact, that he first caught the eye of the Funk brothers, seed corn growers of Illinois, and in 1901 he became the first manager of the Funk Brothers Seed Company, devoting special attention to the scientific breeding of corn.

Another move came within a few months. After hearing Holden speak at a short course, President W. M. Beardshear of the Iowa college resolved to attract him to Ames. Funds were temporarily inadequate, but agricultural leaders in the state agreed to contribute.

[21] "Famous Hybrid Corn Scientist Dies at 94," *Michigan Extension News*, XXX (October–December, 1959), pp. 1, 4; *The Record* (East Lansing, Mich.), LIII (November, 1948), p. 10; P. G. Holden Memoirs (Michigan State University Library, East Lansing); Everett G. Ritland, "The Educational Activities of P. G. Holden in Iowa" (M.A. Thesis, Iowa State College, 1941), pp. 6–8; Herbert K. Hayes, *A Professor's Story of Hybrid Corn* (Minneapolis, 1963), p. 29; Lester S. Ivins and A. E. Winship, *Fifty Famous Farmers* (New York, 1925), p. 86.

As a result, in August, 1902, the board of trustees appointed Holden professor of agronomy and vice-dean at a salary of $2,600.

Holden remained in Iowa for a decade. During that time he won a reputation unequaled among agricultural educators; he also encountered considerable opposition, arising especially from his tremendous popularity with ordinary farmers and his unusual teaching methods. For a variety of reasons he resigned in January, 1912, to seek the Republican nomination for governor. A progressive, Holden had the support of the influential Wallace interests but he found it impossible to overcome the power of the party organization. Soon he returned to agricultural education as an employee of International Harvester Company.[22]

Holden made an indelible impression on extension teaching techniques and ideas. A dynamic, opinionated man, he dominated all that he touched. Kenyon L. Butterfield, who at times differed with Holden, said that he was a "terror" when aroused.[23] In his teaching methods Holden was never limited to conventional practices; the classroom had little appeal for him, he preferred mass meetings. There he could speak in his evangelical way, urging farmers to adopt new methods. For subordinates he demanded men of similar temperament and talents. College training was desirable but not essential; ability to speak clearly and effectively and to win the confidence of farmers was far more important. During his Iowa years Holden developed or significantly modified and systematized several extension techniques. Most important were county farm experiment work, traveling short courses, and educational trains.[24]

The use of small test plots at points removed from the college campuses was by no means new when Holden went to Ames. The

[22] P. G. Holden Memoirs; Ritland, "Educational Activities of P. G. Holden," pp. 8–9; *Breeder's Gazette,* XXXVII (March 14, 1900), p. 336; Ross, *Iowa State College,* pp. 258–259; Morgan, *Extension Service of Iowa State College,* p. 23; *Breeder's Gazette,* LXI (January 10, 1912), p. 82; *Country Gentleman,* LXXVII (February 3, 1912), p. 1.

[23] K. L. Butterfield to L. H. Bailey, February 15, 1911, Bailey Papers.

[24] Morgan, *Extension Service of Iowa State College,* pp. 23, 29; Earle D. Ross, "The New Agriculture," *Iowa Journal of History,* XLVII (April, 1949), p. 128.

Connecticut experiment station started cooperative experiments with farmers in 1877, and stations and colleges in the other New England states soon took up the work. These early efforts were usually limited to testing the value of fertilizers and different crops on various soils. Participating farmers normally supplied land, labor, and ordinary tools and retained the crops; the colleges or stations provided seed, fertilizer, and supervision.

Cooperative experiments took on an extension role in Ohio in 1895. Graduates of the state's land-grant college established the Agricultural Students' Union and agreed to test on their farms ideas developed at the experiment station and to report the results. They also agreed to help disseminate any new information by making themselves available as lecturers at institutes and other farm meetings.[25]

After 1900 the cooperative experiment technique was further modified and came to be used in some form by a majority of the colleges and experiment stations. In Rhode Island the station used cooperative arrangements to teach farmers proper spraying and stock fumigation methods, to suggest means for improving soil fertility, and to expose farmers to other practices already in use on college lands. Mississippi utilized the same technique in its effort to increase dairying and cattle raising in the state. By 1909 the North Dakota station had in operation twenty-one experimental plots in as many counties. Varying in size from five to twenty acres, the plots were located on the properties of farmers who agreed to follow explicitly the instructions given them. The next year Minnesota instituted a similar program.[26]

All such work suffered from certain problems and limitations. Insofar as projects were experimental in nature, they too often were

[25] *American Agriculturist*, XL (March, 1881), pp. 93, 96–97; AAACES, *Proceedings*, 1901, p. 103; Day, *Farming in Maine*, p. 246; Mendenhall, *Ohio State*, II, 146–148.

[26] AAACES, *Proceedings*, 1909, p. 33; *Progressive Farmer and Southern Farm Gazette*, XXVIII (March 1, 1913), p. 299; *Breeder's Gazette*, LVI (December 29, 1909), p. 1436; *ibid.*, LVII (February 2, 1910), p. 276.

superficial in character and suffered from the farmer's impatience and from his tendency to abandon them before completion. Proper supervision was costly. For that reason the North Carolina station dropped cooperative experiments after two years of effort in 1889 and 1890.

On the other hand the extension aspects of the work seemed to have significant possibilities. According to John Craig of New York, the most useful purpose of cooperative experiments was to "demonstrate and illustrate facts already discovered . . . [and] to bring these facts . . . before the farmer."[27]

From these failures and successes Holden drew ideas for a successful extension technique. The immediate catalyst came in 1903 when farmers attending an institute at Hull, in Sioux County, pointed out that results in experiments conducted at Ames might well have little validity elsewhere. Since Holden was well aware of the tremendous differences in the yield qualities of seed corn, he decided to establish a series of experimental plots to show farmers the basic steps to better corn production.

As the plan developed it involved the establishment of test plots on the Sioux County farm near Orange City. Businessmen and farmers induced the county board of supervisors to appropriate funds for the work; the college provided leadership and guidance. The cooperation between county and state authorities suggested the arrangement adopted over a decade later in the Smith-Lever Act. At the outset the programs were aimed primarily at the testing of various varieties of seed corn; later other experiments were undertaken. Throughout the growing season farmers were encouraged to visit the plots. The program culminated in fall field meetings at the county farm where results were studied and discussed.

From the beginning in Sioux County, the technique spread across the state. The second year five counties had comparable programs; the number increased to ten in 1906. That year the state legislature authorized county authorities to give up to $300 annually for such

[27] AAACES, *Proceedings*, 1894, p. 51; *ibid.*, 1901, pp. 102–103.

projects. By 1912 a total of thirty-two counties had maintained plots, in some instances for periods of several years. In later years the average number of participating counties was fifteen, roughly the maximum that Holden's staff could handle.

Although the work was partly experimental in that the tests showed which varieties of seed performed best under given conditions, the extension feature was most important. Holden's plots enabled ordinary farmers to see the results of better practices under conditions that made it difficult to avoid being impressed and influenced. The fall meetings were especially effective. As many as 3,500 persons appeared at them. Tours of the plots, lectures, and the inevitable basket lunch made the gatherings both instructional and enjoyable.[28]

Holden also used traveling short courses in a new, highly successful way. These itinerant schools were outgrowths of campus short courses that were themselves considered by many agricultural educators to be a form of extension.

Campus short courses were well established by 1900. The idea of a special course on campus for ordinary farmers first appeared in 1867 when the state agricultural society in Michigan urged the college to institute such a program. Two years later the Pennsylvania college inaugurated a four-day course that was repeated for several years devoted to trials of agricultural implements and lectures by faculty members. Between 1882 and 1885 the institution sponsored an institute on campus that sought to draw farmers from all parts of the state. Meanwhile, from 1874 to 1899 the agricultural college in Illinois experimented with special courses for farmers that ranged in length from three months to two years. In the winter of 1877–78, Ohio State launched a project that, although premature, established patterns followed later. At the outset it was a noncredit, ten-week

[28] Epsilon Sigma Phi, *Extension Work*, pp. 46–47; Martin L. Mosher, *Early Iowa Corn Field Tests and Related Later Programs* (Ames, Iowa, 1962), pp. 15–20; *Corn*, I (January, 1902), p. 17; Morgan, *Extension Service of Iowa State College*, p. 32; *Wallaces' Farmer*, XXXVI (November 3, 1911), p. 1521; A. H. Snyder, "Extension Work in the West," *Addresses Delivered at the University of Virginia Summer School in Connection with the Conference for the Study of the Problems of Rural Life*, 1909 (n.p., n.d.), pp. 28–29.

course open to all farmers. Response was poor, so the course was cut to four weeks and later abandoned.[29]

Wisconsin was the leader in developing the campus short course in its modern form. Under the capable leadership of W. A. Henry, the first session opened at Madison in January, 1886, with 19 students. It proved to be a continuing success, and in 1890 authorities added a comparable course emphasizing dairying. These were still in existence in 1909, with an added course designed especially for women. Altogether they attracted 1,262 students.[30]

Other states soon took up the work. Short courses began at Purdue University in 1888 and with the exception of one year proved to be an annual event. The North Dakota Agricultural College held its first short course in January, 1891, only a few months after the institution opened its doors. Illinois, Pennsylvania, and Nebraska began the work in 1892, and the Kansas State Agricultural College and Cornell commenced the next year, the latter school modeling its program after the one in Wisconsin. As early as the 1896–97 academic year, the agricultural college in Michigan was holding four short courses dealing with as many branches of agriculture.[31] North Carolina was probably the first state in the South to have a short course, beginning in 1895, but it was soon joined by Mississippi. Iowa held its pioneer short course in 1901, and as late as 1907 such states as

[29] Vernon C. Larson, "The Development of Short Courses at the Land Grant Institutions," *Agricultural History*, XXXI (April, 1957), p. 31; Fletcher, *Pennsylvania Agriculture and Country Life*, p. 460; Robert R. Hudelson and Anna C. Glover, History of Agricultural Education of Less than College Grade: Contributions of the University of Illinois, manuscript (University of Illinois Library, Urbana, Ill.); Eddy, *Colleges for Our Land and Time*, p. 79.

[30] Larson, "Short Courses," p. 31; D. J. Crosby to C. E. Adams, February 10, 1909, Extension Service Records; *Hoard's Dairyman*, XL (March 5, 1909), p. 148; "Fifty Years of Cooperative Extension in Wisconsin," p. 22.

[31] William M. Hepburn and L. M. Sears, *Purdue University: Fifty Years of Progress* (Indianapolis, 1925), p. 114; Hunter, *North Dakota's Land-Grant College*, p. 23; *Breeder's Gazette*, XX (November 18, 1891), p. 369; *ibid.*, XXI (January 20, 1892), p. 48; *ibid.*, XXIII (February 22, 1893), p. 142; *Cultivator and Country Gentleman*, LVII (September 29, 1892), p. 724; *ibid.*, LIX (November 15, 1894), p. 825; Kansas State Agricultural College, *Ninth Biennial Report*, 1893–94, p. 31; Colman, "History of Agricultural Education," p. 111; Michigan Agricultural College, *President's Report*, 1898 (Lansing, 1898), p. 2.

Virginia, Arkansas, and New Jersey were launching their programs.[32]

Despite false starts in a number of instances, short courses grew in popularity and by 1914 they had become a fixture on most agricultural college campuses. Varying greatly in length, content, and subject matter, their purpose was to reach and instruct rural adults who could not attend the full four-year course. By 1907 the length of the various courses ranged from one to fourteen weeks. Instruction was by lectures, practical demonstrations, and assigned readings in Department of Agriculture and experiment station publications. At Wisconsin in 1904, for example, faculty members dissected cattle and other livestock to show students the effects of tuberculosis and other diseases, conducted spraying operations, taught farmers to perform the Babcock milk test, and presented lectures. According to Dean W. A. Henry, these short courses were "the most inspiring thing in the way of agricultural education work that has come to my experience."[33]

The reaction of farmers as reflected in their attendance and their comments became steadily more favorable. The first short course offered in Pennsylvania in 1891 attracted only 3 farmers, but in 1906 a total of 1,130 students attended an annual corn school at Purdue. Enrollees came from seventy-nine counties and from seven other states. Such attendance was unusual, but Missouri reported over 600 participants at a short course in 1910. A farmer who attended the short course at Mississippi Agricultural and Mechanical College in 1907 observed that it was almost "incredible the amount of instruction given in so short a time." In Iowa a group of farmer-students

[32] David A. Lockmiller, *History of the North Carolina State College* (Raleigh, 1939), p. 50; Mississippi Agricultural and Mechanical College, *Annual Catalogue,* 1897–98, p. 28; Morgan, *Extension Service of Iowa State College,* p. 23; *Western Fruit Grower,* XVIII (February, 1907), p. 85; John P. Cochran, "The Virginia Agricultural and Mechanical College" (Ph.D. Thesis, University of Alabama, 1961), p. 259; *Breeder's Gazette,* LI (April 24, 1907), p. 971.

[33] *Breeder's Gazette,* LI (January 2, 1907), p. 14; *Southern Farm Gazette,* XII (October 15, 1907), p. 6; True, "Popular Education for Farmers in the United States," p. 281; Mississippi Agricultural and Mechanical College, *Annual Catalogue,* 1907–08, p. 64; *Progressive Farmer and Southern Farm Gazette,* XV (August 27, 1910), p. 601; W. A. Henry to L. H. Bailey, February 19, 1904, Bailey Papers.

were so eager for additional instruction that they agreed to meet their lecturer by lantern light during the early hours of the morning.[34]

Traveling short courses or itinerant schools of agriculture combined the methods and purposes of campus short courses with the flexibility of institutes. The former gave instruction of a desirable, concentrated type but was limited to those persons who could leave their farms for extended periods of time; the latter reached large numbers of farmers but tended to provide only superficial instruction. A traveling short course offered the best features of both types of instruction.

As early as 1892 George E. Morrow of the University of Illinois suggested that land-grant colleges offer four to five-day courses at different localities. In the winter of 1893–94 the university gave courses at Dixon and Mount Vernon that consisted of four lectures in each of five general areas of agriculture. No fees were charged, but attendance was less than satisfactory, and the experiment was discontinued.[35]

Other early efforts came in Maine, Michigan, New Jersey, Wyoming, and New York. The agricultural college in Maine held ten such courses in 1893, with the 257 enrollees paying all costs. New Jersey began its work in 1891. In 1897 Wyoming sent out instructors to conduct schools in several towns with local people paying the traveling expenses of the instructors, and the same year Michigan's agricultural college sent lecturers into fruit growing districts where they held schools of one week's duration. Cornell's early instructional work in horticulture and dairying included similar efforts.[36]

[34] *Cultivator and Country Gentleman,* LIX (November 15, 1894), p. 825; *Wallaces' Farmer,* XXXI (February 9, 1906), p. 175; Frederick B. Mumford, Press Release, January 1, 1910, Frederick B. Mumford Papers (University of Missouri Library, Columbia, Mo.); Mississippi Agricultural and Mechanical College, President's Report to the Board of Trustees, June 3, 1907, manuscript (Department of History, Mississippi State University, State College, Miss.); P. G. Holden, Extension Work at Iowa State College, Holden Papers.

[35] *Breeder's Gazette,* XXII (November 30, 1892), pp. 392–393; *ibid.,* XXIV (December 13, 1893), p. 396; *ibid.,* XXV (January 24, 1894), p. 55.

[36] Fernald, *University of Maine,* p. 262; *Cultivator and Country Gentleman,* LIX (February 1, 1894), p. 97; AAACES, *Proceedings,* 1897, pp. 57–58; Woodward and Waller, *New Jersey's Agricultural Experiment Station,* pp. 493–494.

But Perry G. Holden in Iowa developed the most comprehensive program of traveling short courses. The first suggestion came from a group of farmers near Red Oak; they had attended one of the short courses held on the campus at Ames and they suggested that a similar program might be held at their town. Holden promptly seized the idea as a promising technique for reaching additional farmers, and the first of the Iowa traveling short courses was held at Red Oak in January, 1905.

The program was well received, and Holden increased the number of courses as rapidly as possible. In the winter of 1906–07 he held 6, including 2 in domestic science for farm women. The number increased gradually, and in 1910–11 there were 32. Some 120 programs had attracted 35,000 enrollees by the end of 1912. Requests from communities, in fact, far exceeded the capability of Holden's staff, and the demand for them contributed materially to the increase in legislative appropriations for extension.[37]

The Holden short course system was thoroughly planned and left little to chance. To make certain of continuing interest, Holden held short courses only in response to requests for them; he demanded that at least 200 people, including both farmers and townspeople, sign a guaranty pledge to protect against any deficit arising from the school; and he always charged a fee, ranging from $1.50 to $3.00 per enrollee, since he was firmly convinced that people would be more conscientious if they paid for the instruction. In addition, he required local communities to provide appropriate facilities, all necessary advertising, and the programs for some of the evening sessions. The college paid the salaries of the instructors and provided much of the demonstration material, which was moved from point to point by rail, but all other expenses were defrayed by the fees and by the guaranty pledge.

In the courses Holden and his men endeavored to "give much more sequence to the program than is ordinarily the case at a farmers'

[37] Morgan, *Extension Service of Iowa State College*, pp. 24, 27, 31; Ritland, "Educational Activities of P. G. Holden," p. 19; *Wallaces' Farmer*, XXXVI (January 6, 1911), p. 25; *ibid.*, (January 27, 1911), p. 122; *Corn*, I (November, 1912), p. 187; *Breeder's Gazette*, LXV (February 5, 1914), p. 282.

institute, and to give the meeting decidedly more of the atmosphere of a school." Instruction dealt with crops, soils, animal husbandry, and dairying and, for the women, domestic science. Subject matter was made as practical as possible, and demonstrations were used extensively, especially in judging of livestock and grains, the identification of weeds, and the care of animals. Students were given no outside reading, but they were kept busy throughout the day. At the end of the week, all students took examinations and those who passed received certificates. Evening sessions were left primarily to the local people, and generally they combined instruction with entertainment.[38]

Soon after Iowa began its traveling short course program, other states took up the work, using essentially the Holden plan. Indiana had its first itinerant short course in 1907, and the pioneer school of that type in Ohio opened the next year. Authorities planned to have schools in eighty of the state's eighty-eight counties in 1910–11. Nebraska inaugurated a program in 1908 with a week-long meeting at Pawnee City, followed by courses the next year at Broken Bow and Hebron. Virginia Polytechnic Institute held its first school late in 1910, while across the nation Washington State College launched its traveling short courses with a six-day dairy school at Lyden in December, 1911.[39] The agricultural college in Idaho moved into the new field in 1911, and North Dakota followed in 1912. In the winter of 1913–14 Tennessee's college of agriculture arranged for eight courses in as many communities; Missouri had eleven "branch" short courses in 1913.[40] By that time, in fact, most of the agricultural col-

[38] A. H. Snyder, "Traveling Schools," *Addresses Delivered at the University of Virginia Summer School in Connection with the Conference for the Study of the Problems of Rural Life,* 1909 (n.p., n.d.), pp. 48–52; P. G. Holden, Extension Work at Iowa State College, Holden Papers; John Hamilton, "Progress in Agricultural Education Extension," U.S. Office of Experiment Stations, *Circular* 98 (Washington, 1910), pp. 7–8.

[39] Latta, *Outline History of Indiana Agriculture,* p. 301; Mendenhall, *Ohio State,* II, 152; *Breeder's Gazette,* LIX (March 22, 1911), p. 757; Crawford, *University of Nebraska,* p. 115; *Southern Planter,* LXXI (December, 1910), p. 1226; *Western Fruit Grower,* XXIII (February, 1912), p. 128.

[40] *Breeder's Gazette,* LIX (March 1, 1911), p. 564; Hunter, *North Dakota's Land-Grant College,* p. 57; *Progressive Farmer and Southern Farm Gazette,*

leges had some sort of off-campus short course program in operation.

In some states the colleges of agriculture cooperated with other agencies in the holding of traveling short courses. In 1907, for example, the Missouri State Board of Agriculture, which managed the institutes in that state, concluded that farmers needed a more intensive type of instruction than the institutes could provide. The result was four week-long courses offered at the state's normal schools. University faculty members constituted a part of the lecture force at those affairs, an arrangement that continued until the college began holding its own off-campus short courses. Similarly, the first short course in Minnesota, at Tracy in 1909, was sponsored by the farmers' institute management with encouragement and aid from the university. In 1910 the Illinois Farmers' Institute arranged a four-day course at Charleston with university men prominent among the speakers.[41]

Almost everywhere that traveling short courses were used, agricultural educators reported good results. Enrollment in most instances was satisfactory, and instructors usually reported that they experienced little difficulty in retaining the interest of the students. In 1913 a course at Sauk Center, Minnesota, attracted an average of 250 to its sessions; a few weeks earlier a school held in Boone County, Arkansas, had an enrollment of nearly 200. In a few areas, such as Montana, the sparseness of population and the lack of facilities in many of the small towns severely limited the influence of courses there.[42]

Most of the traveling short courses offered instruction in several areas of agriculture, such as dairying, fruit growing, stock raising, and soils. But some schools were more limited, especially those in

XXVIII (December 6, 1913), p. 1263; *College Farmer*, XI (September, 1914), p. 150.

[41] Missouri State Board of Agriculture, *Annual Report*, 1907, p. 1035; *Farm, Stock, and Home*, XXV (February 15, 1909), p. 107; *Breeder's Gazette*, LVIII (November 16, 1910), p. 1035.

[42] *Ibid.*, LXIII (February 12, 1913), p. 404d; *Arkansas Farmer and Homestead*, XIV (February 1, 1913), p. 8; C. H. Tuck to L. H. Bailey, July 2, 1912, Bailey Papers; Montana Farmers' Institutes, *Eleventh and Twelfth Annual Reports, 1912–14* (Bozeman, Mont., 1914), p. 18.

areas where a high degree of specialization existed. For instance, schools in Montana in 1912 and 1913 concentrated on fruit growing and among other things attempted to teach farmers to adopt a uniform system of grading and packing to facilitate the marketing of the product. A school held by Washington State College at Cashmere in December, 1911, dealt with horticulture, while one at Lyden the same month emphasized dairying. In Mississippi, dairying schools were popular as that state attempted to diversify its agriculture in the face of the boll weevil menace. Home economics courses were held in most states, and a few schools appealed especially for rural youth. In a number of instances the annual state fair provided a suitable setting for such schools.[43]

Arrangements for the courses tended to follow very closely those developed in Iowa, modified by particular circumstances. In Washington, for instance, the Great Northern Railway donated a railroad car for the transportation of demonstration materials. Virginia Polytechnic Institute required that at least fifty farmers petition for a course and agree to attend regularly as well as meet all traveling expenses of the instructors. Mississippi Agricultural and Mechanical College set the minimum enrollment at twenty-five.[44]

These methods by no means exhausted the list of techniques used by land-grant colleges during the first years of the twentieth century to take information to the farmers in the field. In fact, the institutions displayed an amazing ability to formulate new means of instructing the rural masses.

Almost as soon as the colleges undertook extension work, for instance, they began to issue publications of various types that were

[43] *Arkansas Farmer and Homestead,* XIV (February 1, 1913), p. 8; *Farm, Stock, and Home,* XXV (February 15, 1909), p. 107; Montana Farmers' Institutes, *Eleventh and Twelfth Annual Reports,* 1912–14, p. 6; *Western Fruit Grower,* XXIII (February, 1912), p. 128; Mississippi Agricultural and Mechanical College, *Biennial Report,* 1908–09, p. 95; *Breeder's Gazette,* LXI (February 21, 1912), p. 466.

[44] *Western Fruit Grower,* XXIII (February, 1912), p. 128; *Southern Planter,* LXXII (June, 1911), pp. 698–699; Mississippi Agricultural and Mechanical College, *Annual Catalogue,* 1910–11, p. 74.

expected to contribute to the education of farmers. In 1899 Professor Edwin F. Ladd of the North Dakota Agricultural College began publication of the *Sanitary Home,* a periodical that for three years served as an extension tool. Later, in 1908, *The Extension* appeared. That monthly went to North Dakota farmers who requested it. From 1907 to 1915 the agricultural college in Maine was another of the schools that issued a popular monthly bulletin.[45] As early as 1905 the University of Illinois extension office had available for distribution pamphlets on sugar beet production, corn growing, and milk testing. By 1911 the extension department of Ohio State prepared each month four columns of educational matter for the county weeklies. The Maryland agricultural college in 1907 issued a quarterly, three issues of which were directed primarily toward those public-school teachers who might be called upon to teach agriculture in the rural schools.[46]

Rural youth work attracted the attention of the early extension personnel in the land-grant colleges. Cornell was the leader with its nature study, but Illinois was not far behind. In its first days Fred H. Rankin's extension office at Urbana did little else; its purpose was to "interest rural young folk in agricultural education and induce them to go where it may be acquired." Within a few years, however, the office ceased to be an "advertising medium," and turned to extension in the broader sense.[47] Youth work at Ohio State also went through a transformation. In 1905 it attempted to induce schools to take up the teaching of agriculture by contacting teachers, pupils, and farmers generally, and by providing some of the supplies necessary as well as a bulletin that gave basic suggestions. Two years later the school was busily engaged in the organization of girls' and boys' clubs in the rural schools. A bulletin with a circulation of 10,000 provided some direction and continuity to

[45] Hunter, *North Dakota's Land-Grant College,* p. 56; *Breeder's Gazette,* LXI (March 27, 1912), p. 776; Fernald, *University of Maine,* p. 263.

[46] *Hoard's Dairyman,* XXXVI (August 11, 1905), p. 718; *Breeder's Gazette,* LIX (March 22, 1911), p. 758; AAACES, *Proceedings,* 1907, p. 28.

[47] *Breeder's Gazette,* XLIII (May 21, 1903), p. 1018.

the work. By the same time Kansas had in operation a statewide boys' corn growing contest and a boys' demonstration project.[48]

In a few states, such as Minnesota, extension personnel undertook to expand and strengthen the farm club movement. These local groups, outgrowths in many instances of nineteenth-century granges, alliances, and independent clubs, had changed little over the years. Still, according to Archie D. Wilson, such groups were useful because they encouraged the study of local conditions, tended to develop local leadership, and stimulated a desire for progress. The clubs also provided a means through which university extension employees could reach large numbers of people. In any event, beginning in 1911 the Division of Agricultural Extension and Farmers' Institutes in Minnesota announced its willingness to give all possible assistance to the organization of the clubs and to their maintenance. Speakers were sent to their meetings and large amounts of literature were made available to members. Wisconsin, too, sought to encourage and make more effective the farm club movement.[49]

In some states extension personnel spent no little time in working with other types of farm groups. Men from the Connecticut agricultural college at Storrs were prominent at the annual meetings of the state dairymen's, pomological, and poultry associations and of the state board of agriculture. Holden's associates in Iowa found that grange meetings and similar gatherings offered excellent opportunities for reaching farmers.[50]

Ordinary county fairs provided another means through which the emerging extension departments could reach the people. By 1910 Mississippi Agricultural and Mechanical College was regularly supplying judges for crop and livestock shows held at county fairs. Soon

[48] AAACES, *Proceedings*, 1907, p. 32; *Breeder's Gazette*, XLVIII (December 20, 1905), p. 1318; *ibid.*, LII (August 21, 1907), p. 333.

[49] *Farm, Stock, and Home*, XXVIII (November 15, 1912), pp. 796–797; Minnesota Farmers' Institutes, *Annual 25*, p. 7; *Wallaces' Farmer*, XXXVI (June 30, 1911), p. 988; Oscar B. Jesness and others, *Andrew Boss: Agricultural Pioneer and Builder* (St. Paul, 1950), p. 26; James W. Witham, *Fifty Years on the Firing Line* (Chicago, 1924), pp. 74–76.

[50] AAACES, *Proceedings*, 1907, p. 25; *Wallaces' Farmer*, XXXV (December 23, 1910), p. 1724.

its extension people recognized that they could perform other types of teaching at such gatherings and at the state fair as well. In 1909 Ohio State held exhibits and demonstrations at three county fairs, increasing the number to twelve in 1910. The demonstrations showed methods of testing milk and soils. As early as 1907 the Iowa agriculture college furnished stock and crop judges for over half the county fairs in the state. Kansas, Indiana, and Pennsylvania were only three of the other states that carried on similar programs.[51]

In the years preceeding 1914, some land-grant schools started programs which indicated that they were approaching the concept of the county agent. In 1907, for instance, the Rhode Island College of Agriculture had what it called its carpetbag campaign. The college sent out agents who went from house to house, engaging farmers in conversation, and holding neighborhood meetings for discussion of rural problems. Six years later Mississippi Agricultural and Mechanical College provided a recent graduate with an automobile and various kinds of equipment and sent him to various towns where he gave demonstrations in spraying, pruning, and milk testing. The Massachusetts agricultural college had a similar program early in 1914. At approximately the same time the agricultural colleges in Minnesota and Kansas announced their willingness to send horticultural specialists to any community where a dozen farmers requested assistance.[52]

As early as 1907 at least three states—New York, Pennsylvania, and Michigan—employed agents who devoted their full time to the promotion of dairy interests. In the summer of 1906 a staff member of Cornell's department of dairy industry traveled throughout the state, visiting farms and suggesting improvements. Occasionally he

[51] Mississippi Agricultural and Mechanical College, *Annual Catalogue*, 1910–11, p. 74; *ibid.*, 1913–14, p. 179; Mendenhall, *Ohio State*, II, 155; *Wallaces' Farmer*, XXXII (November 29, 1907), p. 1401; *Breeder's Gazette*, LVI (July 28, 1909), p. 139; Hamilton, "Progress in Agricultural Education Extension," p. 9; Wayland F. Dunaway, *History of the Pennsylvania State College* (n.p., 1946), p. 395.

[52] AAACES, *Proceedings*, 1907, p. 33; *Experiment Station Record*, XXIX, p. 300; *Banker-Farmer*, I (January, 1914), p. 11; *Wallaces' Farmer*, XXXIX (April 3, 1914), p. 609; *Fruit Grower and Farmer*, XXIII (October, 1912), p. 509.

stayed in a community long enough to supervise the carrying out of his proposals and in a very real sense to act as a resident advisor. The agent in Pennsylvania functioned in roughly the same manner. Meanwhile, in 1907 W. F. Raven was appointed by the college in Michigan to work with dairymen, concentrating especially on the organization of purebred sire associations.[53]

Certainly one of the more unique extension techniques was used in Missouri in 1910 and 1911. On February 17, 1910, in conjunction with the Frisco Railroad, the college launched a night school of agriculture aimed at providing practical information to city men who wanted to go into farming. The school consisted of weekly lectures given by faculty members, the series to continue as long as interest was maintained. A month later a similar school, consisting of just six lectures, opened in Kansas City. It proved to be so popular that the program was repeated in 1911.[54]

Almost all of the land-grant colleges continued to aid to some degree the farmers' institute programs in their states. As late as 1910, for instance, the college in Iowa kept from six to eight members of its extension staff busy during the winter months attending and participating in institutes.[55] Increasingly, however, the college men were coming to the conclusion that their time could be better spent in other extension programs.

After the first years' work, officials recognized that extension activities differed in important particulars from resident instruction or experiment station duties and that new distinct departments were needed to manage the work. In Iowa Henry Wallace suggested that Holden himself draft a bill that would provide the needed funds and organization. Two years later the governor approved a measure that

[53] AAACES, *Proceedings,* 1907, pp. 30, 32; "History of Cooperative Extension Work in Michigan," p. 5.

[54] *Missouri Agricultural College Farmer,* VI (February, 1910), p. 5; F. B. Mumford to L. M. Harris, March 2, 1910, Harris to Mumford, March 3, 1910, and Mumford to K. L. Butterfield, April 22, 1910, Missouri Agricultural College Papers (University of Missouri Library, Columbia, Mo.); *University Missourian,* March 2, 1910, p. 1; *ibid.,* February 12, 1911, p. 1; U.S. Commissioner of Education, *Report,* 1910, I, 270.

[55] *Wallaces' Farmer,* XXXV (December 23, 1910), p. 1724.

provided $15,000 for "disseminating information and for carrying instruction to parts of the state remote from the college in the form of lectures and demonstrations, demonstration experiments, assistance in short courses, and other forms of agricultural education."

To implement the new law, college authorities created an extension department within the agricultural division of the institution. The new department had its own staff, members of which were given academic rank and promoted in the same way as other faculty members. All extension personnel were expected to confine their activities exclusively to off-campus instruction except in the case of such short courses conducted at Ames. As head of the department Holden was given the title of superintendent, responsible to the dean of agriculture. Later the title was changed to director, and that official was made directly responsible to the president of the college, thereby placing extension on the same level as resident instruction and experiment station work.

Under the new arrangement extension was a year-round job. Peak work loads came in December, January, and February when the staff was increased through the addition of part-time workers. In the fall Holden's men participated in agricultural fairs, appeared at farmers' picnics, harvest festivals, and grange and other rural organization meetings. The winter season found the extension personnel engaged in short courses, farmers' institutes, and corn shows. In the spring the staff joined in the operation of special trains, supervised county farm-demonstration work, and conducted a variety of other demonstrations. Some programs, such as the dairy testing work, proceeded around the calendar. Even when not engaged in such labors, Holden's staff was busy preparing outlines, charts, and circulars, gathering data and materials, answering correspondence and inquiries, and discussing problems with the large number of farmers who visited the campus.[56]

[56] Ritland, "Educational Activities of P. G. Holden," pp. 21–27; AAACES, *Proceedings,* 1907, p. 27; Morgan, *Extension Service of Iowa State College,* pp. 24–26; Holden, Extension Work at Iowa State College, Holden Papers; Hamilton, "Progress in Agricultural Education Extension," p. 7; Ross, *Iowa State College,* p. 288.

Elsewhere the expansion of extension programs caused a growing number of land-grant colleges to establish formal departments to conduct the work. Although Rutgers may have been the first to organize an extension department, its work does not appear to have been continuous, so Cornell must be considered the leader in agricultural extension with its pioneer office established in 1894. Seven years later the University of Illinois appointed Fred H. Rankin to a rather vague position, but he served as the university's agent in institute and other extension activities. In June, 1902, he was given the title of superintendent of agricultural college extension. Effective July 1, 1905, A. B. Graham, who had been active in early rural youth work in Ohio, became superintendent of agricultural extension at Ohio State. During his first four years Graham spent most of his time promoting the formation of rural youth clubs, but he also had an adult program in which he spoke at institutes and contributed to educational and farm papers. At the outset Graham's office was attached to the university's department of rural economics, but after July, 1909, he headed a distinct department of agricultural extension.[57]

In some states extension departments grew out of earlier institute organizations. Such was the case in Kansas where the agricultural college created a distinct department of farmers' institutes headed by a superintendent, John H. Miller. He was given charge of all institute work as well as such other extension methods as he might devise. At the outset he limited himself to the managing of the institute program, but he soon launched a series of corn contests in several counties, organized farmers' clubs, arranged spring excursions to the college and to the branch experiment station at Hays, and participated in work involving educational trains. In 1907 his office became the department of farmers' institutes and college extension, suggesting the changing nature of his efforts. An increase in financial support in 1909 permitted an increase in his staff and a further broadening of the program. Finally, in 1911 Miller became the

[57] Eddy, *Colleges for Our Land and Time,* p. 104; Hudelson and Glover, *History of Agricultural Education,* p. 8; Mendenhall, *Ohio State,* II, 150–153.

166

director of college extension, and a year later his department became the division of college extension, with four departments. A similar evolution occurred in Minnesota following the retirement of Oren C. Gregg, long-time director of institutes.[58]

The Association of Agricultural Colleges and Experiment Stations played a major role in the establishment of extension departments. At the 1904 meeting of the association, Kenyon L. Butterfield called for the appointment of a committee on extension. The next year the association complied, making Butterfield chairman. In 1906 his committee reported that everywhere the concept of extension was arousing interest, that most colleges as well as numerous other agencies were engaged in a variety of extension techniques, and that farmers were becoming increasingly receptive. Still, extension programs tended to be disorganized, and there was little system to the work. The committee recommended that each college establish a separate department to take over all of the extension teaching in the state and to systematize the different programs.[59]

Two years later the committee asserted flatly "that the present scope of dissemination work among farmers is entirely inadequate. There are tens of thousands of farmers who do not take agricultural papers; probably not one farmer in 25 ever attends a farmers' institute; there is a comparatively small amount of . . . study of agricultural literature among farmers. . . . As a plain matter of fact, we are not today, either directly or indirectly, reaching the great masses of the tillers of the soil with educational processes that may be regarded as even fairly efficient." Moreover, according to the committee, it was clear that the colleges "must go to the farmers in their homes and communities—they will not come to us."[60]

To fill the needs the committee recommended again that each college create a department of extension, that the Association of

[58] Willard, *Kansas State College*, pp. 480–481; AAACES, *Proceedings*, 1907, p. 27; Scott, "Early Agricultural Education in Minnesota," pp. 31–33.

[59] Epsilon Sigma Phi, *Extension Work*, p. 27; AAACES, *Proceedings*, 1906, p. 72; John Hamilton to A. C. True, November 27, 1906, Office of Experiment Stations Records.

[60] AAACES, *Proceedings*, 1908, p. 39.

167

American Agricultural Colleges and Experiment Stations establish within its organization a section on extension similar to those on resident instruction and experiment station work, and that federal appropriations be sought to stimulate and assist the states and the colleges to greater efforts. The latter proposal aroused fears of federal domination, although the committee was careful to point out that the principle involved would be no different from that under which the federal government contributed financially to resident instruction and to experiment station work.

At its 1909 convention the association accepted in the main the recommendations made a year earlier by its committee on extension. A section of extension work was established within the association, and the body went on record as favoring the introduction into Congress of a bill providing for an appropriation to the colleges for extension work. Finally, the association placed its weight behind the movement to establish at each land-grant college a formal extension department.[61]

Pushed by the position taken by their association and by the growing interest of farmers for education, the colleges moved forward rapidly to establish extension departments. Only five states had such offices by the end of 1905, but by February, 1910, there were twenty-seven, and in the middle of 1912 the United States Commissioner of Education reported that forty-three colleges had directors of agricultural extension. As early as 1910 these men headed departments with from one to fourteen regular employees and with appropriations ranging downward from the $50,000 provided in New York. Total outlay for extension in fiscal 1911–12 amounted to $560,000. Regionally, the Middle West was best covered, both in terms of annual appropriations and in number of extension offices. The South tended to establish extension departments but failed to provide adequate funds, while the Pacific and mountain states, with the exception of California, Washington, and Utah, did very little.[62]

[61] *Ibid.*, pp. 40–42; *ibid.*, 1909, p. 45; John Hamilton to L. H. Bailey, October 20, 1909, Bailey Papers.

[62] John Hamilton, ed., "College Extension in Agriculture: Discussions before

By 1911, at least, most of the nation's land-grant colleges were energetically pushing extension work, and collectively they were making a significant contribution to the education of the country-side. As early as 1909 authorities in Kansas claimed that the college was in contact with one out of every three farmers in the state.[63]

Still, many agricultural educators were doubtful that they had found a thoroughly effective teaching device. Many of the methods in use performed useful functions and some of them obviously produced results. On the other hand no one could claim that the colleges had succeeded in reaching every farmer in need of instruction; even more important was the fact that most of the methods in use by the colleges rested on the technique of telling the farmer the route to better agriculture, instead of showing him on his own land and under normal conditions how it might be done. As early as 1900 Isaac P. Roberts had asked, "Will personal visits by trained men break down the farmer's prejudice and insure a better appreciation of specialized education . . . or are there other ways . . . ?" Seven years later, after the colleges had moved aggressively to expand their programs, *Hoard's Dairyman* observed that it knew of but two techniques holding great promise: the cooperative field tests conducted by many colleges and the demonstration work in the South, then just emerging as a regional movement under the dynamic leadership of Seaman A. Knapp.[64] In time, experience would prove the wisdom of that observation.

the Graduate School of Agriculture, at the Iowa State College, Ames, Iowa, July 4–27, 1910," U.S. Office of Experiment Stations, *Bulletin* 231 (Washington, 1910), p. 22; U.S. Commissioner of Education, *Report,* 1912, I, 266.

[63] A. C. True to James Wilson, January 4, 1911, Office of Experiment Stations Records; H. J. Waters, "The Duty of the Agricultural College," *Science,* XXX (December 3, 1909), pp. 777–789.

[64] Epsilon Sigma Phi, *Extension Work,* p. 32; *Hoard's Dairyman,* XXXVIII (August 2, 1907), pp. 684–685.

VI. RAILROADS AND EDUCATION
FOR THE FARMER

IN THEIR EFFORTS to take knowledge to farmers, the land-grant colleges and other agencies enjoyed significant assistance from a wide variety of business firms. Most important were the railroads. They cooperated with the colleges and other groups engaged in the work and in several important cases they were educational innovators, formulating and implementing teaching techniques that were later adopted by traditional educational agencies. Moreover, railroad men often were powerful voices in state legislatures, using their influence on behalf of new or expanded programs.[1]

That railroads should assume such roles is not surprising. The colonization activities of the great transcontinental and other lines stemmed not only from the carriers' desire to dispose of land grants but also from their recognition that they needed to become traffic-generating as well as traffic-moving agencies. For similar reasons, long before 1900 railroad managers found it advantageous to undertake other types of developmental programs in their territories in order to expand business or to reduce seasonal or other fluctuations in the flow of traffic.[2] Perhaps the best-known example of such ac-

[1] This chapter appeared in somewhat different form in *Business History Review* XXXIX (Spring, 1965), pp. 74–98.

[2] For outstanding studies of colonization activities of single railroads see Paul

tivity was James J. Hill's distribution of purebred stock in the early 1880s and the Burlington's work with the development of dry farming techniques on the western prairies.[3] Consequently, it was perfectly logical that railroad managers would watch with considerable interest the rise of agricultural extension and would be potent forces in its development.

Among the oldest and most common forms of railroad assistance to agricultural education was the granting of free transportation or reduced rates to persons engaged in that work. A few carriers gave lowered rates to participants in the first sessions of the Association of American Agricultural Colleges and Experiment Stations, and twenty years later such lines as the Louisville and Nashville gave reduced rates to persons attending the meetings of the Southern Educational Conference. Until the practice was prohibited by law, many carriers regularly provided annual passes for presidents and other officials of land-grant colleges.[4]

Railroads also provided cheap transportation for farmers attending educational gatherings. By the 1890s some carriers had discovered that the agricultural colleges had something to offer adult farmers and that it was to the advantage of the railroads that agriculturists come into contact with the institutions. Late in the decade James J. Hill gave free transportation to farmers visiting the North Dakota college at Fargo, while several Iowa raidroads conducted

W. Gates, *The Illinois Central Railroad and Its Colonization Work* (Cambridge, Mass., 1934) and Richard C. Overton, *Burlington West: A Colonization History of the Burlington Railroad* (Cambridge, Mass., 1941). Other works include Ira G. Clark, *Then Came the Railroads: From Steam to Diesel in the Southwest* (Norman, Okla., 1958), pp. 208–215, 275–280, 300–315; Carlton J. Corliss, *Main Line of Mid-America: The Story of the Illinois Central* (New York, 1950), pp. 292–300, 420–422; James H. Lemly, *The Gulf, Mobile and Ohio* (Homewood, Ill., 1953), pp. 72–76, 259–261.

[3] Joseph G. Pyle, *The Life of James J. Hill* (2 vols., New York, 1917), I, 365–367; Richard C. Overton, *Burlington Route: A History of Burlington Lines* (New York, 1965), p. 285.

[4] AAACES, *Proceedings,* 1889, p. 17; J. E. Davenport to F. B. Mumford, April 1, 1912, Missouri Agricultural College Papers; J. C. Hardy to M. V. Richards, December 23, 1904, Hardy to F. S. White, June 24, 1905, and Hardy to M. Schuler, February 15, 1905, J. C. Hardy Correspondence (Mississippi State University Library, State College, Miss.); Watts, "Agricultural Extension in Tennessee," p. 106.

171

excursions that took farmers from all parts of Iowa to Ames. When the colleges began holding short courses on the campuses, the railroads habitually gave reduced rates to those who attended. In other instances carriers offered lower than ordinary fares to persons attending graduation ceremonies at land-grant schools so that adults might see the strides being taken in the development of scientific agriculture. Reduced fares also helped farmers to attend sessions of such groups as the Iowa State Dairy Association, the National Creamery Buttermakers' Association, and a multitude of similar groups.[5]

In most parts of the United States, railroads were significant contributors to farmers' institute programs. Again the most popular type of aid was free or reduced-rate transportation. In the 1880s the Winona and St. Peter gave such aid to Oren C. Gregg of Minnesota so that the widely-known dairy farmer might lecture at county fairs. In 1903 the New Orleans and Northeastern Railroad Company provided free transportation for Mississippi institute lecturers and directed train crews to make nonscheduled stops to discharge or take on speakers. In Nebraska, where carriers provided a similar service, the savings to the state amounted to as much as $2,500 annually.[6]

Occasionally, railroad men took more active roles in institute work. A railway official appeared as a lecturer at a Louisiana session in 1897, and Thomas Shaw, an employee of the Great Northern, spoke in Montana during the 1911–12 year. The Gulf and Ship

[5] Hunter, *North Dakota's Land-Grant College*, p. 55; Ross, "The New Agriculture," p. 127; W. A. Hopkins to F. B. Mumford, December 26, 1911, and W. S. St. George to Mumford, December 20, 1911, Missouri Agricultural College Papers; C. H. Tuck to H. J. Webber, January 29, 1910, Bailey Papers; R. C. King to General Passenger Agent, Mobile and Ohio Railroad, May 31, 1899, and S. D. Lee to General Passenger Agent, Southern Railway, May 25, 1898, R. C. King Letterbooks (Mississippi State University Library, State College, Miss.); *Hoard's Dairyman*, XXVIII (November 5, 1897), p. 758; *ibid.*, XXXIII (October 31, 1902), p. 784.

[6] Scott, "Oren C. Gregg and Farmers' Institutes," p. 21; Illinois Central General Passenger Agent to J. C. Hardy, July 8, 1903, C. E. Jackson to Hardy, June 24, 1902, G. H. Smith to J. A. Turnipseed, July 31, 1903, and Smith to R. P. Wright, August 3, 1903, Hardy Correspondence; Nebraska Farmers' Institute, *First Report*, p. 27.

Island and the St. Louis and San Francisco cooperated actively with the institute management in Mississippi in holding meetings along their lines. Normally the railroads assumed the responsibility for advertising the gatherings, for making local arrangements, and for providing a part of the lecture corps. Railroad men often appeared at the roundup institutes. M. V. Richards of the Southern and F. S. White of the Frisco, for example, were regular speakers at such gatherings in the South. A few railroads aided the work by subsidizing the publication of institute annuals, the volumes of agricultural information that in many states were distributed free of cost to farmers.[7]

Railroad companies also distributed educational literature. This type of work was not new since the western transcontinental and other lines had made use of the printed word in their colonization efforts, but the literature broadcast after 1900 was less promotional and more educational in character. The Southern was one of several lines that obtained bulletins from the United States Department of Agriculture and distributed them throughout its territory. The Rock Island placed advertisements in local newspapers, calling the attention of farmers to bulletins that it had available. Other railway agents prepared their own literature, usually basing it on materials obtained from the Department of Agriculture or from the agricultural colleges and experiment stations. As early as 1903 J. F. Merry of the Illinois Central produced one-page circulars for enclosure in farm journals and local newspapers published in the carrier's territory. In other instances railroads employed academic men to prepare scholarly but popularly written bulletins for them. In 1903 the Chicago and Northwestern distributed materials of that type, while in 1912 the Southern reported heavy demand for a bulletin on beef production that had been written by professors at the agricultural colleges in North

[7] Louisiana Farmers' Institute, *Bulletin* 1, p. 70; Montana Farmers' Institutes, *Ninth and Tenth Annual Reports,* 1910–12, p. 17; J. H. Bouslog to J. C. Hardy, June 29 and July 9, 1903, Bouslog to J. A. Turnipseed, July 2, 1903, Hardy to F. S. White, May 6, 1905, White to Hardy, February 14, May 23, 1904, M. V. Richards to Turnipseed, August 1, 1904, and Hardy to J. F. Merry, May 2 and July 31, 1902, Hardy Correspondence; *Breeder's Gazette,* XLIV (September 16, 1903), p. 445; Hunter, *North Dakota's Land-Grant College,* p. 54.

Carolina and Alabama. The Soo Line was perhaps unique when in 1905 it gave to a large number of farmers along its line subscriptions to *Hoard's Dairyman*, the influential Wisconsin farm paper.[8]

There can be little doubt that much of the literature distributed was of genuine value and that it had some impact on those farmers willing to read to improve their operations. A Frisco bulletin on dairying in the Ozarks was of such high quality that it was used for a time in the agricultural classes at the University of Missouri. The quantity of literature, alone, was impressive. In 1912 the Frisco had a half a dozen different publications available for distribution, all written by experts, while the Pennsylvania claimed that in a few weeks it had distributed 220,000 copies of pamphlets dealing with eight crops.[9]

In some cases railroad men induced experiment stations or other agencies to undertake studies and to produce educational literature on subjects needing treatment. The Northern Pacific, for example, encouraged the Montana farmers' institute management to prepare and make available for distribution a 25,000-copy edition of a study on dry farming.[10]

Railroads were interested in strengthening the agricultural experiment stations, the sources of much of the information being made available to farmers. In the 1890s James J. Hill gave 480 acres for a substation at Crookston, Minnesota, while in the same decade the Florida East Coast line leased a tract of land at Boca Raton to the Florida experiment station for citrus research. The Southern Railroad aided the South Carolina station in the same way, and the

[8] *Southern Farm Gazette*, VIII (February 1, 1903), p. 1; M. V. Richards to James Wilson, February 1, 1904, George H. Lee to Wilson, June 20, 1905, and W. B. Kniskner to Wilson, December 1, 1903, Records of the Office of the Secretary of Agriculture; *Breeder's Gazette*, LXII (August 28, 1912), p. 372; F. B. Mumford to C. H. Eckles, January 11, 1913, Missouri Agricultural College Papers; *Corn*, I (June, 1912), pp. 108, 114; *Hoard's Dairyman*, XXXV (January 6, 1905), p. 1197.

[9] F. B. Mumford to B. W. Redfearn, October 28, 1910, Missouri Agricultural College Papers; *Missouri Farmer*, IV (June, 1912), p. 20; *Railway Age Gazette*, n.s. LII (January 5, 1912), p. 29.

[10] Montana Farmers' Institutes, *Eleventh and Twelfth Annual Reports*, 1912–14, p. 10.

Burlington contributed considerable sums to the stations in Colorado and Wyoming. Railroads also used their political influence to obtain greater state support for experiment stations or for the establishment of branch stations. The Rock Island was active in Arkansas in 1905, drumming up support for larger appropriations. The Illinois Central played no small role in the creation and location of Mississippi's Delta Branch station at Stoneville.[11]

Since railroad men recognized the educational value of county fairs and similar gatherings, they began to encourage their establishment as early as the 1850s. In that decade the Illinois Central aided the pioneer state fairs in Illinois, while two and three decades later the Burlington was only one of several railroads that cooperated with sponsoring agencies to make local fairs more attractive. These efforts continued into the twentieth century with railway agents urging farmers to attend, offering reduced rates for fair visitors, and awarding prizes for best exhibits. Occasionally railroads promoted agricultural festivals or held the affairs themselves.[12]

In a diversity of other ways the carriers encouraged the development of agricultural knowledge useful in extension teaching. Early in the twentieth century railroads urged the United States Department of Agriculture to expand its soil survey program, especially in the southern states. In 1901 the Southern Pacific distributed new varieties of seed wheat, oats, and barley to farmers in western Ore-

[11] Jesness, *Andrew Boss,* p. 24; G. H. Aull, "The South Carolina Agricultural Experiment Station: A Brief History, 1887–1930," South Carolina Agricultural Experiment Station, *Circular* 44 (Clemson, 1931), pp. 26–27; C. Clyde Jones, "A Survey of the Agricultural Development Program of the Chicago, Burlington and Quincy Railroad," *Nebraska History,* XXX (September, 1949), p. 239; W. P. Fletcher to James Wilson, May 25, 1905, Records of the Office of the Secretary of Agriculture; J. A. Turnipseed to J. F. Merry, November 18, 1901, and J. C. Hardy to Merry, December 20, 1904, Hardy Correspondence.

[12] Corliss, *Main Line of Mid-America,* pp. 292, 299; Jones, "Survey of the Agricultural Development Program of the Chicago, Burlington and Quincy Railroad," pp. 236–237; Jesse C. Burt, "History of the Nashville, Chattanooga and St. Louis Railway, 1873–1916" (Ph.D. Thesis, Vanderbilt University, 1950), pp. 117–118; E. D. Comstock to James Wilson, March 2, July 7, 1903, L. Trice to Wilson, June 8, 1903, and J. F. Merry to Wilson, February 12, 1903, Records of the Office of the Secretary of Agriculture; *Southern Farm Magazine,* VII (March, 1899), pp. 16–17.

gon, and thirteen years later the Chicago and Northwestern leased right-of-way land in Nebraska for the testing of alfalfa. The Chesapeake and Ohio sought to disseminate forage crop information by forming alfalfa clubs among farmers in its territory. In 1912 the Pennsylvania employed an agent who devoted his full time to the establishment of general farmers' clubs, similar to those that had existed for more than a half a century. When the Santa Fe placed certain lands near Dodge City on sale in 1911, the railroad stipulated that buyers were to follow a cropping plan worked out by the state agricultural college. As a final example of miscellaneous activity, in 1913 the Santa Fe sponsored and provided speakers for a series of meetings in eastern Colorado where the value of silage and the methods for constructing silos were explained to farmers.[13]

Of all the extension techniques employed by railroads, however, none was more popular or more widely used during the decade before 1914 than the educational or demonstration train. Like most extension techniques educational trains had diverse and obscure origins. As early as 1891 the Ontario Agricultural College sent two lecturers into the countryside in a wagon loaded with butter-making equipment and other tools of the well-operated dairy as a means of instructing farmers in the methods of successful dairying. Later in the decade James J. Hill operated over Great Northern lines in Minnesota an excursion train that stopped at several stations where the exhibits that it carried were viewed by rural residents. In 1901 the National Good Roads Association, in conjunction with the United States Bureau of Public Roads Inquiries, road machinery manufacturers, and others, sent what was called a Good Roads Train over the Illinois Central from New Orleans to Chicago. Loaded with machinery and men to operate it, the train stopped at numerous

[13] G. A. Parks to James Wilson, January 13, October 28, 1903, and John Sebastian to Wilson, June 7, 1905, Records of the Office of the Secretary of Agriculture; *Railway Age,* XXXII (November 1, 1901), p. 496; *Railway Age Gazette,* n.s. LVI (January 16, 1914), p. 134; *Southern Planter,* LXXII (June, 1911), p. 703; *American Fertilizer,* XXXVI (April 20, 1912), p. 41; Rich, "Railroads and Agricultural Interests in Kansas," pp. 118–119; *Breeder's Gazette,* LXIV (November 6, 1913), p. 896.

points en route where demonstrations illustrated road-building techniques.[14]

A year later John T. Stinson of the Missouri fruit experiment station used a Frisco boxcar to transport spraying equipment and other materials to various points in southern Missouri where he held demonstrations for fruit growers. When reception was favorable the Missouri State Board of Agriculture adopted the technique as a means of making its institute programs more attractive. From 1902 to 1904 the Missouri Pacific provided a standard passenger car to transport display materials for institutes, and in 1904 the Missouri-Kansas-Texas made a similar contribution.[15]

Whether these widely separated episodes had any relationship to the launching of later educational train work is a matter for conjecture; certainly more significant was an experiment by the Minneapolis and St. Louis Railroad. In the winter of 1896–97 prices for grain were so low that farmers hesitated to ship it, and railway revenues fell. E. F. Farmer, a freight agent of the road, suggested to Henry Wallace that the carrier and Wallace's farm paper combine to hold a series of creamery-promotion meetings in northern Iowa. A bargain was struck and a number of meetings was scheduled in February, 1897. The railroad advertised the gatherings, located suitable halls in the towns, paid all expenses, and provided transportation for the speakers via the company's passenger service. Henry Wallace served as a lecturer and only James Wilson's appointment as secretary of agriculture prevented his appearance. A few weeks later officials of the Des Moines and Fort Dodge branch of the Rock Island sponsored a similar series of institute-like meetings.[16]

[14] A. J. Madill, *History of Agricultural Education in Canada* (Toronto, 1930), p. 151; Erickson, *Rural Youth,* pp. 44–45; Earl Mayo, "The Good Roads Train," *World's Work,* II (July, 1901), pp. 956–960.

[15] Roy V. Scott, "Railroads and Farmers: Educational Trains in Missouri, 1902–1914," *Agricultural History,* XXXVI (January, 1962), pp. 5–6.

[16] *Wallaces' Farmer,* XXX (February 10, 1905), p. 176; *ibid.,* XXXIII (December 25, 1908), p. 1599; Ritland, "Educational Activities of P. G. Holden," pp. 13–14; Morgan, *Extension Service of Iowa State College,* p. 24n; Henry Wallace, *Uncle Henry's Own Story of His Life* (3 vols., Des Moines, 1917–19), III, 79–81.

These programs were forgotten as prosperity returned, but when a new problem arose in 1904 they were recalled and the precedent was put to use. A cold, wet summer in 1903 combined with an early fall freeze reduced the Iowa corn crop of that year; even more important, the seed that farmers selected for the next year's planting was certain to be of even poorer quality than usual. Speaking before the Iowa Grain Dealers' Association, Perry G. Holden of Iowa's agricultural college warned the grainmen of the impending danger and pointed out that careful seed selection and other simple procedures could increase corn production by at least 40 percent. The problem was one of getting this rather elementary information into the hands of farmers.

In the audience as Holden spoke was W. H. Given, a Rock Island official who recalled the lecture programs of 1897. A conference including Holden, Given, Henry Wallace, and others followed, and the first true educational train took shape. The promoters of the enterprise decided to operate a special train consisting of three coaches and two private cars, the latter to accommodate the speakers who would accompany the train. Each coach served as an auditorium and was equipped with a speaker's platform and an array of lecture charts and other demonstration materials. The lecture crew consisted of Holden, a second staff member from the Iowa agricultural college, and representatives of the railroad, the rural press, and the grain dealers' association. The carrier made all local arrangements and advertised the meetings through the local press and by means of posters.

The three-day tour of the Seed Corn Special began April 18, 1904, covered more than 400 miles, and visited fifty communities in fifteen counties. Holden's message was simple: farmers should select the best ears from any strong home-grown corn, test the seed with a germination box, grade it by shape and size of the grain, and adjust their planters so that the machines would drop three kernels into each hill. When the tour was completed, promoters estimated that more than 3,000 farmers had been contacted. Response was favorable enough to induce the Chicago, Burlington and Quincy Railroad to arrange for a similar tour over its lines in southern Iowa during the last week in April.[17]

Following the appearance of educational trains in Iowa, the movement spread rapidly, becoming a popular means of presenting basic agricultural knowledge to large numbers of farmers. The Burlington was an early leader in the work, sending demonstration trains into Nebraska, Illinois, Missouri, and Iowa during a three-month period beginning in December, 1904.[18] Other lines were not far behind and educational trains began to appear in other parts of the United States. Early in March, 1905, the Minneapolis, St. Paul and Sault Ste. Marie operated a special in North Dakota, the Frisco dispatched one over its lines from Missouri to Texas, and a month later the St. Joseph and Grand Island Railway operated the first demonstration train in Kansas. In December, 1905, the first in Indiana appeared on the tracks of the Lake Erie and Western, and a year later the pioneer venture in Ohio traveled over the Big Four lines. The Lake Shore and Michigan Southern introduced the technique to Michigan farmers in April, 1906.[19]

The Maryland and Pennsylvania Railroad operated the first educational train in the East and the South in April, 1905. New England farmers saw their first trains in 1906 when the Boston and Maine operated one in Massachusetts, New Hampshire, and Vermont; the same year the Maine Central and the Bangor and Aroostook railway companies sent specials over their lines in Maine. The Illinois Central managed the first demonstration train in the Deep South when its Diversified Farming Special toured Mississippi and Louisi-

[17] How the Seed Corn Specials Got Started, Holden Papers; *Review of Reviews,* XXX (November, 1904), pp. 563–565; *Wallaces' Farmer,* XXIX (April 29, 1904), p. 605; *ibid.* (May 6, 1904), p. 644; *ibid.,* XXX (February 10, 1905), p. 176; Mosher, *Early Iowa Corn Field Tests,* p. 116; Wallace, *Uncle Henry's Own Story,* III, 95–96. The early specials in Iowa are also discussed in Jacob A. Swisher, "The Corn Gospel Trains," *The Palimpsest,* XXVIII (November, 1947), pp. 321–333.

[18] *Breeder's Gazette,* XLVII (February 8, 1905), p. 264.

[19] North Dakota Farmers' Institute, *Annual,* 1905, p. 3; *Wallaces' Farmer,* XXX (December 26, 1905), p. 708; Willard, *Kansas State College,* p. 165; Hepburn and Sears, *Purdue University,* p. 115; Mendenhall, *Ohio State,* II, 153; *Breeder's Gazette,* XLVIII (December 20, 1905), p. 1316; *ibid.,* LI (January 9, 1907), p. 76; Beal, *Michigan Agricultural College,* p. 166.

179

ana in October, 1906. In 1907 at least three lines operated specials in South Carolina and several specials appeared in Virginia.[20] The Southern Pacific operated the first special in California, beginning in November, 1908. In the same year the Pennsylvania Railroad opened educational train work in that state and the Erie operated the first special in New York. Few worlds remained to be conquered by 1912 when a special appeared in northern Michigan on the tracks of the Duluth, South Shore and Michigan Railroad.[21]

The original demonstration trains in Iowa in 1904 were devoted exclusively to seed-corn improvement, but later trains took up other agricultural topics so that by 1914 all aspects of farming had been discussed from their lecture platforms. Programs tended to become more elaborate and more sophisticated, with greater emphasis on demonstration materials and exhibits and with sufficient lecturers to cover a wide array of subjects in more than superficial detail. Displays of prize-winning livestock and modern farm machinery enabled lecturers to improve their presentations, while charts and graphs of various types showed farmers the results of better feeding and tillage practices. Several trains in the South dealt specifically with the boll weevil problem and with the need for diversification. Nor was the farm wife ignored, since some of the later efforts included in their programs lectures by trained home economists. In at least one area, southern Louisiana, speakers found it necessary to talk to their audiences in a language other than English.[22]

Educational trains were most often operated in cooperation with

[20] *Southern Farm Magazine,* XIII (May, 1905), p. 3; Caswell, *Massachusetts Agricultural College,* p. 43; Fernald, *University of Maine,* p. 262; *Railway Age,* XLII (September 21, 1906), p. 363; *Southern Farm Gazette,* XI (October 1, 1906), p. 9; *Southern Farm Magazine,* XV (April, 1907), pp. 3, 7.

[21] *Breeder's Gazette,* LV (June 30, 1909), pp. 1452–1453; *ibid.,* LXII (December 6, 1912), p. 988; "Agricultural Demonstration Trains of the University of California," *University of California Chronicle,* XI (April, 1909), pp. 186–187; Fletcher, *Pennsylvania Agriculture and Country Life,* p. 470; *Cornell Countryman,* VI (January, 1909), pp. 116–119.

[22] *Breeder's Gazette,* LV (March 17, 1909), p. 654; *ibid.,* LIX (March 15, 1911), p. 705; *Kimball's Dairy Farmer,* XI (April 15, 1913), p. 242; *Wallaces' Farmer,* XXXVIII (October 24, 1913), p. 1459; Illinois Central Railroad, *Organization and Traffic of the Illinois Central System* (Chicago, 1938), p. 244; *Railway*

the agricultural colleges which provided most of the lecturers. State boards or departments of agriculture, a variety of farmers' groups, a few other business concerns, and the United States Department of Agriculture occasionally joined in the work. Insofar as the railway equipment was concerned most educational trains followed patterns established by the pioneers. The railroad company furnished motive power, the necessary number of cars of different types, and operating crews. A few specials were unique. The Illinois Traction System ran a special trolley through central Illinois in 1911, a year earlier the Kansas City, Clinton and Springfield hooked together three gasoline-powered cars to form a train, and in June, 1909, the Delaware and Virginia Railway took to the water when it operated the Farmers' Special Instruction Steamboat on the Rappahannock.[23]

As an extension technique the demonstration train was spectacular, attracting farmers in droves. In one month in 1910 lecturers from the University of Missouri contacted 40,000 people, far more than could have been reached by any other means. A Norfolk and Western special distributed 100,000 pieces of scientific literature among Virginia farmers during a twelve-day tour.

Such successes encouraged sponsoring agencies to redouble their efforts. From December, 1908, through March, 1909, five different specials appeared in Mississippi, and in 1911 fourteen trains operated in Ohio alone. During the year ending June 30, 1911, a total of sixty-two educational trains carried 740 lecturers more than 35,000 miles and brought them into contact with almost one million people. Thereafter, the number of specials declined, but attendance continued to increase until 1914.[24]

For almost a decade after 1904 most spokesmen were well pleased

---

Age Gazette, n.s. LIII (November 29, 1912), p. 1058. For a discussion of trains operated by the Rock Island, see Victor H. Schoffelmayer, *Southwest Trails to New Horizons* (San Antonio, Tex., 1960), pp. 27–29. Educational trains in Tennessee are mentioned in Watts, "Agricultural Extension in Tennessee," pp. 122–125.

[23] *Kimball's Dairy Farmer,* XI (May 15, 1913), p. 292; *Southern Farm Gazette,* XI (October 1, 1906), p. 9; *Railway Age Gazette,* n.s. XLVI (June 11, 1909), p. 1224; *Breeder's Gazette,* LIX (February 22, 1911), p. 502; Scott, "Educational Trains in Missouri," p. 10.

[24] F. B. Mumford to W. J. Stone, March 28, 1910, Missouri Agricultural Col-

with the educational contributions of the trains. Educational trains, said Dean Frederick B. Mumford of the Missouri College of Agriculture in 1910, constituted an excellent means of "taking the University out to the people instead of having them come to the University." Continuing in that vein, the veteran agricultural educator claimed that never before had "so many people heard about the College of Agriculture . . . in so brief a time." James Withycombe of the Oregon Experiment Station, President J. C. Hardy of Mississippi Agricultural and Mechanical College, and President E. A. Bryan of the State College of Washington were only three of the other educators who echoed Mumford's sentiments.[25] Railroad men disclaimed any philanthropic motives and stated quite frankly that they were engaged in agricultural extension because better farming "means more business, hence more money for us." Educational train work was expensive, ranging up to $30,000 for an extensive campaign by the Southern Pacific in California, but few railroad men doubted that their companies would be repaid for their expenditures.[26]

Still, long before the educational trains had reached a peak of popularity, some railroad and academic officials were coming to the conclusion that as an effective teaching device the trains left much to

lege Papers; *Southern Planter,* LXXII (June, 1911), p. 700; Mississippi Agricultural and Mechanical College, *Biennial Report,* 1908–09, p. 93; *Railway Age Gazette,* n.s. LII (March 8, 1912), p. 443; *Mississippi Agricultural Student,* I (November 1, 1913), p. 1; *Missouri Ruralist,* February 3, 1912, p. 4; Hamilton, "Farmers' Institute and Agricultural Extension Work," p. 2; *Farm, Stock, and Home,* XXIX (July 1, 1913), p. 491; *Kimball's Dairy Farmer,* XI (May 15, 1913), p. 292.

[25] *Missouri Agricultural College Farmer,* VII (April 1910), p. 17; F. B. Mumford to A. W. Douglas, March 25, 1910, Missouri Agricultural College Papers; James Withycombe to E. W. Allen, February 9, 1909, Office of Experiments Stations Records; J. C. Hardy to J. C. Clair, October 15, 1906, Hardy Correspondence; Bryan, *State College of Washington,* p. 535.

[26] *Missouri Agricultural College Farmer,* VII (April, 1910), p. 16; B. W. Redfearn to C. R. Gray, February 28, 1911, Missouri Agricultural College Papers; *Railway Age Gazette,* n.s. LII (March 22, 1912), p. 695; Council of North American Grain Exchanges, Committee on Crop Improvement, *Proceedings of a Meeting in Chicago, February 8, 1911* (n.p., n.d.), p. 102; *University Missourian,* March 21, 1910, p. 1.

be desired. There was, after all, no means to insure that a farmer would practice what he had heard in the course of an educational-train lecture. One prominent agricultural educator expressed the sentiments of many by 1912 when he observed that "the present form of educational train has about reached its limit of efficiency." James J. Hill of the Great Northern maintained as early as 1911 that the talks given from the ordinary educational train were of doubtful value.[27] Such critics agreed that the educational train was a highly successful propagandizing and popularizing agent, useful in dissolving rural prejudices and in awakening interest in scientific agriculture. But for most farmers practical teaching required the creation of an example so they could actually see the results of proper methods.

Some railroad men recognized the value of the object-lesson teaching device before the turn of the century and, in fact, before most academic men grasped its significance. As early as 1860 a few railroad leaders maintained private estates which they hoped would serve as examples for neighbors, but true demonstration farms proved to be far more useful. In the main these were established in the late nineteenth century by companies whose tracks extended through underdeveloped territory or through districts suffering from sparse rainfall or other problems. Fairly typical of such efforts was the work of the Burlington in the development of dry farming techniques in Nebraska.[28] Generally these farms had a dual purpose: their promoters hoped to improve the practices of farmers already in the area and at the same time to attract new settlers by demonstrating the potential of the region.

An early example of the dual-purpose demonstration farm was the Cycloneta-Model Stock and Agricultural Farm, established by the Georgia Southern and Florida Railroad in 1889 near Tifton, Georgia. The carrier sought to show that cutover land could be used

[27] AAACES, *Proceedings,* 1912, pp. 217, 222; L. H. Bailey to I. H. Shoemaker, March 29, 1912, Bailey Papers; *Railway Age Gazette,* n.s. LI (December 1, 1911), p. 1143; Railway Development Association, *Proceedings,* 1913 (Roanoke, Va., n.d.), p. 7; *ibid.,* 1914, p. 38.
[28] O. O. Waggener, *Western Agriculture and the Burlington* (Chicago, 1938), pp. 21–23.

profitably. At the outset the property was managed by a former staff member of Mississippi Agricultural and Mechanical College.[29]

Later, in 1905, the Long Island Railroad undertook a similar project to demonstrate that large undeveloped areas of Long Island could be productive. Using only techniques available to the ordinary farmer, the carrier showed that scrub oak wasteland was suitable for growing of grasses, fruit, and garden truck. As a result, long-time residents improved their operations and new arrivals moved into the area to farm the land in accordance with the plans worked out by the company's agriculturists.[30]

Demonstration farms were also established by a number of other lines. In 1907 the Norfolk and Western acquired 300 acres of pine and oak forest near Ivor, Virginia, and transformed the tract into a flourishing farm. The promotion of immigration and the forwarding of agricultural education were both prominent in the company's plans. Two years later the Pennsylvania purchased a run-down farm of 50 acres near Bacon, Delaware, and employed H. S. Lippincott, a graduate of Cornell University, to manage the property. Lippincott also appeared at farmers' institutes and otherwise helped the general development of agricultural education in the area.[31]

In 1910 New York Central took title to three farms in New York and placed them in the hands of competent managers to demonstrate what could be done on rough, run-down farms with new scientific and business methods. A few months later the Norfolk and Southern Railroad Company obtained a farm at Wenona, North Carolina, while the St. Louis–Southwestern had one at Morrill, Tex-

[29] B. Irby to W. O. Atwater, January 8 and November 21, 1890, Office of Experiments Station Records; Range, *Georgia Agriculture*, pp. 114, 132.

[30] Railway Development Association, *Proceedings*, 1914, pp. 38–40; Edith L. Fullerton, *The Lure of the Land* (New York, 1906), pp. 5–8; *Railway Age Gazette*, n.s. XLVII (October 29, 1909), p. 789; *Fruit Grower and Farmer*, XXIII (October, 1912), p. 466.

[31] *Southern Planter*, LXVIII (June, 1907), pp. 561–562; *ibid.*, LXXI (June, 1910), pp. 661–662; *Railway Age Gazette*, n.s. XLVII (October, 1909), p. 606; *Breeder's Gazette*, LVII (April 27, 1910), p. 1036; *American Fertilizer*, XXXI (October, 1909), p. 17.

as. In 1911 the Toledo, Peoria and Western Railway had a ten-acre farm at Crugar, Illinois, under the management of a prominent farmer living in the area. Even some minor companies took up the work. The Waycross and Western was reported in 1913 to be considering the establishment of a model farm near Kings, Georgia.[32]

The single demonstration or model farm, however, suffered from distinct and, in fact, fatal weaknesses as a teaching device. While farmers would visit a model farm to see what was being done, they invariably left with the firm conviction that the techniques used were economically unsound on their properties or were unsuitable because of variations in soil or other elements.[33] Such spokesmen as G. J. Ryan of the Great Northern maintained by 1911 that a series of small plots would be more valuable since they would allow for local soil and other variations while permitting a larger number of farmers to observe better methods. The same year, a leading southern farm journal advised the carriers that "demonstrations, such as Dr. [Seaman A.] Knapp is making, will . . . pay the railroads a great deal."[34]

Precedents had been set for this type of work prior to 1900. In 1899 the Seaboard Air Line reported that it had launched twenty-two "experimental farms" along its tracks. More effective, apparently, was early work of the Oregon Railroad and Navigation Company. Late in the 1890s that carrier became concerned with the decline in carrying capacity of grazing lands in its territory. It established on railroad land a series of experimental plots to determine what grasses

[32] *Experiment Station Record*, XXII, p. 598; *Wallaces' Farmer*, XXXV (May 6, 1910), p. 745; *Breeder's Gazette*, LVII (June 22, 1910), p. 1414; B. E. Rice to C. L. Goodrich, February 28, 1913, Bureau of Plant Industry Records (National Archives, Washington); G. I. Stewart to F. B. Mumford, January 12, 1912, Missouri Agricultural College Papers; *Railway Age Gazette*, n.s. LV (December 5, 1913), p. 1090.

[33] Bradford Knapp, "Education through Farm Demonstration," The American Academy of Political and Social Science, *The Annals*, LXVII (September, 1916), pp. 225–226; *Progressive Farmer and Southern Farm Gazette*, XXVIII (February 1, 1913), pp. 138–139.

[34] Council of North American Grain Exchanges, Committee on Crop Improvement, *Proceedings of a Meeting in Chicago, February 8, 1911*, p. 122; *Progressive Farmer and Southern Farm Gazette*, XVI (February 25, 1911), p. 192.

185

could best be grown in semi-arid districts. In all more than 100 varieties were planted in May, 1898; alfalfa and bromegrass were found to offer the greatest hope. The Oregon experiment station considered the experiments to be of sufficient validity to cooperate with the railroad, and later tests were made to determine the most suitable planting time. Rather similar were the experimental plots established in 1897 along the Burlington line to demonstrate Hardy W. Campbell's dry farming methods.[35]

Little more than a decade later several railroads were using the demonstration plot technique to instruct farmers. By 1913, for example, the Illinois Central had a dozen forty-acre plots in Mississippi and Louisiana. The tracts were located on privately owned farms whose proprietors signed contracts with the railroad, binding them to operate the tracts in accordance with instructions given by trained agents and to use such seed, fertilizer, and special equipment as were provided. In return the railroad agreed to protect the farmer against any loss incurred as a result of the operation. Management of the plots, which were located with an eye toward their accessibility for farmers, aimed at showing the advantages of agricultural diversification with emphasis on the production of foodstuffs for humans and animals, the growing of soil building crops, and the raising of livestock.[36]

The Missouri Pacific–Iron Mountain System, the Frisco, and the Santa Fe used similar arrangements to instruct farmers. By 1913 the Missouri Pacific had plots in Louisiana, Arkansas, and Missouri, with a manager assigned to each plot. These officials also functioned as demonstrator-instructors for farmers in the surrounding areas. The

[35] *Southern Farm Magazine*, VI (January, 1899), p. 25; *Railway Age*, XXXII (December 27, 1901), p. 761; *ibid.*, XXXIII (January 31, 1902), p. 142; H. T. French to A. C. True, February 12, 1898; B. Campbell to H. T. Newcomb, February 7, 1898, True to Campbell, February 16, 1898, and Campbell to True, March 30, 1898, Office of Experiment Stations Records; Waggener, *Western Agriculture and the Burlington*, pp. 21–22.

[36] *Railway Age Gazette*, n.s. LI (December 22, 1911), p. 1290; *ibid.*, LVII (November 13, 1914), p. 910; Railway Development Association, *Proceedings*, 1913, p. 6; Illinois Central Railroad, *Mississippi, A Wonderful Agricultural State*

Frisco used essentially the Illinois Central plan but tried to inspire farmers under contract with it to excel by offering an annual prize to the most proficient. The Santa Fe also utilized plots, but emphasized dry land techniques and crops.[37]

In the final analysis the Hill lines had the most comprehensive and sophisticated system of demonstration plots. In 1906 the Great Northern and the Northern Pacific provided funds to permit the North Dakota experiment station to establish a half dozen twenty-acre plots at various points in the state, an action that stimulated the state to provide adequate funds for these and other demonstration tracts. Meanwhile, in Montana where grazing land was being transformed by the plow, the Great Northern by 1911 had several demonstration or experiment farms where the suitability of various crops to soil and climate was tested.[38] Officials of the company, however, resolved to put into operation a more comprehensive teaching program. Recognizing that answers to all the problems of agriculture were not available and knowing that public agencies lacked the funds to take existing knowledge to the farmer, Great Northern's management decided to assume a portion of the burden.

Late in 1911 the Great Northern employed Fred R. Crane, a soils expert from the University of Wisconsin, and placed him in charge of a program that in its operation was not greatly different from that instituted in the South by Seaman A. Knapp. Farmers interested in better methods were asked to agree to till, fertilize, and plant a five-acre plot on their properties in accordance with in-

(Chicago, n.d.), p. 20; Railway Development Association, *Proceedings,* 1914, pp. 33–34; *Illinois Central Magazine,* XIV (January, 1926), p. 32.

[37] *Arkansas Farmer and Homestead,* XIV (January 18, 1913), p. 13; *Fruit Grower and Farmer,* XXIV (March, 1913), p. 218; (Nevada, Mo.) *Southwest Mail,* January 31, 1917; *Breeder's Gazette,* LXIII (June 3, 1912), p. 27; Railway Development Association, *Proceedings,* 1915, p. 39; *ibid., Proceedings,* 1914, pp. 30–32; *Railway Age Gazette,* n.s. LI (November 10, 1911), p. 969.

[38] North Dakota Agricultural Experiment Station, *Seventh Annual Report,* 1907 (Fargo, 1907), pp. 13–14; North Dakota Agricultural Experiment Station, *Bulletin* 104 (Fargo, 1913), p. 68; Hunter, *North Dakota's Land-Grant College,* p. 53; *Breeder's Gazette,* LXVI (November 5, 1914), p. 773; *ibid.,* LX (November 8, 1911), p. 948.

structions given by Crane and a corps of assistants. Cooperators paid all costs but they retained the crop and received a rental of eight dollars an acre from the railroad. In order to determine the type and quantity of fertilizer needed, soil samples from each of the plots were tested in a laboratory established by the Great Northern in St. Paul. At the outset the testing work was conducted in greenhouses adjacent to James J. Hill's residence. The first year 151 Minnesota and North Dakota farmers cooperated in the project; by 1914 the number had grown to 780. Observers claimed that the Great Northern effort constituted the largest private system of agricultural extension in the nation.

Results of the work were often dramatic. Yields of wheat on the test plots averaged almost twice those on unfertilized land, and gains in oats and barley were comparable. A cooperator from Grafton, North Dakota, reported that "I am very well pleased with the experimenting of the five-acre plot. I see that it would readily pay to farm in that way, as I can see the difference between the other wheat and that of the plot, and I wish that all of my wheat had been as good as that." Moreover, since the cooperators were located at least five miles apart, a major portion of the company's territory in Minnesota and North Dakota was touched, and few rural residents failed to hear and see the results of scientific farming.[39]

While the vigor and enthusiasm with which railroad men took up agricultural extension work amazed many farmers and disturbed a few of them, agricultural educators in the main were pleased with the efforts of the carriers. The United States Department of Agriculture "appreciates the great value of the service the railroads are rendering to agriculture through their extension departments," wrote John Hamilton in 1913, and J. C. Hardy of Mississippi Agricultural and Mechanical College was lavish in his praise of rail-

[39] *Breeder's Gazette*, LXVI (November 5, 1914), pp. 773–774; *Literary Digest*, LXVI (February 22, 1913), p. 429; Railway Development Association, *Proceedings*, 1915, p. 73; *American Fertilizer*, XXXVIII (May 17, 1913), pp. 25–27. The quotation is from Joseph G. Pyle, "A Farm Revolution That Began in a Greenhouse," *World's Work*, XXV (April, 1913), p. 670.

road contributions to the work. Most other educators were only slightly less enthusiastic.[40]

Spokesmen for the railroads regularly admitted that the carriers' interest in improved agriculture was a selfish one, but they maintained that their goals were identical with those of rural educators and the more enlightened farmers. Certainly the public relations aspect of such railway programs was not lost on astute railroad men.[41] Still, any objective observer could see the relationship between better farming methods and heavier traffic, and in balance the work of the rail carriers in forwarding the education of the countryside was of major significance in the development of modern agricultural extension.

[40] John Hamilton to Railroad Presidents and Railroad Industrial Agents, April 3, 1913, Office of Experiments Stations Records; J. C. Hardy to F. S. White, May 25, 1905, Hardy Correspondence; AAACES, *Proceedings,* 1912, p. 216.

[41] Railway Development Association, *Proceedings,* 1913, p. 25; *Progressive Farmer and Southern Farm Gazette,* XXVII (February 17, 1912), p. 209; Watts, "Agricultural Extension in Tennessee," pp. 107–108.

WHILE RAILROADS were the business group most active in the promotion of agriculture and in the development of extension methods, other elements in the business world devoted varying amounts of attention to rural problems. Included in this group were bankers, fertilizer concerns, grain exchanges, and a variety of other businesses. Most important were the farm machinery manufacturers. In some instances their programs were little more than the type of sales propaganda that such concerns had furnished for decades and in no case was their work as comprehensive or of as long a duration as that of the railroads. But in the aggregate their efforts had a significant impact.

Such concerns as John Deere and Company and J. I. Case, both major manufacturers of farm equipment, had programs designed to improve agriculture. Early in 1911 President William Butterworth of John Deere reported that his company expended annually $150,000 for crop improvement. As a part of its program Deere and Company employed a trained agricultural scientist and prepared and distributed literature dealing with a wide variety of farm topics. The company scientist, who in 1913 was W. E. Taylor, spoke at numerous rural gatherings ranging from ordinary farmers' institutes to sizeable conventions.[1] In 1913, also, the company purchased a farm in

[1] Council of North American Grain Exchanges, Committee on Crop Improve-

190

Prince George's County, Maryland, where it expected to carry on demonstrations showing the value of intensive farming and better cultivation. In the same period J. I. Case employed F. H. Demaree as agronomist and advertising manager. His duties also included speaking to farm groups. A number of less well-known firms also found methods for helping farmers. The Mitchell Implement Company of Fort Dodge, Iowa, for instance, tested seed corn for farmers who sent samples to it.[2]

It was the International Harvester Company, however, that entered the work most aggressively. Company spokesmen explained the expenditure of several million dollars in rural improvement on the grounds of service to the community. In 1913 an officer wrote, "One of the greatest responsibilities of our . . . corporations is that of citizenship. . . . Every business concern doing business in a community is under obligation to do many things . . . which are not directly connected with or even directly beneficial to it."[3]

Some critics were quick to point out that International Harvester was well aware of the effect that rural improvement programs would have on the public image of the company, while others noted that rural advancement would lead directly to heavier purchases of International Harvester products. No doubt the firm's activities could be explained on the grounds of simple self-interest, but according to one observer this "selfishness contains within it a resulting philanthropy."[4]

International Harvester Company began work that was essentially extension in character in August, 1910, when it established its ser-

ment, *Proceedings of a Meeting in Chicago, February 8, 1911*, p. 124; *Farm Implement News*, XXXIV (January 30, 1913), p. 21; *ibid.* (February 13, 1913), p. 17; *Wallaces' Farmer*, XXXV (March 18, 1910), p. 481; Agricultural Extension Committee of the National Implement and Vehicle Association, *Agricultural Extension* (Chicago, 1916), p. 15.

[2] *American Fertilizer*, XXXVIII (June 14, 1913), p. 31; *Farm Implement News*, XXXI (March 24, 1910), p. 23; *ibid.*, XXXIV (March 13, 1913), p. 14.

[3] Edgar W. Cooley, "Service, the Duty of Citizenship," *Farming*, XIX (January, 1921), pp. 210–211; J. G. Merrison, "Extension of Agriculture to the American Farmer," *Journal of the Board of Agriculture*, XXVI (December, 1919), p. 881; *Country Gentleman*, LXXXIV (February 8, 1919), p. 46; *Breeder's Gazette*, LXIV (December 18, 1913), p. 1236.

[4] *Farm Implement News*, XXXIV (March 20, 1913), p. 25.

vice bureau. Headed at the outset by J. E. Waggoner, a graduate of
Iowa State College and a former faculty member at Mississippi
Agricultural and Mechanical College, the bureau was expected to
aid farmers by acting as an information office, by distributing literature, and by presenting to audiences of farmers and agricultural
leaders a variety of motion picture and lecture programs. To launch
the program the company asked farmers to send their questions and
problems to the bureau, and company officials sought permission of
college authorities to appear before various groups meeting on the
campuses. One of the first and most successful of the motion picture-lecture programs dealt with the evolution of the grain harvester.
It was given at Farmers' Week held on the University of Missouri
campus in January, 1911, and at other gatherings at the Universities
of Illinois and Wisconsin.[5] Meanwhile, the bureau began to produce
pamphlets and bulletins that were distributed to farmers, often
through bankers and seed merchants. For factual data, personnel of
the bureau were expected not only to rely upon their own knowledge and experience but also to call upon the specialists in the
agricultural colleges and the United States Department of Agriculture.[6]

Less than two years after the service bureau had begun operations, International Harvestor found that farmer reaction was so
favorable and that demands placed on the bureau were so heavy
that a new agency, one with more manpower, better financial support, and capable of conducting more active programs, was needed.
In 1912 the company created its agricultural extension department
and employed Perry G. Holden to manage it. Having failed in his
bid for the Republican nomination for governor of Iowa in the

[5] International Harvester Company, *The Golden Stream* (Chicago, 1910),
p. 66; *Breeder's Gazette*, LVIII (November 9, 1910), p. 976; E. L. Barker to L.
H. Bailey, September 30, 1910, and January 24, 1911, Bailey Papers; J. E. Wag-
goner to A. C. True, September 6, 1910, Office of Experiment Stations Records;
E. L. Barker to A. R. Hill, September 30, 1910, and F. B. Mumford to Barker,
October 7, 1910, Missouri Agricultural College Papers.

[6] J. E. Waggoner to W. H. Beal, November 28, 1910, Office of Experiment
Stations Records; Merrison, "Extension of Agriculture to the American Farmer,"
p. 886.

spring of 1912, Holden was more than receptive when Cyrus H. McCormick called him to Chicago, stated that the company was prepared to spend a million dollars in rural improvement, and asked him to take charge of the work.[7]

Under Holden's leadership International Harvester's agricultural extension department expanded rapidly, both in terms of manpower and functions. At times the Department employed as many as forty persons, with others serving for short periods in specific programs. Several of the staff members were widely known. Holden, of course, was recognized wherever corn growing or extension teaching was discussed. J. E. Waggoner had been a professor of agricultural engineering before he joined International Harvester. J. G. Haney was the former superintendent of a branch experiment station at Hays, Kansas. Others of Holden's subordinates were practical farmers with more than local reputations. Included in this group were A. P. Grout of Illinois, Joseph E. Wing of Ohio, and W. R. Baughman of Texas.[8] Meanwhile, functions expanded and increased. Correspondence with individual farmers and motion picture-lecture program work at rural gatherings continued, apparently at an increased pace. The distribution of literature was carried forward, and evidence suggests that the pamphlets and bulletins became more sophisticated and useful. New was the establishment of model or demonstration farms in the South and the Middle West and the management of alfalfa campaigns in various parts of the country, campaigns that were without doubt the most concentrated rural education programs ever undertaken by businessmen.[9]

[7] P. G. Holden, "Why and How I Came to Join the International Harvester Company," clipping from *IHC World*, October, 1946, Holden Papers; *Fruit Grower and Farmer*, XXIII (December, 1912), p. 603; International Harvester Company, *Annual Report*, 1912 (n.p., n.d.), p. 18; Ivins and Winship, *Fifty Famous Farmers*, p. 90; *Breeder's Gazette*, LXII (November 20, 1912), p. 1102b.

[8] *Fruit Grower and Farmer*, XXIII (December, 1912), p. 603; *Kimball's Dairy Farmer*, XI (August 15, 1915), p. 447; *Farm Implement News*, XXXIV (March 20, 1913), p. 21; Andrew S. Wing, "'Alfalfa Joe' Wing, Exponent of Beardless Barley, Brome Grass, and Other Forage Crops," *Agricultural History*, XXXVIII (April, 1964), pp. 96–101.

[9] George H. Alford, *Diversified Farming in the Cotton Belt* (Chicago, n.d.), n.p.; Ivins and Winship, *Fifty Famous Farmers*, p. 91; International Harvester

In all, by 1913 the International Harvester Company operated five demonstration farms. Two were 300-acre or larger units, located on the northern plains at Grand Forks, North Dakota, and Aberdeen, South Dakota. The primary purpose was to illustrate the possibilities and advantages of diversified farming in those areas by growing corn, alfalfa, and clover and by raising livestock on land that previously had been used only for grain production. Three smaller farms were in the cotton belt, at Brookhaven, Mississippi; Marion, Alabama; and Trimble, Georgia. The Brookhaven and Marion farms were typical southern properties, suffering from the results of single cropping, erosion, and low fertility. There the objective was the rebuilding of the soil through proper tillage, cropping, and fertilizer application and the demonstration of other techniques that might be expected to improve the position of farmers utilizing them. The Trimble farm was used to carry on experiments with varieties of cotton, corn, and oat seed, seeking high-producing types.[10]

These kinds of work, useful though they were, were overshadowed in the popular mind by Holden's dramatic alfalfa campaigns, programs carried forward with a concentration of manpower and energy rarely seen before in extension teaching. When Holden joined International Harvester, he was faced with the problem of allocating the resources and energies of his department to achieve the maximum benefits in both the long and the short run for the farmer and for the company. The basic objective of his work, of course, was to encourage greater production of farm commodities through the rebuilding of soil fertility, the raising of livestock, and the diversification of crops. On the basis of his experience and knowledge Holden was firmly convinced that the key element in achieving these goals,

Company, *For Better Crops in the South* (Chicago, 1913), n.p.; Cooley, "Service, the Duty of Citizenship," p. 201; Merrison, "Extension of Agriculture to the American Farmer," p. 881.

[10] J. G. Haney, *IHC Demonstration Farms in the North* (Chicago, 1917), pp. 5, 13, 27; George H. Alford, *Autobiography* (Progress, Miss., 1949), pp. 70–71; International Harvester Company, *IHC Demonstration Farms in the South* (Chicago, 1917), p. 4; *American Fertilizer*, XXXIV (June 3, 1911), p. 36; *Progressive Farmer and Southern Farm Gazette*, XXVIII (December 13, 1913), p. 1272.

at least on midwestern and southern farms, was the production of alfalfa. That crop adapted itself to most soil and climatic conditions; it was a soil builder that produced heavy yields of hay with high feed values; and its production led directly to increased livestock farming.

The problem was one of taking this information to farmers and inducing as many as possible to adopt alfalfa as a regular crop in their rotation systems. Few men were as well acquainted as Holden with the value of farmers' institutes and educational trains, but he recognized their limitations as teaching techniques. Holden's alfalfa campaigns featured meetings in the fields and homes of farmers, concentration of work in relatively small areas in order that practically every farmer could be reached, the use of automobiles to increase the mobility of lecturers, and the extensive utilization of demonstration materials.[11]

The first alfalfa campaign was held in Kent County, Michigan, April 29 to May 3, 1913, when the Grand Rapids Chamber of Commerce and the Kent County Alfalfa Growers' Association invited Holden to bring his traveling school to that area. In the course of the five-day affair Holden and a dozen other experts moved about the county by automobile, stopping at prearranged spots for meetings with farmers and often for demonstrations in the fields. Night sessions were held in the country towns. Everywhere the lesson was the same. Alfalfa was the ideal crop; it could be grown easily; and farmers growing it could expect to see their operations become more profitable. In all, over 6,000 Kent County farmers heard the message in one or more of the thirty meetings held during the campaign.[12]

The Kent County campaign proved to be little more than a dress rehearsal for more elaborate programs elsewhere. After terminating operations in the Grand Rapids area, Holden and his staff moved to Allegan and Barry counties, Michigan; Williams County, Ohio; and Sangamon County, Illinois. In late summer, August 11–14,

[11] Clipping from *Weekly Implement Trade Journal*, May 17, 1913, Holden Papers; *Breeder's Gazette*, LXIV (December 18, 1913), p. 1236; *Kimball's Dairy Farmer*, XI (April 15, 1913), p. 248.

[12] *Corn*, II (June, 1913), p. 94; *Kimball's Dairy Farmer*, XI (August 15, 1913), p. 447; *Breeder's Gazette*, LXIII (May 21, 1913), pp. 1205–1206; *Farm Implement News*, XXXIV (May 15, 1913), p. 18; Unidentified clippings, Holden Papers.

came the Mississippi and Alabama black belt alfalfa campaign in which twenty-eight lecturers talked to farmers in Noxubee County, Mississippi, and Marengo and Perry counties, Alabama. Broader in scope and of longer duration was the Inland Empire campaign in the Pacific Northwest, lasting from September 23 to October 29. Carried on in conjunction with farmers' groups, agricultural colleges, state educational departments, the Spokane Chamber of Commerce, and five railroads, the campaign featured an educational train, but nearly 600 of the 960 lectures were given at off-line points, accessible only by automobile. Some 86,000 people heard the speakers call for diversified farming in the Palouse country and elsewhere where one-crop systems prevailed.[13]

Perhaps as comprehensive as any of the 1913 projects was an alfalfa campaign conducted in cooperation with the Chicago, Burlington and Quincy Railroad in southwestern Iowa and northwestern Missouri between July 28 and August 8. In this instance the Burlington furnished a special train that carried International Harvester men, railway development agents, and personnel from the two states' agricultural colleges to fifty-five towns. From these points automobiles took the lecturers to as many as five meetings held simultaneously in the surrounding countryside. In this way the speakers came into contact with a maximum number of farmers; at the same time the farmers were not called from their work for very long. In the preparation for the campaign the promoters of the project devoted full attention to community cooperation and participation. Prior to the appearance of the special train in a community two groups of company representatives visited the town, called on prominent businessmen and farmers, and arranged for local planning and advertising. Such advance work made good crowds more likely, and in fact Holden noted that farmers "stopped their threshing in order that they might hear the lectures." At the final accounting he reported that 12,649 persons attended the 281 meetings and were instructed

[13] *Kimball's Dairy Farmer,* XI (August 15, 1913), p. 447; Unidentified clippings, Holden Papers; *Progressive Farmer and Southern Farm Gazette,* XXVIII (August 23, 1913), pp. 895, 899; P. G. Holden, Records of the Holden, Wilson, and Other Related Families (Belleville, Mich., 1944), Holden Papers.

through demonstrations and lectures. They also took more than 100,000 pieces of literature home with them.[14]

The first full year of Holden's management of International Harvester's agricultural extension department was more than successful if statistics alone were taken as a guide. Holden had directed twenty-seven separate campaigns touching sixteen states and involving 2,000 meetings, the distribution of a million and a half pieces of literature, and the traveling by International Harvester men of more than 340,000 miles by rail and auto. Everywhere alfalfa had received the major emphasis, but at times Holden gave farmers advice on other topics. In Missouri, for instance, he objected to the practice of successful farmers moving to towns. "A retired farmer is a nuisance in town, when he moves there simply to die cheap," he told his audience. Still, in every instance, according to Holden the people received the aid and advice in the same broad spirit that it was given. Nowhere was the sincerity of the work questioned. In fact, "each campaign has been a campaign of the people, the extension department merely furnishing the plan and assisting in the direction of the work."[15]

These programs continued in 1914, with little slackening of the pace. Early in June, Holden's entourage visited Blackhawk County, Iowa, for a four-day campaign, subsequently moving into South Dakota where the alfalfa, corn, and silo campaign continued until June 27. In that effort Holden's men and agents of the Milwaukee and Northwestern railroads talked with 23,000 persons and distributed over 300,000 pieces of literature. Later, the Oklahoma alfalfa campaign covered eleven counties in the eastern portion of that state. Between November 10 and December 12, Holden managed the Arkansas profitable farming campaign, a joint enterprise with the Little Rock Chamber of Commerce, the Arkansas State Bankers

[14] Unidentified clippings, Holden Papers; *Hopkins* (Mo.) *Journal*, July 14 and 31 and August 7, 1913; (Maryville, Mo.) *Weekly Democratic Forum*, July 10 and 24 and August 7, 1913; *Missouri Farmer*, V (August, 1913), p. 4; Jones, "A Survey of the Agricultural Development Program of the Chicago, Burlington and Quincy Railroad," p. 242.

[15] *Breeder's Gazette*, LXIV (December 18, 1913), p. 1236, 1242b; *Hoard's Dairyman*, XLVI (September 12, 1913), p. 170.

Association, the Iron Mountain Railroad, the agricultural college, and agents of the United States Bureau of Plant Industry.[16]

Despite the success of Holden's better farming campaigns, the extension department participated in various other projects aimed at the same end. In 1913, for instance, Holden and members of his staff served on a number of educational trains operated by the Union Pacific and other lines. He and other company employees were in attendance at a multitude of farmers' meetings of various types, including a gathering called an institute at the industrial school for Negroes at Braxton, Mississippi. The department's charts, lantern slides, and movie reels were regularly available to farmers' groups requesting them. Later the company had a Safe Farming Newspaper Service that each week provided 300 newspapers with articles dealing with farm safety. In 1912 International Harvester awarded a prize bull as a part of a cattle-tick eradication program in Mississippi, Alabama, and Tennessee.[17] Finally, International Harvester agents everywhere added their support to the county agent movement and encouraged any group looking toward rural improvement.[18]

[16] *Wallaces' Farmer*, XXXIX (June 12, 1944), p. 918; *ibid.* (June 19, 1914), pp. 923, 928; Unidentified clippings, Holden Papers; International Harvester Company, *Every Farm a Factory* (Chicago, 1916), pp. 9–10; *Progressive Farmer and Southern Farm Gazette*, XXIX (November 14, 1914), p. 1165; Merrison, "Extension of Agriculture to the American Farmer," pp. 882, 884. Campaigns of a similar nature continued into 1915 and 1916, even though the enactment of the Smith-Lever Act suggested that such work would in the future lose some of its importance. Among these efforts were the Texas Will Feed Herself Campaign, January 11–24, 1915; the Alabama Crop Diversification Campaign, February 3–March 24, 1915; the Memphis Profitable Farming Campaign, November 9–December 11, 1915; and the Western Nebraska Safe Farming Campaign, May 1–May 5, 1916.

[17] *Breeder's Gazette*, LXIV (December 18, 1913), p. 1242b; *Farm Implement News*, XXXIV (April 24, 1913), p. 20; *ibid.* (February 13, 1913), p. 12; *Banker-Farmer*, I (June, 1914), p. 11; Cooley, "Service, the Duty of Citizenship," p. 210; Unidentified clippings, Holden Papers; *Progressive Farmer and Southern Farm Gazette*, XXVII (November 30, 1912), p. 1261.

[18] *Farm Implement News*, XXXIV (March 20, 1913), p. 25; *ibid.* (April 24, 1913), pp. 36, 38. International Harvester's agricultural extension department continued to function until 1933 when it was abolished. Holden retired in 1932. Clipping from *International Harvester Company World*, October, 1946, Holden Papers; Jackson E. Towne, Perry Greeley Holden, 1865–1959, paper read before

Most other concerns were more inclined than International Harvester to work through industry-wide associations and to devote major efforts to the development of sentiment in Congress and elsewhere that would lead to adequate public support for those teaching techniques deemed worthwhile. The banking industry, for instance, threw its strength into the struggle to obtain greater federal and state aid for agricultural education, especially for off-campus instruction.[19] Still, even bankers undertook what might be called action programs, not dramatically different from some carried on by other industries and concerns.

In 1909 at least two state bankers' associations established committees on agricultural development, steps that marked the formal beginnings of banker interest in agricultural extension. Four years later thirty-four states had such committees. Meanwhile, in 1911 the American Bankers' Association established its Committee on Agricultural Development and Education, an agency later known as the Agricultural Commission of the American Bankers' Association, and appointed as chairman Joseph Chapman of the Northwestern National Bank of Minneapolis. During its first months the functions of the Agricultural Commission were limited to encouraging the establishment of bankers' agricultural development committees in those states not yet having them and to promoting Congressional bills pertaining to agricultural education. To carry its message to local bankers and others, in 1913 the commission established in Champaign, Illinois, the *Banker-Farmer*, a monthly that for all practical purposes was managed during its first year of publication by Benjamin F. Harris, an official of the First National Bank of Champaign.[20]

Michigan Historical Society meeting, October 8, 1960, copy provided by the author.

[19] *Breeder's Gazette*, LXII (November 20, 1912), pp. 1115–1116. For a general statement of the role of bankers, see Grant McConnell, *The Decline of Agrarian Democracy* (Berkeley and Los Angeles, 1953), pp. 30–31.

[20] American Bankers' Association, *Proceedings,* 1913 (New York, 1913), p. 224; *ibid.*, 1911, pp. 386, 395; *ibid.*, 1912, pp. 27, 213; American Bankers' Association, *Journal*, VII (December, 1914), p. 362; *ibid.* (April, 1915), p. 749; *American Fertilizer*, XXXIX (December 13, 1913), p. 35. For information concerning the life of Harris, see Mrs. Frank V. Harris, ed., "The Autobiography of Benjamin

More specific in objective were the efforts of many of the state bankers' associations and a number of individuals in the industry. Throughout the Middle West, in fact, bankers could be found promoting or participating in planning for institutes and short courses. In Minnesota, bankers were foremost in the struggle to introduce agriculture as a subject in the public schools, while in neighboring Wisconsin they sponsored corn contests for rural youths, awarded prizes for essays by high school students in agriculture, and participated in various better-seed distribution programs. For a time the state bankers' group there issued an informative bulletin to farmers. As an example of action by individual firms, the local bank in Colfax, Wisconsin, provided the necessary funds for three neighborhood farmers to attend the annual short course in agriculture at Madison, while the bankers of Aberdeen, South Dakota, contributed $7,000 to stimulate the production of dry-land alfalfa in that area. A bank in Jerseyville, Illinois, offered to loan money at no interest to farmers who wanted to apply commercial fertilizer or lime to their land.[21] Such policies, no doubt, were as much developmental as they were educational, but at least bankers showed that they were acutely aware of the value of the field demonstration as a teaching technique.

In fact, by 1912 most bankers had concluded that the county agent represented the best answer to the needs of the countryside. According to the *Journal* of the American Bankers' Association, "Existing educational institutions and methods logically developed promise to provide suitable agricultural instruction for coming generations, but the present problem . . . is to bring individually to farmers of the present generation some of the abundance of scientific knowledge possessed by educated specialists. This can only be done effectively

---

Franklin Harris," Illinois State Historical Society, *Transactions*, 1923 (Springfield, Ill., 1923), pp. 72–101.

[21] *Wallaces' Farmer*, XXXV (June 16, 1911), p. 933; *Country Gentleman*, LXXVII (February 3, 1912), p. 1; *Hoard's Dairyman*, XLII (August 11, 1911), p. 868; *ibid.*, XLVI (September 19, 1913), p. 205; *ibid.* (August 29, 1913), p. 122; *ibid.*, XLIII (June 14, 1912), p. 738; Arlan Helgeson, *Farms in the Cutover: Agricultural Settlement in Northern Wisconsin* (Madison, 1962), pp. 79–80; *Kimball's Dairy Farmer*, XI (December 15, 1913), p. 739.

through the location of agricultural experts in all rural counties...."[22] Consequently, bankers not only were powerful voices in Washington during the months that preceded the enactment of the Smith-Lever Act, but in numerous areas they participated in the movements that placed agents in the fields prior to May, 1914.

Meanwhile, the National Fertilizer Association, a trade organization that had been formed in 1894, turned its attention to agricultural extension. The first step was the formation on January 7, 1911, of the association's educational bureau headed by John D. Toll. A second step came on May 12, 1911, when a regional agency, the Middle West Soil Improvement Committee, was established in Chicago and a fund of nearly $48,000 was subscribed to support its operations for three years. Henry G. Bell, a graduate of Ontario Agricultural College and a former professor of agronomy at the University of Maine, became head of the agency, taking his post on September 1.[23] Later, southern fertilizer concerns created a second regional group that functioned under the title of Southern Soil Improvement Committee.[24]

The objectives of these groups were "to encourage better methods of building up and maintaining soil fertility, by rotation of crops and the use of barn yard manure and legumes, supplemented by fertilizers bearing nitrogen, phosphoric acid and potash; by drainage; and by the scientific use of lime where required." To promote such goals, the educational bureau and the regional groups relied especially upon the preparation of literature that was distributed by fertilizer concerns, bankers, millers, feed dealers, and directly to farmers by means of a mailing list compiled for that purpose. Some railroads used the publications on their educational trains. Between Septem-

---

[22] American Bankers' Association, *Journal*, V (November, 1912), p. 287.

[23] *American Fertilizer*, XL (June 27, 1914), pp. 78–79; *ibid.*, XXXV (July 29, 1911), pp. 51, 90–91; *ibid.* (August 12, 1911), p. 27; *ibid.* (September 9, 1911), p. 44; Railway Industrial Association, *Proceedings,* 1911 (St. Louis, n.d.), pp. 7–8. As early as 1892 some fertilizer concerns awarded prizes to farmers who produced huge crops through the application of fertilizers. See *Cotton Plant*, IX (February 13, 1892, p. 7.

[24] *American Fertilizer*, XL (January 24, 1914), p. 29; *ibid.* (March 7, 1914), pp. 35–36.

ber, 1911, and March, 1914, the Middle West Soil Improvement Committee alone prepared twenty-three different publications and released more than a million and a half copies of them. In addition staff members spoke before farm groups whenever possible, while the Chicago-based committee, at least, supplied informative articles, including many based on experiment station bulletins, to a number of farm periodicals and local newspapers.[25]

Businessmen involved in the grain trade assumed a role in the educational movement through their organization, the Council of North American Grain Exchanges. In September, 1910, at a meeting of council members in New York, an official of the St. Louis Merchants' Exchange presented an address, calling attention to the poor quality of grain and the generally low yields achieved by farmers. The upshot of the ensuing discussion was the formation of the National Crop Improvement Committee, with Bert Ball of the Chicago grain exchange as secretary. Later, farm machinery and fertilizer manufacturers, railroads, and other business concerns made financial donations and other contributions to its work.

At the outset the Crop Improvement Committee devoted its energies to the arousing of interest in agricultural improvement through education among other business groups and by the distribution of literature to farmers. Under Ball's energetic leadership, meetings with business interests were held across the country during the winter of 1910–11. As one result of these efforts the National Federation of Millers placed in the hands of farmers over a million copies of a pamphlet written by a Purdue University professor dealing with the selection of seed grain. In addition the federation circularized local millers, asking them to take steps to improve the quality of grain sold as seed and to offer premiums for grain of exceptional quality. The Crop Improvement Committee also encouraged local commercial clubs and similar groups to sponsor seed improvement programs and to hold crop contests for farmers. Finally, Bert Ball prepared editorials on seed grain improvement which were repro-

[25] Ibid., XXXVII (July 27, 1912), pp. 69–70; ibid., XL (March 7, 1914), pp. 35–36; ibid. (March 21, 1914), p. 43; ibid., XLI (July 25, 1914), pp. 44–45.

duced in almost 450 farm and other papers. The favorable reception of this work led to the creation within the Crop Improvement Committee of a publicity bureau that regularly provided farm periodicals with articles on better grain-growing methods.[26]

Increasingly, however, the Crop Improvement Committee turned its attention to the expert in the field as the best means of enlightening the farmer. The grain men joined with the bankers, fertilizer manufacturers, and others in urging the enactment of legislation that would place an agricultural agent in every county where farming was important. Prior to the enactment of such a law the committee took an active role in the installation of agents in counties where before 1914 the work was financed in large measure by private and local funds. This phase of the committee's activities received a boost on May 10, 1912, when Julius Rosenwald of Sears, Roebuck and Company offered to donate $1,000 to any county able to raise locally the remainder needed to meet an agent's salary and expenses.[27]

A great number of other businesses and industries displayed more than casual interest in agricultural education, and many carried on active programs during the years before the enactment of the Smith-Lever Act. In 1910 packing interests in Texas, for instance, operated three demonstration farms to show Texans that swine would do well in that state. The Grasselli Chemical Company, a fertilizer company, had an experimental farm near Birmingham in 1911.[28] Many other concerns distributed literature of one nature or another. A reasonably representative list would include the American Steel and Wire Company, the Armour Company of Chicago, a number of the larger

[26] Railway Industrial Association, *Proceedings*, 1911, pp. 5–6; Council of North American Grain Exchanges, Committee on Crop Improvement, *Proceedings of a Meeting in Chicago, February 8, 1911*, pp. 3–7, 9; McConnell, *Agrarian Democracy*, p. 32; *American Fertilizer*, XXXV (July 29, 1911), p. 49; *Wallaces' Farmer*, XXXVIII (May 16, 1913), p. 827.

[27] *Railway Age Gazette*, n.s. LII (May 24, 1912), p. 1172; *Breeder's Gazette*, LXII (November 6, 1912), p. 964; *Farm Implement News*, XXXIV (October 23, 1913), p. 34; M. R. Werner, *Julius Rosenwald: The Life of a Practical Humanitarian* (New York, 1939), pp. 62–63.

[28] *Breeder's Gazette*, LVIII (July 27, 1910), p. 138; *American Fertilizer*, XXXV (July 1, 1911), p. 32.

and more successful implement dealers, and the Association of American Flax Seed Buyers and Crushers. Several publishers, including those of *Hoard's Dairyman, Breeder's Gazette,* and the Memphis *Commercial Appeal,* sponsored contests for farmers, a practice also adopted by such diverse groups as the Association of Manufacturers of Linseed Oil and the organized businessmen of Joliet, Illinois.[29]

Elsewhere businessmen participated in the rural club movement and sponsored exhibits with distinct educational overtones. In 1912 the Tampa, Florida, Board of Trade took upon itself the responsibility for organizing in the neighborhood a club for farm boys, and two years later the Commercial Club of Franklin, Tennessee, opened its doors to farmers and then proceeded to organize a number of commodity growers' associations in the area. A similar businessmen's group in Richmond, Indiana, in 1911 raised the funds necessary to send fifty boys to Purdue University for a one-week short course in agriculture. As early as 1906 fruit sprayer manufacturers participated in exhibits and demonstrations in Kansas, and a few years later the E. I. DuPont Powder Company gave lessons to farmers interested in land clearing, through both lectures and literature. The Quaker Oats Company regularly distributed educational literature and placed exhibits at agricultural fairs across the country.[30]

The diverse and extensive role of business in the promotion of agricultural extension prior to 1914 raised no small number of questions in the minds of farmers and presented perplexing problems for agricultural educators who were often asked to cooperate in

[29] *Banker-Farmer,* I (January, 1914), p. 10; *American Fertilizer,* XXXV (December 2, 1911), p. 31; *Farm Implement News,* XXXI (March 31, 1910), p. 16; Montana Farmers' Institutes, *Ninth and Tenth Annual Reports,* 1910–12, pp. 14–15; *Hoard's Dairyman,* XLVI (September 12, 1913), p. 167; *Breeder's Gazette,* LX (October 25, 1911), p. 824a; *Wallaces' Farmer,* XXXV (March 18, 1910), p. 487.

[30] *American Fertilizer,* XXXVI (April 6, 1912), p. 35; *Progressive Farmer and Southern Farm Gazette,* XXIX (March 21, 1914), p. 366; Council of North American Grain Exchanges, Committee on Crop Improvement, *Proceedings of a Meeting in Chicago, February 8, 1911,* pp. 13, 77–78, 121; *Western Fruit Grower,* XVII (March, 1906), p. 144; *Southern Planter,* LXXI (June, 1912), pp. 661–662.

business-sponsored programs. To many farmers who remembered well the agitation of the late nineteenth century, the interest of bankers, grain dealers, and machinery manufacturers could only be suspicious. "Can any good come out of Nazareth?" asked an Illinois farmer in 1913, and Thomas C. Atkeson, prominent agricultural educator and granger from West Virginia, said categorically that the interest of bankers in agriculture "is the same interest that the shepherd has in his sheep; he takes care of them in order that he may secure more wool at shearing time."[31] For the agricultural colleges and experiment stations, the activities of business presented a sharp dilemma. Business funds were desperately needed and gladly welcomed, but many professors felt that they must not appear before farmers as the allies of business. In addition the academic men and commercial interests at times differed as to goals, the latter being more interested in immediate results while the professors looked for long-run gains. Academic men also feared that business interests, through misguided enthusiasm or ignorance, might spread absolute falsehoods among farmers.

In a majority of cases, however, most college men and common farmers were able to meet the problems posed by business participation in educational projects and to convince themselves that businessmen were sincerely interested in rural improvement. Academic personnel no doubt found their task easier; at any rate, they generally concluded that business participation was welcome as long as there was no commercialism involved and as long as business interests did not attempt to dictate to public agencies.[32] Farmers reasoned that, while they may have opposed business in the past, there was no reason that cooperation could not be possible in those cases where agricultural and commercial interests had common objectives. Many farmers, in fact, came to be appreciative of business efforts and to recognize their contribution to the development of better methods in agriculture.

[31] *Missouri Farmer*, V (May, 1913), p. 10; *Wallaces' Farmer*, XXXVIII (May 16, 1913), p. 827; Bertels, "National Grange," pp. 177–178.
[32] AAACES, *Proceedings*, 1913, pp. 254–255, 258–259.

VIII. SOUTHERN

DEMONSTRATION WORK

IT WAS APPROPRIATE that an effective teaching device would be discovered in the South; more than any other region of the nation that area was in need of rural rejuvenation. Agricultural methods were poor, characterized by one-horse farming, shallow plowing, inadequate seed-bed preparation, dwindling humus content, and defective drainage. The soil lacked those moisture-retention qualities desirable during droughts and it washed badly after heavy rains. Livestock raising was in no more than its infancy, due partly to the failure of southern farmers to produce good pastures. Instead, for livestock feed, they tended to rely on fodder and ear corn, expensive commodities that caused many farmers to try to operate with too few work animals and with too much hand labor. Meanwhile, the shortage of natural manures forced southern landowners to make large outlays for commercial fertilizers in vain efforts to maintain production levels and gross income.

The sharecropper and crop-lien systems, unknown in the North, further complicated matters. "The credit system . . . might have been a necessary evil 40 years ago, but it prospered and became dominant, oppressive, and insolent. . . . It substituted involuntary for voluntary servitude; ownership by agreement and poverty by contract under fear of the sheriff for the ownership of birthright and a government by proprietary right," said one authority. There was, in fact, no doubt that these unique characteristics of southern farming placed severe restrictions on agricultural change and progress.[1]

Nor did the immediate past suggest that changes were in the foreseeable future. According to Seaman A. Knapp, who probably knew as much about southern agriculture as any man, "Rural conditions . . . in the Southern States have changed little in thirty years. The houses are a little more dilapidated; the fences give evidence of more decay; the highways carry more water in the wet season, and are somewhat less easily traveled in the dry; but the environments are about the same; no paint, and slight evidence of thrift. The same old mule stands at the door with his rope end on the ground, hitched to a plow that Adam rejected as not up-to-date; the same old bushes are in the fields, and the same old weeds in the fence corners; no strange sights disturb the serenity of Rip Van Winkle; wages are about the same, and the conditions of farm life are almost exactly as they were thirty years ago."[2]

In the first years of the twentieth century the United States Department of Agriculture was acutely aware of the problems in southern agriculture. By 1902 various bureaus and offices within the department, most notably the Bureau of Plant Industry headed by Beverly T. Galloway, launched a program to promote agriculture in the South. A many-sided project, it envisioned among other features the establishment of a number of government model farms in specific problem areas. The department hoped to demonstrate that existing difficulties could be overcome by better management and the use of good seed, crop rotation, diversification, and other improved methods. Five farms were established in Texas and Louisiana. Arrangements varied, but, in the main, local businessmen and other landowners leased the land and provided tools and teams, the gov-

[1] U.S. Department of Agriculture, *Yearbook*, 1909, pp. 155–157; Seaman A. Knapp, "Farmers' Cooperative Demonstration Work in Its Relation to Rural Improvement," U.S. Bureau of Plant Industry, *Circular* 21 (Washington, 1908), pp. 3–7; Epsilon Sigma Phi, *Extension Work*, p. 40; *Progressive Farmer and Southern Farm Gazette*, XVI (July 1, 1911), p. 592.

[2] Seaman A. Knapp, "Farmers' Co-Operative Demonstration Work," *Southern Educational Review*, III (October, 1906), pp. 50–51.

207

ernment contributed seed, fertilizer, and the cost of necessary labor as well as the all-important supervision.

Placed in charge of the project was one of the more significant figures in American agricultural history. Seaman A. Knapp was born in New York State in 1833 and educated at a preparatory school in Vermont and at Union College where he graduated in 1856 with a Phi Beta Kappa key. For a decade Knapp and his wife served as preparatory school teachers and administrators in the East. Then in 1865 an injury to Knapp's knee and subsequent poor health induced the family to migrate to Iowa the next year. There followed a busy life for Knapp as a progressive farmer, minister, superintendent of a school for the blind, and editor of a farm paper. It was the latter work, especially, that brought Knapp into contact with Henry Wallace and James Wilson, two of the dominant figures in Iowa agriculture. It also helped to make Knapp professor of agriculture at the Iowa agricultural college in 1879 and for a few months president of that institution.

His academic years were important ones for Knapp and for his subsequent career. He learned much of the potential as well as the limitations of the land-grant colleges of that day. Aware that the agricultural schools suffered from a lack of scientific knowledge, Knapp drafted the bill introduced in Congress in 1882 by C. C. Carpenter providing for the establishment of experiment stations. The Hatch Act of 1887, which accomplished the purpose, differed but little from the earlier bill. Meanwhile, Knapp came into contact with academic types, men with whom he disagreed bitterly. As a result he developed a suspicion and distrust of college men that he would carry for the remainder of his life.[3]

When an opportunity came to leave the academic world, Knapp

[3] Joseph C. Bailey, *Seaman A. Knapp: Schoolmaster of American Agriculture* (New York, 1945), pp. 137–138, 147, chaps. 1–4; Frederick W. Williamson, *Yesterday and Today in Louisiana Agriculture* (Baton Rouge, 1940), pp. 64–71; Ross, *Democracy's College*, pp. 59–60; *Wallaces' Farmer*, XXXVI (April 14, 1911), p. 668; Texas Agricultural Extension Service, *Silver Anniversary: Cooperative Demonstration Work, 1903–1928* (College Station, Tex., n.d.), pp. 44–46.

jumped at it. A predominantly English syndicate had acquired more than a million acres in southwestern Louisiana, a region inhabited by easygoing Acadians or Cajuns who lived by raising a poor grade of cattle and by hunting, fishing, and trapping. The syndicate hoped to populate its land with diligent, industrious farmers using modern methods, and in 1885 it asked Knapp to join in the enterprise. Unfortunately, few northern farmers had much interest in Louisiana, and what interest they did have quickly disappeared after they visited the area and talked with the natives who were never hesitant to discuss the region's limitations. To meet this situation Knapp suggested that the syndicate place in each township a subsidized farmer who would demonstrate what could be done. The technique succeeded; within ten years Louisiana was the leading rice-growing state in the union. "We learned then the philosophy and power of demonstration," recalled Knapp.[4]

In 1898 Knapp began a new career with the Department of Agriculture that would engage him for the rest of his life. His experiences with rice growing in Louisiana, combined with his long acquaintance with Secretary of Agriculture James Wilson, made him a logical choice for the post of special agent when the department decided to seek a better milling variety of rice in the Far East. In 1900 the department gave Knapp another task, sending him to Puerto Rico to study the agriculture of that island. Later, in 1901–02 he went again to the Far East to search for additional varieties of rice and for plants suitable for use in the Gulf states. Upon his return to the United States he expected to supervise a government experiment farm where his discoveries might be tested under varying soil and climatic conditions. Thus it was that when the Department of Agriculture undertook to rejuvenate southern agriculture, Knapp was well equipped

[4] Bailey, *Knapp*, chap. 5; Oscar B. Martin, *The Demonstration Work: Dr. Knapp's Contribution to Civilization* (San Antonio, 1941), pp. 1–2; *Experiment Station Record*, XXIV, pp. 497–498; Seaman A. Knapp, "An Agricultural Revolution," *World's Work*, XII (July, 1906), p. 7734; Raymond B. Fosdick, *Adventure in Giving: The Story of the General Education Board* (New York, 1962), pp. 41–42; B. T. Galloway to S. A. Knapp, April 27, 1901, Bureau of Plant Industry Records.

to head the program. In the summer of 1902 he became special agent for the promotion of agriculture in the South.[5]

Knapp's long experience in agriculture, combined with his personality, made him almost the ideal man for the task at hand. Described by one writer as a "bucolic Franklin," Knapp knew the farmers and had a rare rapport with them. Wearing a frock coat and an old-fashioned square-topped derby, he spoke the language of the soil, and farmers instinctively accepted him as one of their own.[6]

Knapp took up his new duties in the fall of 1902. There was an urgency to the work, induced in part by the insistent demands of men like E. H. R. Green of the Texas Midland Railroad that something be done to revive business. Besides planning and placing in operation the demonstration farms, Knapp represented the Department of Agriculture at important farm and other meetings in the state, learning all that he could of the people and conditions.

Within a few weeks Knapp concluded that government demonstration farms such as he was supervising would not do the job. Similar farms had served well in Louisiana, but there the task had been simply to convince prospective immigrants that the coastal plains of Louisiana were fertile. In Texas the objective was to induce established farmers to adopt new methods on land that they had long tilled. Such men were quick to assume that techniques proven successful on a government farm would not succeed on their property; after all, they said, they lacked the resources of the government. Knapp recognized that a new approach was necessary.[7]

An opportunity to test a new teaching device came in 1903. Early that year in New York Knapp chanced to meet E. H. R. Green. The men discussed problems in Texas, and Green suggested that Knapp

[5] Bailey, *Knapp,* pp. 133–137.

[6] Fosdick, *General Education Board,* p. 42; *Progressive Farmer and Southern Farm Gazette,* XXVII (April 20, 1912), p. 529; Bay, *County Agricultural Agents,* p. 11.

[7] B. T. Galloway to S. A. Knapp, September 3, 1902, E. H. R. Green to James Wilson, August 27, 1902, and Knapp to A. J. Pieters, November 16, 1903, Bureau of Plant Industry Records; J. A. Evans, "Recollections of Extension History," North Carolina Extension Service, *Circular* 224 (Raleigh, 1938), p. 7.

visit Terrell, Texas, to meet with local businessmen and farmers. At that gathering, held February 25, 1903, Knapp proposed that a new type of demonstration farm be established. He asked that community leaders select a suitable farm and raise a sum of money that could be used to compensate the owner for any loss that might be incurred in the tests to be made. The United States Department of Agriculture, through Knapp, would provide nothing but supervision and technical guidance. The end result, hopefully, would be to demonstrate what an ordinary farmer, using improved methods, could accomplish on his own land.

In this fashion the famous Porter demonstration farm was launched. A committee of eight Terrell businessmen and farmers selected the 800-acre Walter C. Porter property as the test site; Porter agreed to follow explicitly all instructions given him. The 70-acre plot actually used was moderately rolling, light sandy loam upland, with a clay pan at a depth of about thirty inches. It had been planted with cotton or corn for twenty-eight years, no fertilizer had been applied during that period, and humus was almost nonexistent. Knapp prescribed only basic innovations. A number of different varieties of properly selected corn and cotton were planted, fertilizer was applied in large amounts, and more thorough cultivation and better planting techniques were utilized. The fertilizer applications produced especially dramatic results, and, despite an attack by the boll weevil, Porter claimed at the end of the season that he had cleared $700 more than he would have by the old methods. The next year, he announced, he expected to apply the lessons learned to his entire property.

Few farmers who saw the results of the Terrell experiment could argue with them. Porter had performed the work on his own land; the government had provided nothing but information and guidance. Presumably, any farmer could do as well. Therein rested the significance of the Porter farm; it was nothing less than an experiment in which Knapp discovered that a community demonstration farm conducted by the farmer himself was a highly effective teaching device.[8]

[8] Evans, "Texas Agriculture," p. 70; Evans, "Recollections of Extension History,"

If farmers in the vicinity of Terrell were convinced that Knapp's methods could produce results, the United States Department of Agriculture was less certain. In fact, officials in Washington were quite uninterested until rising demands for new means to combat the boll weevil brought important government figures to Texas in the fall of 1903.

In 1892 the Mexican boll weevil crossed the Rio Grande into Texas, spreading devastation in its wake. By 1903 the plague had moved across the southern half of Texas. Businessmen and farmers alike were in a panic; 500 of the latter decamped from one county in a single year. In areas where the crop-lien system was widely used, the economic disruption was especially dramatic. Fearing that a crop could not be made, merchants refused to give credit, and sharecroppers had no choice but to leave the district. Scientists believed that the weevil was the most dangerous pest ever to attack a crop in the United States.[9]

The boll weevil came to the attention of the Department of Agriculture soon after the pest appeared in Texas. Various insecticides proved to be of little value, and it seemed to be impossible to produce a weevil-resistant variety of cotton. Some authorities ultimately concluded that the inroads of the weevil could at least be limited by changes in farming methods. In the so-called cultural methods they suggested that cotton be planted as early as possible and that an early maturing variety be used in an effort to produce a crop before the weevil became most destructive. In addition, they urged thorough cultivation and weed control and suggested that farmers burn old stalks promptly and protect the wildlife, such as quail, that fed on the weevil.[10]

---

p. 7; Epsilon Sigma Phi, *Extension Work,* p. 36; Seaman A. Knapp, "The Work of the Community Demonstration Farm at Terrell, Texas," U.S. Bureau of Plant Industry, *Bulletin* 51 (Washington, 1905), part 2, pp. 9–12; S. A. Knapp to A. J. Pieters, November 16, 1903, Bureau of Plant Industry Records.

[9] B. T. Galloway to S. A. Knapp, December 1, 1903, Bureau of Plant Industry Records; Evans, "Recollections of Extension History," p. 8.

[10] Work of the Bureau of Plant Industry Relative to Cotton, memo dated October, 1903, Bureau of Plant Industry Records; Evans, "Texas Agriculture," pp. 63–64.

The wave of fear that swept Texas in 1903 produced in the fall of the year a mass meeting in Dallas attended by Texas congressmen, local businessmen and farmers, Secretary of Agriculture James Wilson, and Beverly T. Galloway. While visiting in the state Wilson and Galloway toured the Porter property with Knapp, and there they concluded that the community demonstration farm could be used in the fight against the boll weevil. Perhaps it would do little to arrest the spread of that pest but it was hoped that it would help to keep farmers in the area by giving them faith that something was being done.

Accordingly, Wilson and Galloway agreed to provide federal funds for the support of community demonstration work. Late in 1903 Congress appropriated $250,000 for boll weevil control programs, the amount to be divided equally between Galloway's Bureau of Plant Industry and the Bureau of Entomology. Of the funds given Galloway $40,000 was set aside for Knapp's use in 1904.[11]

Knapp lost little time in getting his project under way. Two days after opening an office in Houston on January 25, 1904, he called a meeting of the agricultural agents of the railroads serving Texas. Knapp valued the advice of these men since most of them had worked in farmers' institutes and had a thorough knowledge of rural people and conditions. He asked them to devote the next two months to working with him, carrying his message to communities served by their companies' tracks. Eager to improve the economy of the territory the railroad men agreed. To supplement their efforts and to reach other points, Knapp appointed the first of his special agents. W. M. Bamberg led the list, with an appointment dated January 27, and four others—W. F. Procter, J. A. Evans, J. L. Quicksall, and W. D. Bentley—were added in February. These men were to serve in Texas, but early in March J. E. Wemple and J. E. Adger were appointed for work in western Louisiana. Within a few months more than twenty special agents were at work.

At the outset appointments as special agents were temporary, usually for sixty days. Salaries were sixty dollars a month plus ex-

[11] Evans, "Recollections of Extension History," p. 8; Martin, *Demonstration Work,* p. 4.

213

penses. Although Knapp sought the cooperation of the agricultural colleges in Texas and Louisiana and of the commissioner of agriculture in Louisiana, he chose as special agents men who were or had been practical farmers. College graduates, Knapp felt, would not be satisfactory for the type of work he had in mind.[12]

The agents had no easy task. Their duty was to go into communities and urge individuals to cooperate in the Knapp program. Too often farmers were suspicious and apathetic; so in the first years the usual technique called for the agent to contact the leading businessmen in a community, especially the merchants and bankers who controlled local credit sources, and induce them to arrange a community meeting where the cooperative demonstration work could be explained. There an attempt would be made to raise a guaranty fund and to form a committee that would select one or more demonstrators from among the better farmers in the neighborhood. Then on a plot of approximately ten acres of their land, a program would be carried out that was roughly similar to the one conducted on the Porter farm in 1903.

During the first years these community demonstration farms were the basic elements of Knapp's instructional technique. The demonstrator signed an agreement pledging that he would follow all instructions given him and that he would render such reports as might be requested. The agent personally visited the farms in his territory at least once a month to note progress and to give more detailed instructions. The farms were to be simply object lessons; it was hoped that one could be established in each county and that neighbors could easily see the results. Field meetings held during the growing and harvesting seasons provided other opportunities for instruction. In addition, as Knapp saw them, the community demonstration farms would also be useful for gathering further information on purely local conditions.

[12] S. A. Knapp to B. T. Galloway, January 31, 1904, and Knapp to James Wilson, March 10, 1904, Bureau of Plant Industry Records; Evans, "Recollections of Extension History," pp. 8–9; Kate A. Hill, *Home Demonstration Work in Texas* (San Antonio, 1958), p. 194; Williamson, *Agricultural Extension in Louisiana*, p. 52.

214

Besides establishing community demonstration farms, agents were directed to secure as many as possible of a second class of demonstrators; Knapp called them general cooperators. These farmers agreed simply to follow such instructions as might be given them, usually by mail, and to report the results. They received no direct assistance, and agents visited them rarely, if at all. Their demonstration plots might be as little as one acre or they might include the entire farm. Knapp hoped to enlist from fifteen to twenty general cooperators in each county. Their successes, he believed, would constitute an important teaching tool.

The announced purpose of Knapp's work was the control of the boll weevil, and his agents placed in use the cultural method developed by the Department of Agriculture. Farmers were directed to burn all stalks and other matter that might serve as hibernation quarters for weevils and to plow cotton land in the fall, followed by periodical cultivation during the winter. In its studies of the life cycle of the weevil, the Bureau of Entomology had found that the pest did not begin to multiply until the cotton squares began to mature and that they usually were not numerous enough to damage a crop severely until late in July. With these facts in view Knapp's men urged early planting, use of an early maturing variety, and the application of phosphate and potash to hasten the growing process. Intensive cultivation during the growing season and the topping of plants also contributed to early maturation of the crop. If weevils appeared, little could be done, but the agents recommended that growers use mechanical devices to agitate the stalks during cultivation to shake off some of the larvae. They also suggested that all infected squares be gathered and burned and that cooperators try insecticides of various types and report the results.[13]

But Knapp wanted to do more than simply retard the spread of

[13] S. A. Knapp to B. T. Galloway, December 28, 1905 and July 29, 1906, Bureau of Plant Industry Records; Outline of Cooperative Demonstration Work to Be Carried on during the Fiscal Year, 1905–1906, Suggestions to Special Agents in Regard to Business Methods Relative to Special Farms, memos, *ibid.*, Knapp to Cooperators, July 14, 1905, *ibid.*; Evans, "Recollections of Extension History," pp. 10–11; U.S. Department of Agriculture, *Yearbook*, 1909, pp. 153–157.

the boll weevil, dangerous though it was. He was firmly convinced that the poverty of the rural South was due in large part to the backwardness of its farmers; he was equally convinced that in the demonstration method he had a technique that could revitalize the entire section. Indeed, the boll weevil campaign, in so far as it dealt with fertility, cropping systems, and erosion control, pointed to the basic needs of southern agriculture. Consequently, Knapp soon was engaged in a general crusade against prevailing conditions. The demonstrators and cooperators were the instruments through which Knapp hoped to reach ultimately all southern farmers, showing them the most satisfactory methods of producing the standard farm crops in their areas.[14]

Even with this broader purpose in mind, agents' instructions tended to be simple and basic, consisting of what Knapp called his ten commandments. First, they attempted to show by demonstration how to prepare a suitable seed bed. It should be plowed to a depth of eight to ten inches, preferably in the fall, well pulverized, and adequately drained. Then farmers were instructed in selecting, testing, and storing seed as essential steps toward obtaining high germination and a sturdy, heavy producing crop. Seed should be planted with adequate spacing between plants and rows. Frequent but shallow cultivation during the growing season was a necessity. Knapp's men proposed the use of a simple rotation system—cotton, corn, and cowpeas, followed by a winter cover crop. Manures and commercial fertilizers should be used to rebuild the soil in terms of fertility and humus content. Livestock production received attention; the agents showed how to develop adequate pastures, prepare desirable rations for farm animals, and otherwise achieve the optimum production of livestock on given farms. Finally, agents tried to impress on farmers the need for reducing operating costs through the more efficient use of labor, the utilization of more horse power and better machinery, the keeping of accurate records, and the achievement of self-sufficiency

[14] Evans, "Texas Agriculture," p. 71; *Wallaces' Farmer,* XXXV (June 17, 1910), p. 895; Southern States Association of Commissioners of Agriculture and Other Agriculture Workers, *Proceedings,* 1907 (Raleigh, 1908), pp. 13–14.

through the production on the farm of all food needed for the family and its livestock.[15]

Quite obviously Knapp's men taught little that was original. "The work is in no sense experimental; no experiments are tried; the instructions are not new nor doubtful; everything recommended has been fully tested by practical farmers." The important thing was not what was taught, but how it was taught. "The object of the Farmers' Co-operative Demonstration Work is not only to place a practical object lesson illustrating the best methods of producing standard farm crops before the farm masses, but to secure their active participation in such demonstrations to an extent that will prove that the average farmer can do better work. . . ."[16]

In their teaching, agents were instructed to use the characteristics of farmers, which Knapp knew so well, to achieve their goals. "It is easy to enlist the masses in the army of reform, if wisely managed; but impossible if undertaken along the lines generally pursued," he said. Individual pride and local rivalry, Knapp believed, could be potent factors in bringing about change. Farmers watched their neighbors carefully, and most were quick to note successful innovations and to attempt to exceed the results achieved by the innovator.[17] On the other hand, farmers were suspicious of "experts" and those considered to be outsiders, and they placed great pride in their own knowledge, however inadequate it might be. Agents should not present themselves as final authorities on all agricultural matters and they should disturb prevailing ideas no more than necessary. "Sometimes farmers have peculiar views about agriculture," Knapp noted. "They farm by the moon. Never try to disillusion them. Let them believe in farming by the moon or the stars, if they will faithfully try our methods." Agents were further instructed to "avoid discussing

[15] S. A. Knapp to B. T. Galloway, July 29, 1906, Bureau of Plant Industry Records; B. Knapp, "Education through Farm Demonstration," p. 229; U.S. Department of Agriculture, *Yearbook,* 1909, pp. 155–157.

[16] *Southern Planter,* LXIX (March, 1908), pp. 226; B. Knapp, "Education through Farm Demonstration," p. 226.

[17] S. A. Knapp, "Agricultural Revolution," p. 7735.

politics or churches. Never put on airs. Be a plain man, with an abundance of good practical sense."[18]

Knapp's approaches to the problems of the boll weevil and of the generally inadequate southern farming practices aroused considerable question and no little opposition among men of more scientific temperament. Jealousy, too, motivated some of his critics. At the outset the Bureau of Entomology of the Department of Agriculture objected bitterly to the development of an anti-boll weevil campaign outside its jurisdiction. An official of the bureau claimed that since his agency had been studying the weevil for years it should be given the responsibility for carrying out any control program. Unfortunately, the bureau had failed to act except to issue publications pertaining to various restrictive techniques. Its officers, according to the Bureau of Plant Industry's Beverly T. Galloway, ignored the fact that it "requires something more than a mere circular to induce the ordinary farmer to change his practices."[19]

Later, the Bureau of Entomology accused Knapp of failing to give it sufficient credit for developing the cultural method of weevil control. Knapp contended that the cultural method was merely the good farming practices that speakers had been discussing from institute platforms for twenty years. He was willing to credit the bureau only with significant studies of the life cycle of the weevil and with the suggestions that cotton be matured as soon as possible, that stalks be burned in the fall, and that infected squares be gathered and burned.[20]

Other controversies arose over technical questions of weevil control. Knapp, for instance, was convinced that direct sunlight killed the larvae, an issue that he claimed "anyone can settle with two good eyes." When entomologists protested vigorously, Knapp displayed both his grasp of bureaucratic realities and his own stubbornness. "I have decided not to discuss this question," he wrote,

[18] Quoted in Epsilon Sigma Phi, *Extension Work*, p. 39.
[19] S. A. Knapp to B. T. Galloway, December 1, 1903, Galloway to Knapp, October 7, 1904, and Knapp to A. J. Pieters, February 3, 1905, Bureau of Plant Industry Records.
[20] *Progressive Farmer and Southern Farm Gazette*, XV (March 19, 1910), p. 199; W. D. Hunter to S. A. Knapp, October 24, 1904, and Knapp to Hunter, November 2, 1904, Bureau of Plant Industry Records.

"and if the Bureau of Entomology decides that the boll weevil larvae turns into a monkey and climbs up a tree, I shall say Amen . . . but I will not print it or talk about it."[21]

Relations with the agricultural colleges were little better than those with the Bureau of Entomology. College professors and experiment station men tended to be suspicious of a purely federal project, and some of them were more than a little jealous of an educational system that was beyond their control but which proved to be popular with farmers and with their elected representatives. At the same time Knapp was perhaps less than objective in his assessment of the college men, a failing that stemmed in part from his unhappy experiences in Iowa two decades earlier. After a preliminary interview at Texas Agricultural and Mechanical College, Knapp reported that he was "a good bit disappointed with the College people; they are immensely narrow and fault finding." President David F. Houston, later to be secretary of agriculture, was in Knapp's view "largely the inspirer of . . . small potato ideas" that he found at College Station.[22]

Still Knapp made an effort to work with the college men. Within limits he instructed his agents to refrain from expressing views counter to those prevailing on the college campuses in given states. In 1904 and 1905 some funds from Knapp's office went into the farmers' institute programs in Texas and Louisiana in the hopes that the institute lecturers would call the attention of farmers to the procedures and results of demonstration work. In July, 1905, Knapp and Houston worked out an arrangement by which a professor joined Knapp's staff as an agent, primarily for the purpose of doing "what he could to secure the sons of prosperous farmers as students for the College," while three of Knapp's regular agents served at the college for a time to absorb the views there.[23]

[21] S. A. Knapp to B. T. Galloway, February 14, 1905, and Knapp to A. J. Pieters, February 17, 1905, Bureau of Plant Industry Records.

[22] Evans, "Recollections of Extension History," p. 14; Martin, *Demonstration Work,* p. 125; McConnell, *Agrarian Democracy,* p. 25; P. G. Holden to R. K. Bliss, March 15, 1949, Holden Papers; S. A. Knapp to B. T. Galloway, December 17, 1903, Bureau of Plant Industry Records.

[23] S. A. Knapp to A. J. Pieters, February 3, 1905, Knapp to B. T. Galloway, January 4, July 7, and October 3, 1905, and Galloway to Knapp, July 11, 1905,

But the mutual suspicion could not be dispelled. Knapp soon concluded that federal funds placed in institute work in Texas and Louisiana had been wasted. Both states, according to Knapp, used the money to hire lecturers who were special pleaders for truck or livestock farming and who in reality knew very little about the needs of ordinary farmers. In Texas the institute management "paid men $100.00 per month and expenses who would have willingly gone for nothing as an advertisement for the stock they had for sale." C. J. Barrow, in charge of institute work in Louisiana, was characterized as "no farmer and not very level headed." Moreover, Knapp became convinced that his demonstration work in large measure took the place of institutes and to better advantage; there was no need to spend federal money to continue them. Knapp thought he found a better spirit at Mississippi Agricultural and Mechanical College, but he remained unreconciled to the land-grant institutions, and until his death in 1911 the Farmers' Cooperative Demonstration Work was in the main conducted separately from them.[24]

If scientists were skeptical or jealous of Knapp's efforts, businessmen were among his foremost supporters. From the outset the railroads were more than willing to cooperate in every way possible. They advertised the program, gave special rates to farmers attending meetings, detailed their men for service with Knapp, and provided transportation for his agents, saving the government as much as twenty thousand dollars annually. Local bankers and merchants also helped. Many of them were landowners, eager to use the Knapp system on their properties. In addition, they were powerful forces in urging ordinary farmers to follow Knapp's suggestions, occasionally writing such stipulations into leases and liens.[25]

The reaction of farmers, those most concerned, was mixed.

Bureau of Plant Industry Records; Beverly T. Galloway, "Work of the Bureau of Plant Industry in Meeting the Ravages of the Boll Weevil and Some Diseases of Cotton," U.S. Department of Agriculture, *Yearbook*, 1904, p. 508.

[24] S. A. Knapp to B. T. Galloway, November 27 and December 13, 1905, and September 10, 1906, Bureau of Plant Industry Records; Epsilon Sigma Phi, *Extension Work*, p. 24.

[25] S. A. Knapp to B. T. Galloway, July 29, 1906, Bureau of Plant Industry Records.

Knapp's men at the outset met a great deal of apathy and prejudice. Added to the farmers' usual resistance to change was the tendency to suspect anything that came from Washington, as many farmers mistakenly believed Knapp's ideas did. One agent in Arkansas was suspected of being an Internal Revenue officer. In areas where there were "large cotton plantations," according to Knapp, "the owners actually object to having diversification taught to the tenants, because they fear the tenants will become independent and leave their employ."[26]

Still, farmers everywhere were interested in the work, and their prejudices dissolved in time. Such groups as the Grange, the Farmers' Union, and the Southern Cotton Growers' Association tended to be friendly. A Mississippi farmer in the spring of 1906 reported that farmers in his area were watching a neighborhood demonstration "like a hawk watches a spring chicken."[27] Gradually, ordinary farmers became more cooperative and eager to join in the work. Their prejudices were slowly overcome by the realization that the better methods offered a route to economic salvation. As a Negro farmer from Marshall, Texas, expressed it, "If President Roosevelt thinks enough of us . . . to send a white man down here to tell us how to raise cotton, I think we ought to raise cotton."[28]

Often actual results were sufficiently impressive to convince even the habitual doubters. In 1904 the work did something to restore confidence that a cotton crop could be made in spite of the boll weevil. Those farmers who followed Knapp's advice in the main harvested fair crops; others rather generally failed. A Texas demonstrator reported a crop of 250 pounds per acre while his neighbors using traditional methods were fortunate if they got 25 pounds. A year later agents in western Louisiana found a 250- to 300-pound per acre dif-

[26] Martin, *Demonstration Work*, p. 23; Fosdick, *General Education Board*, p. 47; S. A. Knapp to B. T. Galloway, April 26, 1906, Bureau of Plant Industry Records.

[27] S. A. Knapp to B. T. Galloway, July 29, 1906, and G. H. Alford to W. M. Bamberg, April 9, 1906, Bureau of Plant Industry Records.

[28] L. J. Berryman to S. A. Knapp, January 15, 1906, Bureau of Plant Industry Records; Everett W. Smith, "Raising a Crop of Men," *The Outlook*, LXXXIX (July 18, 1908), p. 603.

ference between yields of cooperators and noncooperators. Knapp claimed that other crops improved and farmers displayed a better attitude. "Farmers were more notable in the field and less in town than usual," he observed.[29]

Not only did results suggest the wisdom of expanding the work, but the steady movement eastward of the boll weevil caused Knapp and his associates to seek larger federal appropriations. In 1904 between twenty-seven and forty-one field agents carried the message to farmers in Texas and western Louisiana; in all 7,603 cooperating farmers were enlisted, and Knapp estimated that at least 50,000 farmers followed the instructions that came from his office. In 1905 lack of funds prevented any significant expansion in numbers, but the area covered began to shift gradually to the east. Louisiana got more attention, and in November, 1905, preliminary efforts commenced in Arkansas, Mississippi, and western Tennessee. Still, as late as June, 1906, Texas had ten of Knapp's twenty agents and Louisiana had four, while the remainder were little more than introducing the work into Arkansas, Mississippi, Tennessee, and Oklahoma. Meanwhile, in December, 1905, Knapp moved his office from Houston to Lake Charles, Louisiana, in order to be in a more central location.

By the first months of 1906 it was apparent to most that Knapp's work was successful and that demand for its expansion was growing. The major handicap was the lack of funds. Unfortunately, the Bureau of Plant Industry held out little promise of greater allocations and Knapp faced the prospect of curtailing his program.[30]

When it appeared that Knapp's work was to be severely limited by a shortage of funds, assistance came from an unexpected quarter. Early in January, 1902, John D. Rockefeller, the oil tycoon, undertook a program to improve southern education. For that purpose, on

[29] S. A. Knapp to B. T. Galloway, September 2 and 10, 1904, and December 28, 1905, and J. E. Wemple to Knapp, January 16, 1906, Bureau of Plant Industry Records; Evans, "Recollections of Extension History," p. 11.

[30] S. A. Knapp to A. J. Pieters, November 9, 1904, and Knapp to B. T. Galloway, April 1, September 5, and December 28, 1905, and June 29 and October 6, 1906, Bureau of Plant Industry Records; Evans, "Recollections of Extension History," p. 11.

January 12, 1903, the General Education Board was incorporated by Congress, and by July, 1909, Rockefeller had given $53 million to it. Preliminary studies of southern needs began even before incorporation. In the fall of 1902 the general secretary, Wallace Buttrick, and others began a survey of the nature and problems of southern school systems.

There were problems enough. By northern standards all southern states devoted far too little money to education. Salaries of teachers were low—white teachers in Alabama received an average of $152 a year as compared to $516 nationally—and school terms were short, averaging just over five months in Georgia. Mississippi law required only a four-month term, and it was apparent that many students received no more than the minimum.

Obviously, southern schools suffered from indifference and apathy, but the major difficulty stemmed from the southern economy, from rural poverty. Education could best be improved by improving southern farming. Some spokesmen close to the General Education Board proposed that the board undertake to promote the teaching of agriculture in rural schools, but the board resolved to educate the farmers themselves. In 1905 it set out to ascertain how that might best be done.[31]

In May, 1905, Wallace Buttrick began a tour of leading agricultural colleges, gathering information concerning their extension techniques. At Texas Agricultural and Mechanical College, in the home of President Houston, Buttrick met Knapp and learned of his methods. Buttrick also learned that existing federal appropriations prohibited the use of government funds in areas where the boll weevil was not yet an immediate threat. He became convinced that Knapp's demonstration technique had potential, not only for helping to control the inroads of the weevil but also for the teaching of better farming methods. Moreover, Knapp emphasized that once the value of

[31] *Progressive Farmer and Southern Farm Gazette,* XXVIII (January 18, 1913), p. 68; General Education Board, *The General Education Board: An Account of Its Activities, 1902–1914* (New York, 1915), pp. 3, 12, 15–16, 18–19, 21–22; Fosdick, *General Education Board,* pp. 40–41; Wallace Buttrick, "Seaman A. Knapp's Work as an Agricultural Statesman," *Review of Reviews,* XLIII (June, 1911), p. 684; "General Education Board," U. S. Senate *Document* 453 (1910), p. 1.

his methods was fully recognized, local funds would become available to continue the program. Outside support, such as the General Education Board could provide, might well have a powerful catalytic effect.

In January, 1906, Buttrick and other members of the board went to Washington for additional conferences with Knapp and the Secretary of Agriculture. There it was agreed that the board would finance the expansion of the Knapp demonstration method into states hitherto little touched. Almost immediately, on February 13, the board appropriated the first money, seven thousand dollars, to be used in Mississippi. Two months later on April 20, 1906, the General Education Board and the Department of Agriculture drew up a memorandum of understanding covering the relationship between the two agencies. It provided that the demonstration work financed by the General Education Board would be entirely distinct in territory and monetary support from that carried on by the Department of Agriculture. In its allotted territory the board would pay all costs, direct and indirect, but the Department of Agriculture would have complete supervision, determine all policies, and be solely responsible for the appointment of agents selected to implement the work.[32]

Although the General Education Board gave its financial assistance freely and asked for nothing in return, from the outset the arrangement had its critics. Among others the *Southern Farm Magazine* denounced the cooperative endeavor of the "Educational Trust" and the Department of Agriculture as a means of depriving the farmer of his self-respect and independence while extending federal control over education. The "end, if the scheme is not blocked, will be the Africanizing of the South under the cover of philanthropy for 'education' of the South."[33]

[32] General Education Board, *General Education Board,* pp. 23–25, 27; Fosdick, *General Education Board,* pp. 40–44; David F. Houston, *Eight Years with Wilson's Cabinet* (2 vols., Garden City, N.Y., 1926), I, 203; Memorandum of Understanding between the United States Department of Agriculture and the General Education Board, April 20, 1906, Bureau of Plant Industry Records; S. A. Knapp to Editor of the *Manufacturers' Record,* September 17, 1906, *ibid.*

[33] B. T. Galloway to Wallace Buttrick, May 22, 1906, Bureau of Plant Industry

Nevertheless, under the agreement of April, 1906, General Education Board money flowed into the Farmers' Cooperative Demonstration Work in ever-increasing amounts, playing an important role itself and contributing to the raising of additional funds from the federal government, local governmental units, and private sources. In 1906 the Board gave $7,000, all of which was used in Mississippi, but thereafter allocations rose rapidly, standing at $102,000 in 1910 and $187,000 in 1914.

Federal funds also increased, although at a rate Knapp thought was distressingly low. He had used only $27,316 in his pioneer efforts in 1904 and in 1906 he expended a mere $37,677. Such support he considered entirely inadequate since his demonstration system had proven itself, and in December, 1906, he spoke out strongly on the subject, suggesting to farmers that they let their congressmen know their needs. The effect of his appeal remains unknown, but in 1908 he spent $85,901 in federal money, and by 1912 federal outlays had climbed to $335,856.[34]

The use of federal and General Education Board funds, combined with the decision to keep the expenditure of those funds separate, meant that after 1906 there were in reality two branches of the Farmers' Cooperative Demonstration Work. The general rule was that federal funds would be used in those areas where the boll weevil was an immediate threat, while General Education Board money would be used elsewhere. Knapp began using General Education Board funds in Mississippi in 1906 and in Virginia, Alabama, South Carolina, and Georgia in 1907. The next year he took demonstration work into North Carolina and Florida. After Knapp's death, his son and successor, Bradford Knapp, used General Education Board money to expand the program into Maryland in the last weeks of 1911 and into Kentucky and West Virginia in 1913. Federal funds

Records; *Southern Farm Magazine*, XV (June, 1907), pp. 3–5; *ibid.* (November, 1907), p. 8.

[34] William B. Mercier, "Status and Results of Extension Work in the Southern States, 1903–1921," U.S. Department of Agriculture, *Department Circular* 248 (Washington, 1922), p. 7; Martin, *Demonstration Work,* p. 23.

continued to support the work in Texas and Louisiana and financed its expansion in Arkansas, Oklahoma, and Tennessee.

As the boll weevil moved eastward, federal funds tended to replace those provided by the General Education Board. In 1908 federal money began supporting the program in Mississippi; later, it supplanted that of the board in Alabama, Georgia, and Florida. By early 1914 General Education Board money was being used in Maryland, Virginia, West Virginia, and the Carolinas; elsewhere the program was supported by congressional appropriations.[35]

With the geographical expansion of the work came a rapid increase in the number of agents who implemented the program in the field. As late as 1907, Department of Agriculture figures showed that Knapp operated with only 49 agents; five years later the number had increased to 700. From the outset Texas had the largest number and in 1912 it employed 118, but that year Alabama had 82 and Georgia 78. By contrast, Maryland, where the work had just begun, used only 6.[36]

Regardless of any increase in numbers, the individual agent and his method of operation changed but little until after Knapp's death. The founder of demonstration work never lost his suspicions of agricultural colleges and their graduates and he continued in the main to employ men who had gained their knowledge through long and practical experience. In fact, many were farmers who still lived on their properties. Others were not actual farmers—for example, former railroad development agents continued to be valuable members of his staff—but Knapp insisted that all agents have a farm

[35] General Education Board, *General Education Board*, pp. 35, 40; Fosdick, *General Education Board*, p. 45; Bradford Knapp to B. T. Galloway, February 9 and March 4, 1912, Bureau of Plant Industry Records; B. Knapp, "Education through Farm Demonstration," pp. 228, 236; Seaman A. Knapp, "Demonstration Work in Cooperation with Southern Farmers," U.S. Department of Agriculture, *Farmers' Bulletin* 319 (Washington, 1908), p. 6; Ronald J. Slay, *The Development of the Teaching of Agriculture in Mississippi* (New York, 1928), p. 75; Range, *Georgia Agriculture*, p. 236; Armstrong, "Agricultural Education in North Carolina," p. 92; J. M. Napier, *Guide and Suggestions for South Carolina Agricultural Extension Workers* (Clemson, S.C., 1950), p. 13.

[36] Mercier, "Extension Work in the Southern States, 1903–1921," pp. 9, 15.

background and that they have personalities that would allow them to move easily among ordinary farmers. The early agents averaged forty years of age.

Bradford Knapp was less rigid in his selection of agents. In fact, the rapid expansion of the work, combined with the development of ties with the agricultural colleges, forced him to turn to recent graduates, at times picking them directly from the campuses. The younger Knapp was careful, however, to retain the personality and farm background qualifications.

Nor did an agent's sources of information and his working methods change markedly prior to 1914. Equipped with his own stock of common sense and with material provided by the Department of Agriculture, supplemented later by that from the agricultural colleges, the agent traveled about his territory by horse and buggy, on horseback, or even on foot. Leaving his home on Monday morning, he spent the week visiting farmers, inspecting demonstrations, and launching others. At times, he helped with the chores, assisted a farmer to regain control of an unruly bull, or entertained the baby. In a very real sense, the agent lived with his farmers and was truly an itinerant teacher.

The rapid expansion of the work did produce a number of significant alterations in procedures. Probably most important was the reduction of the area covered by an agent to a single county. At first agents worked six to eight counties or more, the demonstration farms were widely spread, and many farmers had no opportunity to observe the results of good farming practices. For instance, early in 1906 in Smith County, Texas, there were four or five demonstration farms, a number considered inadequate by local businessmen and leading farmers. They petitioned Knapp, asking that one man be assigned permanently to Smith County. Knapp recognized the value of such an arrangement but he was limited by available funds so he suggested that local people contribute a portion of an agent's salary and expenses. Community leaders seized the idea and quickly raised $1,000 for one year. The Department of Agriculture matched that amount and on November 12, 1906, W. C. Stallings was appointed agent in

227

Smith County. Not only was he in reality the first county agent but the cooperative financing of his services set a pattern that in the future would be used throughout the United States.

From Smith County the idea of an agent being assigned to a single county with his salary being met in part by local sources spread across the South. Before the 1907 crop season five counties in Texas and two in Louisiana had agents. As early as 1911 Alabama had at least one agent in every county and the term "county agent" had come into common use.[37]

The rapid increase in agents was aided after 1908 by the additional money raised by the counties, supplementing funds from the Department of Agriculture, the General Education Board, and local private sources. Mississippi took the lead in this regard, in 1908 enacting a law that authorized counties to use tax money in paying part of agents' salaries. Adams County pioneered under the new law, naming W. D. Clayton as agent in May, 1908. North Carolina followed in 1911 and by 1915 every state that was constitutionally permitted to grant such power to counties had done so. According to Bradford Knapp, "When we go before the Boards of County Supervisors . . . we tell them that we are not asking for money but that we would be glad to cooperate in case they desire to contribute. . . ." Arrangements generally called for the drawing up of a memorandum of understanding between the Department of Agriculture and the county, but county supervisors were not required to sign it; most of them would have flatly refused to have so bound themselves. "They are not always easy to get along with," said Knapp.[38]

Continuing demands for the expansion of demonstration work in

[37] Martin, *Demonstration Work*, p. 5; Evans, "Recollections of Extenson History," pp. 12–13; W. C. Stallings to S. A. Knapp, March 23, 1907, and J. A. Evans to Knapp, May 4, 1907, Bureau of Plant Industry Records; Southern States Association of Commissioners of Agriculture and Other Agricultural Workers, *Proceedings*, 1907, pp. 16–17; AAACES, *Proceedings*, 1911, p. 213; Clarence Ousley, *History of the Agricultural and Mechanical College of Texas* (College Station, Tex., 1935), p. 25; B. Knapp, "Education through Farm Demonstration," p. 228.

[38] U.S. Department of Agriculture, *Report on Agricultural Experiment Stations and Cooperative Extension Work in the United States*, 1915 (Washington, 1916), part II, p. 17; Bradford Knapp to B. T. Galloway, August 14, 1912, Bureau of Plant Industry Records; Armstrong, "Agricultural Education in North Carolina,"

due course induced the states, beginning with Alabama in 1911, to appropriate money for the program. In that state a bill that passed the legislature almost unanimously provided $25,000 a year for co-operation with the Department of Agriculture in demonstration work. The fund was to be expended under the direction of a committee of the state board of agriculture. By 1912 Virginia, North Carolina, South Carolina, and Florida were contributing in the same manner with amounts varying from $5,000 to $15,000 a year.[39] By 1913 the total of state and county funds approached the federal appropriation and was almost double the annual contribution from the General Education Board. Very clearly, Knapp, Buttrick, and others had been correct when in 1906 they foresaw the catalytic effect of the board's early donations.[40]

The multiplication of agencies contributing to the work suggested that there should be one central office in each state through which the program in that state could be administered. In many ways the land-grant college was the logical choice; it was carrying on an extension program itself and presumably all types of extension could be combined. But Seaman A. Knapp, with his slight regard for the colleges, refused to join with them or to work through them, except in a very limited way. After his death, however, his successor was less rigid, and late in 1911 the first agreement with a land-grant college was arranged. Effective January 1, 1912, all demonstration work in South Carolina was administered through a cooperatively employed state agent, headquartered at Clemson College. Similar arrangements were reached with Texas late in 1912, with Georgia in 1913, and with Florida early in 1914.[41]

p. 93; R. S. Wilson, Early Days of Extension Work, James E. Tanner Papers (Mississippi State University Library, State College, Miss.).

[39] *Progressive Farmer and Southern Farm Gazette,* XVI (February 25, 1911), p. 199; *ibid.* (July 22, 1911), p. 640; *ibid.,* XVII (June 8, 1912), p. 673; B. Knapp, "Education through Farm Demonstration," p. 229; Bradford Knapp to C. L. Blease, January 8, 1912, Clemson University History Collection.

[40] AAACES, *Proceedings,* 1912, p. 142; U.S. Department of Agriculture, *Report on Agricultural Experiment Stations and Cooperative Extension Work in the United States,* 1915, p. 17.

[41] Epsilon Sigma Phi, *Extension Work,* p. 24; Evans, "Recollections of Extension History," p. 24; AAACES, *Proceedings,* 1912, p. 142; B. Knapp, "Education

Other innovations or changes in operating procedures came after 1906. Increasingly, Knapp's agents were employed and at work not for a few months during the growing season but on a year-round basis. Among other things, this allowed the men to participate in institute programs in late summer and in winter. They could generally advertise the demonstration work and from among the farmers with whom they came into contact select better cooperators. In addition, as the county agent became popular and as the work became more intensive, there was a tendency for the old demonstration farm, such as the pioneer Porter experiment, to be replaced by the individual demonstration, conducted by a farmer without a guaranty of any kind. In certain instances the Department of Agriculture continued to supply improved cotton and corn seed.[42]

Gradually, too, Knapp's demonstration work came to encompass every crop and every type of farming in the South. In theory, at least, cotton was the only crop considered at the outset since federal money provided through the Bureau of Plant Industry was by the terms of the appropriation limited to boll weevil control programs. No such restrictions applied to the General Education Board funds, and agents working in those areas where the board supported the program interested themselves in all crops and types of farming. During the first years after the cooperative agreement between the Department of Agriculture and the board or when demonstration work had just been introduced in an area, agents tended to work primarily with a few staple crops; but as the number of agents increased and as demonstration work became better established, instruction came to include livestock raising, dairying, truck farming, and other specialties. Meanwhile, the language of federal appropriation measures was gradually changed until by 1913 it was broad enough, under even the strictest interpretation, to cover all types of farming.[43]

---

through Farm Demonstration," p. 237; Ousley, *History*, p. 26; Bradford Knapp to C. L. Blease, January 8, 1912, Clemson University History Collection.

[42] S. A. Knapp to B. T. Galloway, July 29, 1906, Bureau of Plant Industry Records; Evans, "Recollections of Extension History," p. 13.

[43] Mercier, "Extension Work in the Southern States, 1903–1921," p. 6; Martin, *Demonstration Work*, p. 158; *Southern Planter*, LXX (February, 1909), p. 136;

The changes in Mississippi were typical. At first agents limited themselves in the main to the staple crops, in line with Knapp's dictum that farmers needed to recognize first the value of better preparation of soils, better seed, better cultivation, laborsaving equipment, and general thrift. But by 1910 agents in the state concluded that demonstration work had progressed to the point where they could introduce new crops and instruct farmers in regard to all problems related to rural development. They began to conduct demonstrations in clover, vetch, oats, lespedeza, and alfalfa. Several introduced terracing and the use of limestone. Others urged the improvement of livestock through the use of purebred sires, and in cooperation with the state veterinarian they inoculated hogs against cholera. In Clay County the agent assigned there concluded that the area was suited to the production of cattle. He began to discuss the topic and when a few farmers showed an interest he located stock for them. Later, he induced local businessmen to purchase stock, bring it into the county, and sell it at auction to farmers.[44]

Organization became more systematic, less urgent in appearance. The Farmers' Cooperative Demonstration Work continued to be administered by the Bureau of Plant Industry, but Knapp's central office, which had been located in St. Charles, Louisiana, since December, 1905, was moved to Washington in the fall of 1907. Seaman A. Knapp served as general agent in charge until his death on April 1, 1911, when he was succeeded by his son Bradford. There was no little opposition to the naming of Bradford Knapp to the post, especially from the scientists in the Bureau of Plant Industry and from some college people, but southern senators and congressmen made it plain that the Knapp name was of major importance in the South. Outside the central office, the field organization consisted of state agents, district agents, and, increasingly, county agents. The structure resembled a pyramid, with a rapidly broadening base.[45]

Mason Snowden, "Demonstration Work in Louisiana," U.S. Bureau of Education, *Bulletin* 30, 1913 (Washington, 1913), pp. 28–30.

[44] *Progressive Farmer and Southern Farm Gazette,* XXIX (June 6, 1914), p. 667; Interviews with P. E. Spinks and J. E. Sides, March 23, 1968.

[45] Evans, "Recollections of Extension History," pp. 13, 20–21; Williamson,

The addition of General Education Board money in 1906 produced still another significant innovation, the first work with Negroes by Negro agents. When Knapp launched his demonstration program in 1904, he was fully aware of the fact that much of the cotton in the South was produced by Negro laborers and tenants, so he directed his agents to enroll Negro farmers as demonstrators and cooperators. J. A. Evans, in his first day at work early in 1904, enrolled two Negro farmers. For the program as a whole, it was estimated that Knapp's first agents spent 25 percent of their time working with Negroes.[46]

Knapp was opposed to the naming of Negro agents, since he feared the effect of southern prejudice. Moreover, he believed that the use of Negro agents would lead to a needless waste of funds. "Colored agents cannot work except in strictly Negro communities but the typical southern white man whom we employ will of his own volition carry on a large amount of work among the colored people," he wrote.[47]

Southern white attitudes on the matter were mixed. Many southerners feared that education of any sort would disrupt prevailing race and economic arrangements. "In this part of Virginia the average farmer likes his farm hand to have as little education as possible," reported one writer from Winchester. Others believed the Negro incapable of absorbing the ideas of scientific agriculture or too lazy to apply them. On the other hand, some southerners recognized that the welfare of the entire section was tied to the Negro and that his improvement was a necessity if the South were to alter materially its economic condition. "Ignorance is a weakness and a

---

Louisiana Agriculture, p. 98; Napier, South Carolina Agricultural Extension Workers, p. 13; Progressive Farmer and Southern Farm Gazette, XXIX (June 6, 1914), p. 667.

[46] General Education Board, General Education Board, p. 54; Evans, "Recollections of Extension History," p. 12; William B. Mercier, "Extension Work among Negroes, 1920," U.S. Department of Agriculture, Department Circular 190 (Washington, 1921), p. 3.

[47] S. A. Knapp to B. T. Galloway, June 2, 1908, quoted in Rodney Cline, The Life and Work of Seaman A. Knapp (Nashville, Tenn., 1936), p. 69.

menace, not only to those whose mind it clouds, but to all about them," editorialized a farm paper.[48]

It was the General Education Board that induced Knapp to change his mind concerning the usefulness of Negro agents. Booker T. Washington, famous head of Alabama's Tuskegee Institute, was deeply interested in reaching the masses of Negro farmers. For years he and George Washington Carver had made it a practice to speak in nearby Negro communities and at churches, often taking farm tools and other materials with them for demonstration purposes. From these experiences there developed in Washington's mind the idea that was realized on May 24, 1906, when the Jesup Wagon began to make its rounds in Macon and surrounding counties. Made possible by a grant from Morris K. Jesup of New York, the wagon carried farm tools, a variety of seeds, samples of fertilizers, a churn, a cream separator, a milk tester, and other equipment for making demonstrations wherever a meeting might be held. When it proved to be an effective teaching device, Washington discussed its usefulness with Knapp and others, and an arrangement was reached which provided for cooperation between Knapp's office and Tuskegee Institute in its continued operation. Thomas M. Campbell, beginning November 12, 1906, operated the wagon and served as an agent to launch demonstration work among the farmers with whom he came into contact. Not only was Campbell the first Negro agent, but Tuskegee Institute was in reality the first college to cooperate directly in Knapp's program.[49]

Only a few days later a similar cooperative arrangement was worked out with Hampton Institute in Virginia. At the request of Hollis B. Frissell, principal of the school, the General Education

[48] *Breeder's Gazette,* LV (April 7, 1909), p. 863; *Progressive Farmer and Southern Farm Gazette,* XV (February 19, 1910), p. 135; *ibid., XXVIII* (January 18, 1913), p. 68.

[49] Thomas M. Campbell, *The Movable School Goes to the Negro* (Tuskegee, Ala., 1936), pp. 90–95; Booker T. Washington, "Farmers' College on Wheels," *World's Work,* XIII (December, 1906), pp. 8352–8354; Lewis W. Jones, "The South's Negro Farm Agent," *Journal of Negro Education,* XXII (Winter, 1953), pp. 38–45.

Board made a special allocation to allow Knapp and the Institute jointly to employ J. B. Pierce for work with Virginia Negroes. Pierce, in fact, became Knapp's first agent in the state, assuming his duties before demonstration work began with white farmers.[50]

By the end of 1907 there were three Negro agents at work, those in Alabama and Virginia and J. A. Booker in Mississippi, stationed at the all-Negro community of Mound Bayou. Five years later, by 1912, there were thirty-two Negro demonstration agents working under Knapp's direction. As early as May of that year, some thirty-five hundred Negro farmers were enrolled as demonstrators directed by Negro agents, and it was estimated that from ten to fifteen thousand additional Negro farmers were enrolled as demonstrators under white agents.

Still, there was no question that the expansion of work with Negroes was limited by prejudice. Knapp and in fact everyone connected with the program knew that Negro demonstration work could grow no faster than local white sentiment allowed. It was especially difficult to obtain local financial support for Negro agents.[51]

Little discernible difference existed between the instruction given Negro and white farmers or in the results achieved with the groups. With Negroes, agents presented the basic ideas of good agriculture, much as they did with whites. In most instances Negro demonstrators proved to be capable and in the main Negro farmers were appreciative of the agents' efforts.[52]

That farmer, in fact, displayed more appreciation for and understanding of Knapp's work than did some whites. When demonstration work was first extended beyond Texas and Louisiana, where Knapp's reputation had preceded his demonstration work, Knapp

[50] Fosdick, *General Education Board*, p. 44; Gladys Baker, *The County Agent* (Chicago, 1939), p. 194; AAACES, *Proceedings,* 1911, pp. 208–212.

[51] General Education Board, *General Education Board*, p. 55; Mercier, "Extension Work among Negroes, 1920," p. 7; Fosdick, *General Education Board,* pp. 48–50; *Country Gentleman*, LXXVII (May 18, 1912), p. 11; Mississippi Extension among Negroes, Mississippi Extension Collection (Mississippi State University Library, State College, Miss.).

[52] Martin, *Demonstration Work*, p. 22; General Education Board, *General Education Board*, p. 56.

found his task more difficult. Not only were large land units, with the owner-sharecropper relationship, more numerous, but farmers in such states as Mississippi, Alabama, and Arkansas seemed to be less prepared for change than were those further west, although their soils were even more depleted. Far too common was a reaction in Arkansas that was recalled many years later by a Crawford County farmer. "When I think of old Brother Phares [D. L. Phares, an agent employed in 1910] I have to smile. I can see him riding his white mule over the country roads seeking sympathetic hearers and trying to get cooperators in his work. . . . The great bulk of the farmers took him and his work as a joke. They resented the idea that an old man riding over the country could, or even had the right to, question their knowledge of how to farm and cultivate crops."[53]

Still, Farmers' Cooperative Demonstration Work came to be a powerful force producing change in the rural South. As early as 1911 there were 1,445 cotton and corn demonstrators in fifty Mississippi counties, while at the same time there were 2,534 in Alabama. A year later, Arkansas had 2,716 demonstrators and 11,351 cooperators in sixty counties. Taking the southern states as a whole in fiscal 1912 the Department of Agriculture counted more than 100,000 farmers as demonstrators and cooperators.[54]

Even more important, agricultural spokesmen throughout the nation began to point to Knapp's demonstration work as constituting, in primitive form perhaps, the universally used extension technique of the future. Without disputing the importance of other forms of extension, many authorities were convinced that demonstration work had done more than any other method to arouse farmers to the advantages of better practices. John C. Hardy, president of Mississippi Agricultural and Mechanical College and a strong advocate of farmers' institutes, expressed the sentiments of many when he wrote that demonstration work "is even better than institute work,

[53] S. A. Knapp to B. T. Galloway, April 26 and October 22, 1906, and May 4, 1907, Bureau of Plant Industry Records; quoted in Hale, *University of Arkansas,* p. 197.

[54] *Progressive Farmer and Southern Farm Gazette,* XXVII (February 17, 1912), p. 213; *ibid.* (March 2, 1912), p. 291; *Arkansas Farmer and Homestead,* XIV (January 18, 1913), p. 4; U.S. Department of Agriculture, *Yearbook,* 1912, p. 142.

for [Knapp's agents] are in touch with the actual farmer for 365 days in the year."[55]

Agricultural leaders in the North also came to praise Knapp's demonstration work. Many of them considered the lessons that he taught as being far too elementary for northern farmers, but they saw in his program methods that could be used in the presentation of more advanced concepts. In fact, such spokesmen as Cornell's Liberty Hyde Bailey were convinced that a truly successful extension system would have to rest upon the agent in the field, men who could work directly with farmers and with their special problems.[56]

[55] *Breeder's Gazette,* LXII (August 21, 1912), pp. 312–313; "Making Rural Life Profitable," *World's Work,* XVI (May, 1908), p. 10178; *Progressive Farmer and Southern Farm Gazette,* XVI (July 1, 1911), p. 592; J. C. Hardy to H. E. Savely, March 24, 1908, Bureau of Plant Industry Records.

[56] *Breeder's Gazette,* LXVI (July 9, 1914), p. 42; *Cornell Countryman,* X (May, 1913), p. 250.

IX. THE ROUNDING OUT
OF KNAPP'S SYSTEM

SEAMAN A. KNAPP knew that the Farmers' Cooperative Demonstration Work, important though it was, constituted no complete answer to rural difficulties. The problems of the South could not all be found in the fields. Malnutrition, depressingly unattractive farm homes, and poverty of spirit were as detrimental to regional progress as were the poor farming methods so much in evidence. Work with adult farmers alone could not strike at these deficiencies in southern agriculture; instead, it was necessary to reach the farm wife and even more important the children of the South.

Knapp's interest in rural youth work stemmed from a variety of considerations. Most important immediately was his belief that adults could be instructed through their children. Some farmers refused to be influenced by adult demonstrations, in spite of all efforts, but Knapp was convinced that few could ignore the results achieved by their sons and daughters using improved methods. At the same time Knapp was interested in educating rural children for their own sake. The boys and girls would be the farmers of tomorrow, and as children they were more receptive to new ideas than they would be later. In fact, greater long-run results might well be achieved through work with them than with adults. Finally, Knapp hoped that through club programs he would be able to instill in rural boys and girls a love and appreciation for farm life that would retard the movement to the city of the type of young people that agriculture needed to retain.[1]

In developing his rural youth program, Knapp drew heavily upon the innovations of other educators. Though unsystematic, the scattered work with boys and girls in northern states as well as the pioneer efforts in Texas and Georgia served as precedents. In Texas in 1903 the Farmer Boys' and Girls' League had been organized as an adjunct to the Texas Farmers' Congress. Some 1,200 children were enrolled within little more than a year. Two years later C. G. Adams, the school commissioner of Newton County, Georgia, organized 100 boys in a corn growing contest modeled after O. J. Kern's club in Illinois. Prizes were awarded at the county fair that fall, entries being judged on the basis of weight per twenty ears of corn. Later, under the leadership of a University of Georgia professor, the program became statewide. School commissioners and teachers were asked to take direction of the program at the county level, to hold corn and cotton growing contests in the fall, and to join with the state fair association in holding a state contest. The first year twenty counties complied. Meanwhile, Hancock County launched a garden contest for girls. The two programs continued with little change until 1909 when they were absorbed by Knapp's work.[2]

Some authorities give credit for organizing the first boys' corn club in the South to William H. Smith. Their contention is not well founded, but Smith did make major contributions to the development of club work. Born in 1866 in Lamar County, Alabama, Smith grew to manhood on a farm in Clay County, Mississippi. After graduating from normal school, he began a teaching career, and in 1903 he was elected superintendent of schools in Holmes County, Mississippi.

There Smith became convinced "that our system of rural schools is not producing . . . satisfactory results . . . because it is not show-

[1] U.S. Department of Agriculture, *Yearbook*, 1909, p. 158; *The Outlook*, CXIV (February 5, 1910), p. 279.
[2] Howe, "Boys' and Girls' Agricultural Clubs," p. 10; Wheeler, *Agricultural Education in Georgia*, pp. 29–36.

ing the relationship between education and farm life." Something was very wrong with "a system of education that loses its hold on the boy as he passes the age of fourteen; that fails to impress the masses with the importance of . . . citizenship; that fails to relate closely the school life of the child with the every day life of the community." What was needed, Smith believed, was "a school system that will produce educated and scientific farmers; that will bring the boy or girl into intelligent and sympathetic relation to the world . . . around him; that will make the home life and farm life of the child minister to its education."

Smith was aware of some of the developments in youth work in other parts of the country. In 1906 Perry G. Holden of Iowa had crossed the state with an educational train; at various points he had mentioned the clubs that existed in Illinois and elsewhere. Moreover, Smith was acquainted with the work of O. J. Kern and had studied his volume *Among Country Schools*. Thus it was that the Mississippian resolved to inaugurate a similar program for pupils in his county.

The first step came in January, 1907. A letter to the teachers in Holmes County explained the program and asked interested pupils to meet February 23 at Lexington, the county seat. There a faculty member from Louisiana State University and A. S. Meharg of the Farmers' Cooperative Demonstration Work discussed with a group of boys and girls the importance of better corn crops and described improved methods for producing higher yields. Some 120 boys agreed to form the Holmes County Corn Club and to plant plots with seed given them. In the course of the summer Smith provided members with literature and personal guidance. About 60 percent of the enrollees produced a crop and exhibited their corn at a show in the fall. Local businessmen supplied prizes for those who produced the biggest yield, the most ears to the stalk, and the largest ears.

Nor did Smith ignore the girls. He organized them into a Home Culture Study Club. Members devoted themselves to sewing and bread and cake baking.[3]

[3] W. H. Smith to P. G. Holden, May 13, 1932, and Holden to Smith, April 27, 1932, Holden Papers; W. H. Smith to J. F. Merry, October 21, 1907, and Smith to

Seaman A. Knapp soon learned of Smith's project. In December, 1907, after the first year of work, Knapp gave Smith the title of collaborator and placed him on the Department of Agriculture payroll at a nominal salary so that Smith might have the franking privilege. A month later Smith and a committee including Knapp's agent in the area drew up plans for extending the work to other counties. Thus it was that the Department of Agriculture in an informal way entered into the corn club work in the South.

Smith pushed the program in 1908, enrolling some 300 boys in Holmes County and stimulating the establishment of clubs in other areas. In the spring of 1908, in fact, there were clubs in twenty-three Mississippi counties, with a total membership of 3,000 boys and 600 girls.[4]

Meanwhile, comparable groups appeared elsewhere in the South. Soon they were found as far away as Virginia. Texas and Louisiana especially developed vigorous pioneer programs. When in the fall of 1907 adult farmers failed to support a corn show in Jack County, Texas, the promoters of the enterprise turned to the youth in the county. The next year Thomas M. Marks, a Knapp agent, organized the boys into a corn club, using seed provided by the Rock Island Railroad, and in the fall 137 boys displayed their results at a highly successful fair held at Jacksboro. The same year Superintendent V. L. Roy of Avoyelles Parish, Louisiana, established a club of 350 members, and by the end of the year at least fifteen parishes had begun organization. A. L. Easterling, superintendent of education in Marlboro County, formed the first boys' club in South Carolina in December, 1908. The next year there were organizations in seven counties, with a total enrollment of 327.[5]

All Schools in Holmes County, n.d., Tanner Papers; D. C. McFarlane, " 'Corn Club' Smith," *Collier's*, LI (May 17, 1913), pp. 19, 24; *Southern Farm Magazine*, XV (June, 1907), pp. 9–10; *ibid.* (December, 1907), p. 14; *Southern Farm Gazette*, XIV (December 19, 1908), p. 3; *Mississippi School Journal*, XII (March, 1908), p. 43.

4 Reck, *4-H Story*, pp. 50–53; *Southern Farm Magazine*, XV (June, 1907), p. 9; *ibid.* (December, 1907), p. 15; Fosdick, *General Education Board*, p. 51; *Southern Farm Gazette*, XIV (December 19, 1908), p. 3.

5 Hill, *Home Demonstration Work in Texas*, p. 3; Ousley, *History*, p. 25; Evans,

Knapp watched with great interest the development of the early, scattered boys' club activities. Late in 1908, after he had visited William H. Smith in Mississippi, Knapp resolved to take general control of the clubs, systematize them, and make them a valuable adjunct to the cooperative demonstration work system throughout the South. The General Education Board agreed to provide financial assistance; Oscar B. Martin, an educator from South Carolina, was installed in Knapp's Washington office in March, 1909, to take charge of the program; and Knapp set out to provide a comprehensive system for reaching southern farm boys.[6]

Knapp almost immediately introduced a series of innovations designed to make the work more effective. Most important was the standardization and economic practicality that he brought to the work. To be effective as a teaching device, Knapp thought, the boys' plots should be of sufficient size to give a meaningful test, so he established one acre as the standard unit for corn and similar crops. Knapp also knew that in practice the cost of production was as important as yield and of far greater significance than the appearance of individual ears. He required each boy to keep accurate records of labor, fertilizer, and other costs and to write a history of the crop. To Knapp learning took many forms. In the awarding of prizes to club boys, he insisted that judges consider yield, profit on the crop, and the quality of the history and the exhibit. Finally, Knapp knew well the value of the economic incentive; he required that all participants be allowed to retain the profits from their work. Parents were expected to provide land, teams, and implements, but after a boy had paid a fair price for their use, any profit was his alone.[7]

When Knapp first assumed control of the boys' club movement, he worked directly with local authorities. County superintendents of schools who agreed to participate were made employees of the De-

"Texas Agriculture," p. 103; Williamson, *Agricultural Extension in Louisiana*, p. 59; Louisiana State Board of Agriculture and Immigration, *Fourteenth Biennial Report*, 1908–09, p. 8; Napier, *South Carolina Agricultural Extension Workers*, p. 13.

[6] Evans, "Recollections of Extension History," pp. 14–15; W. H. Smith to Editor, *Aberdeen Examiner*, January 31, 1912, Tanner Papers.

[7] B. Knapp, "Education through Farm Demonstration," p. 231; Martin, *Demonstration Work*, p. 30.

partment of Agriculture at a nominal salary. They in turn worked with local schoolteachers who assumed the primary responsibility for organizing and conducting the clubs. Local merchants and bankers were asked to provide prizes for countywide contests in the fall, and county weeklies were expected to give the project the widest possible publicity.

Experience soon suggested that a better program could be conducted through cooperation with the agricultural colleges. Those institutions were in a position to give closer supervision and more specific information to club members; it was felt also that the college men would be able to work better with local educators. Knapp swallowed his dislike of the colleges and entered into a cooperative program with them. An agreement with the agricultural college in North Carolina that took effect July 1, 1909, provided that the institution and the Department of Agriculture would jointly employ an agent to manage the club work in the state. The agent's salary and traveling expenses would be provided by the Department of Agriculture from funds appropriated for the Farmers' Cooperative Demonstration Work; the program that he would administer would be determined jointly by the cooperating agencies. Very clearly, those arrangements foreshadowed those provided for in the Smith-Lever Act.

Cooperation was soon established in other states. Alabama followed almost immediately and by the end of 1909 the agricultural colleges in Mississippi, Louisiana, Georgia, and Arkansas signed agreements with the Department of Agriculture.[8]

The task of the state agents, who administered the boys' club movement from their headquarters on the agricultural college campuses, was not a simple one. But according to Cully A. Cobb, state agent in Mississippi from 1910 to 1919, it was a stimulating and

[8] Evans, "Recollections of Extension History," pp. 15–16; Martin, *Demonstration Work*, p. 34; U.S. Department of Agriculture, *Report on Agricultural Experiment Stations and Cooperative Extension Work in the United States*, 1915, part II, p. 86; AAACES, *Proceedings*, 1911, p. 218; Bradford Knapp to B. T. Galloway, January 24, 1912, Bureau of Plant Industry Records; J. E. Swearington to A. F. Lever, January 12, 1910, Asbury F. Lever Papers (Clemson University Library, Clemson, S.C.).

rewarding experience. The first step was to establish contact with county superintendents and local schoolteachers, meeting with them either at teachers' institutes and other gatherings or more informally at their offices and homes. The objective, of course, was to induce them to take up the responsibility of organizing and conducting the clubs. Quite often bankers and other community leaders played a role in recruiting school officials. After a club was organized, the state agent was the medium through which information from the Department of Agriculture and the college flowed to participants. In the fall the agent was busy with the general supervision of the contests that would produce the state winners and with plans for the next year.

In time problems with local school officials produced another change in the organization of the work, an alteration that was not yet completed when the Smith-Lever Act became law. Too often state agents found that local schoolteachers lacked either the interest or the ability to manage clubs properly. Moreover, many of them found it impossible to continue their work with boys during vacation periods. The rapid turnover among county superintendents produced still other problems and quite often led to interruptions in programs. As a result, county agents gradually assumed the responsibility for directing clubs on the local level. Not only were the agents in most instances better prepared to give expert supervision but they were also able to give the continuity that the work needed. The replacement of the local schoolteacher meant that a weak link in the chain that extended from Washington to individual club members had been removed.[9]

Despite such changes the objectives of the work in 1912 or 1914 remained much the same as earlier. Writing in January, 1912, Bradford Knapp noted that the club program continued to serve a dual purpose. On the one hand it was a practical means to interest farm boys in agriculture and to show them the value of improved farming. At the same time it served to stimulate diversified farming in areas where cotton production had reigned supreme and it helped

[9] Interview by the author with C. A. Cobb, October 6, 1966; Fosdick, *General Education Board*, pp. 51–52.

to break down the prejudice among adult farmers against instructions coming from outside agencies. Knapp, in fact, claimed that in many localities it would have been impossible to introduce cooperative demonstration work if the corn club had not opened the way for it.[10]

In 1909 Knapp added a new incentive for the boys when he offered a trip to Washington to the winner of the corn club contest in Mississippi. The state bankers' association in Arkansas and private groups in South Carolina and Virginia offered similar prizes in those states. While in Washington the four boys met Secretary of Agriculture Wilson and received certificates of merit from him. The next year the state prizewinner in South Carolina had already pocketed a ten-dollar county prize and a seventy-five-dollar state prize before he left for the nation's capital. The Washington trip for winners was discontinued early in 1914, since by that time Bradford Knapp and others had concluded that the offer had served its purpose. Instead, Knapp suggested that county winners be given funds to allow them to attend the winter short course given on the campus of the agricultural college in their state, while state winners might be given tuition and expenses for one year at the institution.[11]

As the size and significance of the boys' corn club movement grew, members were invited to participate in the major agricultural fairs and shows in the South. At the National Corn Exposition held in Columbia, South Carolina, in 1913, there was a school to which the first two prizewinners from each county in the South were invited. The boys were allowed to see all the exhibits and to attend instructional sessions arranged for them. At the end of the visit the boys were required to write a report detailing what they had learned. Somewhat similar was an encampment for a hundred corn club boys

[10] Bradford Knapp to B. T. Galloway, January 24, 1912, Bureau of Plant Industry Records; Oscar B. Martin, "Boys' and Girls' Clubs," Conference for Education in the South, *Proceedings,* 1912 (Washington, n.d.), pp. 206–207.

[11] *Progressive Farmer and Southern Farm Gazette,* XV (January 8, 1910), p. 6; *ibid.* (January 15, 1910), p. 32; *The Outlook,* CXIV (February 5, 1910), pp. 279–280; B. Knapp to D. F. Houston, December 6, 1913, Bureau of Plant Industry Records.

from Mississippi, Arkansas, and Tennessee that was held in conjunction with the Tri-State Fair in Memphis in 1913.[12]

From the outset Knapp insisted that Negro children be included in the work. Despite some criticism his demands were met, and state agents and their subordinates dealt with Negro schoolteachers and through them with Negro children in much the same way as they did with white children. Agents reported that they found little difference in the reaction of children, regardless of color. Rural Negro ministers played an especially significant role in recruiting club members. Prize money was less for Negro children because it was more difficult to raise local funds for that purpose. In some instances, the Negro clubs had different names to distinguish them from the white groups.[13]

From the organization of corn clubs it was but a short step to the formation of clubs specializing in other farm products. Pig clubs, for instance, proved to be popular. The first seems to have been established in January, 1910, in Oktibbeha County, Mississippi. Formed by a staff member of the agricultural college and the county superintendent of schools, the club was open to any white child under eighteen. Four prizes were offered to members whose purebred pigs showed the greatest weight gain per day. Entries were weighed and judged at the county fair in October; top prize went to a little girl whose 8½-month-old sow weighed 328 pounds.[14]

Subsequently, pig clubs spread across the South. In June, 1911, one club in Ouachita Parish, Louisiana, had 56 members, while Texas had forty-one clubs with an aggregate of 1,250 members.

---

[12] *Progressive Farmer and Southern Farm Gazette*, XXVIII (August 9, 1913), p. 859; *ibid.* (June 28, 1913), p. 731; *Country Gentleman*, LXXVII (June 22, 1912), p. 11; Helen M. Cavanaugh, *Seed, Soil and Science: The Story of Eugene D. Funk* (Chicago, 1959), p. 194.

[13] Fosdick, *General Education Board*, p. 55; interview by the author with C. A. Cobb, October 6, 1966.

[14] Hugh Critz to W. R. Perkins, May 17, 1938, Tanner Papers; *Progressive Farmer and Southern Farm Gazette*, XV (February 5, 1910), p. 83; *ibid.* (September 3, 1910), p. 616; *ibid.*, XVI (May 6, 1911), p. 450; *Progressive Farmer*, XXX (December 4, 1915), pp. 1140–1141.

As the movement grew, Knapp's office assumed overall direction of it and by 1913, when clubs had appeared in such additional states as Alabama and Georgia, Bradford Knapp planned to extend it into every state in the South.[15]

Other types of clubs also appeared. Early in January, 1912, leaders in Cumberland County, Tennessee, outlined plans for a potato club, hoping to use the club technique as a means of developing a cash crop to replace the rapidly dwindling lumber industry. Similar in purpose were the poultry clubs formed in Mississippi in 1913. Perhaps most successful were the baby beef, mutton, and dairy clubs existing in Texas by January, 1914. The previous year, some 300 rural children were enrolled in baby beef clubs, each agreeing to feed and keep records on the maintenance cost and growth of one calf for a 120-day period preceding stock shows at which winners were announced. The agricultural college provided detailed instructions for participants.[16]

After Knapp's office took direction of boys' club work, it expanded rapidly. In 1908 there were 10,300 enrolled in corn clubs, but the total rose to 45,000 in 1909 and to 70,000 in 1912. By that time the work was well established throughout the lower South, and a few clubs existed as far north as southern Maryland, West Virginia, and Kentucky. One club was being directed by correspondence in Will County, Illinois. Early estimates of the cost of the program in 1912 were placed at $40,000, of which over half came from the Department of Agriculture and $6,500 from the General Education Board. The balance was contributed by local governmental units, agricultural colleges, and private sources.[17]

The impact of the boys' club work was difficult to determine, but there was little doubt that it was substantial. Some critics maintained

---

[15] *Progressive Farmer and Southern Farm Gazette,* XVI (June 17, 1911), pp. 564–565; *Wallaces' Farmer,* XXXVIII (November 28, 1913), p. 1624.

[16] *Breeder's Gazette,* LXI (February 7, 1912), p. 361; *ibid.,* LXV (January 1, 1914), pp. 14–15; *Progressive Farmer and Southern Farm Gazette,* XXVIII (March 8, 1913), p. 336.

[17] General Education Board, *General Education Board,* p. 59; B. Knapp to B. T. Galloway, January 24, 1912, Bureau of Plant Industry Records; *Breeder's Gazette,* LIX (March 8, 1911), p. 627.

246

that the production of up to 230 bushels of corn on a single-acre plot was of little practical importance; after all, the boys lavished care on their acres that would be impossible on larger tracts. Still "it would be difficult to over-estimate the far-reaching effect of the . . . clubs," reported officials at the Mississippi Agricultural and Mechanical College, and that sentiment was found across the South. Certainly the boys were learning better agricultural methods, some of which would surely remain with them as adult farmers. Perhaps even more important, the participants were learning to read, to think, and to reason logically. These were lessons that would not be out of place in the emerging South of the twentieth century. If nothing else, observers noted that corn club activities suggested to white boys that there was nothing demeaning about physical labor, thereby helping to break down one of the harmful heritages of slavery. Meanwhile, the Department of Agriculture continued to insist that the primary importance of the work lay in its value in the effort to reach the adult farmer. Here again Seaman A. Knapp's shrewd knowledge of human nature seems to have been unusually perceptive.[18]

Seaman A. Knapp made a final contribution to rural education through his promotion of girls' tomato clubs. From the outset Knapp had been interested not only in improving southern agriculture but also in uplifting farm life. This obviously required work with farm women because, as James Wilson reportedly reminded Knapp, "You can not elevate a people without elevating its womanhood."[19]

In many ways the task of inducing a farm wife to change her housekeeping methods was more difficult than inducing her husband to adopt better farming practices. No agent could go into a farm home and tell the lady of the house that her home was unsanitary, that her food preparation methods were poor, and that her family's diet was totally inadequate. Moreover, a farmer could usually be

---

[18] Mississippi Agricultural and Mechanical College, *Annual Catalogue*, 1913–14, p. 177; Martin, "Boys' and Girls' Clubs," pp. 209–210; AAACES, *Proceedings*, 1912, p. 142. For figures on corn yields during the early years, see Martin, *Demonstration Work*, p. 31.

[19] Quoted in *Country Gentleman*, LXXVIII (March 22, 1913), p. 447.

converted to better methods by showing him the advantages in dollars and cents; that approach would in all probability be of little value with farm wives. The problem was to find some simple means of getting into the country homes of the South without arousing the suspicion and antagonism of the women. Work with girls constituted an obvious method.[20]

The interest that farm girls across the South had shown in boys' corn club work suggested the direction a project with them might take. Knapp knew that in the South many people would object to girls growing corn, so he and O. B. Martin, agent in charge of boys' work, concluded that the growing and canning of some garden product or the raising of poultry would be most appropriate. Late in 1909, in an address before South Carolina teachers, Martin outlined in general terms Knapp's thoughts on the matter and he challenged the teachers to take up the new work.

In the audience as Martin spoke was Marie Cromer, a teacher from Aiken County. Early in 1910 she resolved to act upon Martin's suggestions. By letters and personal visits she enrolled forty-six girls in a tomato club. She rejected poultry work because in that area of South Carolina raising chickens was considered not fully in keeping with women's work; the growing of chrysanthemums was rejected because of its lack of practical value. Tomatoes, on the other hand, were universally liked, they were easily grown, they had a long growing season, they could be kept for some time for exhibition, and they could be consumed in a multitude of ways. Any girl between nine and twenty could join the club. Members were expected to plant a tenth of an acre and to perform all the work except preparing the soil. Prizes would be offered at the end of the season to those members achieving the biggest yield, producing the best display in glass jars, and compiling the best collection of tomato recipes.[21]

[20] General Education Board, *General Education Board*, p. 62.

[21] Jane S. McKimmon, *When We're Green We Grow* (Chapel Hill, N.C., 1945), pp. 1–4; *Progressive Farmer and Southern Farm Gazette*, XVI (May 27, 1911), p. 502; Napier, *South Carolina Agricultural Extension Workers*, p. 13.

Miss Cromer was away from her teaching post during the summer months, but her club continued its work under the direction of a home economist from Winthrop College. When the crop was ripe in July, 1910, Martin visited the group and arranged for a canning demonstration, and the girls learned how to handle a marketable commodity. Impressed with the club and its work, Martin suggested and Knapp appointed Miss Cromer as agent in two South Carolina counties, effective August 16, 1910.

Even before Miss Cromer's appointment, the work had been introduced into Virginia and it soon spread to other states. In April, 1910, in an address in Virginia, Martin discussed the forming of the pioneer tomato club in South Carolina and suggested that something of a similar nature might be undertaken there. Impressed with his talk, state authorities took steps to launch the work in Virginia, naming Ella G. Agnew as organizer in two counties. On June 3 she received a Department of Agriculture appointment. Before the year was out Susie V. Powell was appointed in Mississippi and Jane S. McKimmon in North Carolina, permitting girls' club work to begin in those states in the spring of 1911. Soon Virginia P. Moore was named in Tennessee.[22]

The movement spread rapidly from 1911 to 1913. Agnes Morris was appointed state agent in Louisiana in November, 1911; work began in Alabama the same year; and the first tomato clubs in Arkansas and Texas were organized in 1912. Total enrollment shot up from 3,000 in 1911 to about 30,000 in 1913, scattered through fourteen states. Alabama, Georgia, Mississippi, Oklahoma, and Texas were the leaders, each having more than 2,000 members. Meanwhile, canning clubs for Negro girls first appeared in 1911 in southern Virginia. They were not managed by Knapp's agents, but were conducted instead by the Jeanes teachers who modeled the program after that for white girls.[23]

[22] *Progressive Farmer and Southern Farm Gazette*, XVI (May 27, 1911), p. 502; Epsilon Sigma Phi, *Extension Work*, p. 66; C. J. Goodell, Press Release, March 31, 1942, Tanner Papers.

[23] Williamson, *Agricultural Extension in Louisiana*, p. 63; Alabama Agricultural

In 1911, the first year of work outside of South Carolina and Virginia, there were 21 agents working in seven states. Only two years later the number had grown to 199, and the agents were scattered through twelve states. Only West Virginia, Maryland, and Kentucky, of the fifteen states lying south of Pennsylvania, the Ohio River, Missouri, and Kansas, had yet to begin the work.[24]

Until the enactment of the Smith-Lever Act in 1914, there was no federal appropriation for girls' club work. Instead, it was supported primarily by the General Education Board, which gave $5,000 in 1911, $25,000 in 1912, and $75,000 in 1913. In a number of counties, however, local contributions helped to maintain girls' club agents who came to be the female counterparts of the men in charge of boys' club work.[25]

Upon appointment, state agents usually began their labors by calling upon the state superintendent of education. Through him they made contact with the county superintendents. There followed meetings at the county level, where the state agents tried to enlist the cooperation of the local teachers, met some of the girls who would be in the clubs, and sought the financial support of local authorities or leaders. If support was forthcoming, the state agent then selected a local woman who took charge of the work in the county.[26]

At the outset, these women were employed for two-month periods. Given the title of collaborator, they received seventy-five dollars a month. If paid from local sources, they were placed on the government payroll at a dollar a year to give them the franking privilege. During the two-month period the collaborators were expected to

Experiment Station, *Circular* 12 (Auburn, 1911), p. 6; Hale, *University of Arkansas*, p. 198; General Education Board, *General Education Board*, p. 64; *Progressive Farmer and Southern Farm Gazette*, XXVIII (April 5, 1913), p. 460; Fosdick, *General Education Board*, p. 56.

[24] Mercier, "Extension Work in the Southern States, 1903–1921," p. 15.

[25] General Education Board, *General Education Board*, pp. 64, 67; McKimmon, *When We're Green*, p. 4; *Progressive Farmer and Southern Farm Gazette*, LVI (October 14, 1911), p. 877; Williamson, *Agricultural Extension in Louisiana*, p. 64.

[26] McKimmon, *When We're Green*, pp. 12–14.

organize the girls into clubs, induce them to plant the required amount of tomatoes, and teach them to can the fruit that the family could not immediately consume.

The collaborators had a difficult task. According to one observer, the women had to "know their subject, know the practical application of it; they must have the wisdom of a serpent and the harmlessness of a dove, be as diplomatic as ambassadors, have unlimited health, and a willingness to give themselves body and soul."[27] In most instances the responsibilities of collaborators required far more than two months, and beyond that time span their work tended to be a labor of love. The women found it necessary to travel by buggy through the rural areas of their counties, spending nights with farm families. In the main they were well received; most of them were experienced schoolteachers, reasonably well educated, and from the better-known families in their communities. As a result, they were generally respected and at least tolerated, even by farm wives who tended to be skeptical of young women traveling around the country and talking about improved household practices.[28]

Prizes and the possibility of profit helped to retain the interest of club members. Quite often the girls who canned part of their crop were able to sell their products through local outlets. In 1913, for instance, a Georgia girl produced over 5,300 pounds of tomatoes which sold for a profit of $132. Meanwhile, club members were expected to exhibit their work at local and state fairs, with prizes going to the winners. In 1913 state victors were also rewarded with a trip to Washington.[29]

By 1913 the use of the 4-H symbol had come into use. Designed originally to give the movement some unity and to facilitate the sale of the girls' canned products, the symbol represented "head,

[27] Henrietta Calvin, "Extension Work," *Journal of Home Economics*, IX (December, 1917), p. 566.

[28] B. Knapp to B. T. Galloway, February 15, 1912, Bureau of Plant Industry Records; Epsilon Sigma Phi, *Extension Work*, pp. 67–68.

[29] *Breeder's Gazette*, LXIV (October 23, 1913), pp. 787, 791; *Progressive Farmer and Southern Farm Gazette*, XXVII (December 21, 1912), p. 1356; Martin, *Demonstration Work*, pp. 47–48.

hand, heart, and health," and was expected to give participants a desirable sense of belonging to a significant movement.[30]

In Mississippi under the leadership of Susie Powell, the tomato clubs had a club yell and a song. The yell proclaimed the girls' objectives: "To make the best better." The song, to the tune of "Dixie," presumably further strengthened the unity and spirit of the group.[31]

After tomato clubs were well established, it was but a short step to the expansion of the work to include not only other crops but also the direct instruction of adult women. From the outset Knapp and his agents had been fully aware that tomato clubs constituted no more than a first step, so they moved forward aggressively whenever the opportunity presented itself. In Virginia the tomato clubs were in no more than their infancy when Ella Agnew began to consider the formation of clubs specializing in cucumbers, other vegetables, flowers, and poultry. By 1913 in Arkansas, South Carolina, and elsewhere, agents were teaching girls to sew and bake. Inevitably, adult women were caught up in the work. In 1913 Jane McKimmon reported that in one cooking demonstration she instructed four generations of North Carolina women. By that time, in fact, it was already accurate to use the term "home demonstration work" to describe the activities being conducted in many parts of the South.[32]

By 1914 instruction of southern women fell short of that conducted for men in terms of extent and scope of the efforts. Still, in a very real sense, southern home demonstration work at least promised to be more effective than methods in use in the North. There farm women continued to be instructed by means of institutes, short

[30] Epsilon Sigma Phi, *Extension Work*, p. 68; Hale, *University of Arkansas*, p. 198; *Breeder's Gazette*, LXIV (October 23, 1913), p. 791; Martin, *Demonstration Work*, pp. 53–54.

[31] *Progressive Farmer and Southern Farm Gazette*, XXVII (February 22, 1913), p. 257.

[32] *Country Gentleman*, LXXVIII (March 22, 1913), p. 448; *Progressive Farmer and Southern Farm Gazette*, XV (September 10, 1910), p. 627; *ibid.*, XXVIII (August 30, 1913), p. 910; Hale, *University of Arkansas*, p. 198; Maurice C. Burritt, *The County Agent and the Farm Bureau* (New York, 1922), pp. 172–173, 176–177; Martin, *Demonstration Work*, p. 56.

courses, correspondence, and similar techniques. There were only a few efforts in the North, except in such states as North Dakota, to reach farm women directly by means of an agent in the community. In fact, most northern specialists in home economics, especially those in the colleges, were highly critical of southern home demonstration work, taking an attitude that was not dramatically different from that of northern extension men toward Knapp's Farmers' Cooperative Demonstration Work.[33] In both instances, time would show the narrowness of such northern views.

[33] *Breeder's Gazette*, LXIV (November 27, 1913), p. 1047; Evans, "Recollections of Extension History," p. 29; Martha Van Rensselaer, "Home Economics at the New York State College of Agriculture," *Cornell Countryman*, XI (May, 1914), p. 261.

X . THE COUNTY AGENT
COMES TO THE NORTH

FARMERS' COOPERATIVE DEMONSTRATION WORK and the programs with rural youth and farm wives in the South achieved results that could not long be ignored elsewhere. Northern farm leaders, editors of agricultural papers, and various business elements soon came to believe that the techniques in use in the South might well prove equally beneficial in the North and the West. The result was a many-sided program that in time culminated with the United States Department of Agriculture assuming a position of leadership in a movement to extend the county agent system to those regions where cotton was not grown.

Certainly the dominant figure in the rise of the county agent system in northern and western states was William J. Spillman. Born in 1863 in Lawrence County, Missouri, he received bachelor's and master's degrees in 1886 and 1889 from the University of Missouri, specializing in the biological and physical sciences. In 1894, after teaching at normal schools and at Vincennes University, Spillman became professor of agriculture and director of the agricultural experiment station at Washington State College. There he made a name for himself. Young, energetic, deeply interested in a variety of agricultural topics, Spillman was a rarity among agricultural college men; not only was he a research worker with a fine scientific mind but he was also a highly successful institute lecturer.

A paper that Spillman presented before the Association of American Agricultural Colleges and Experiment Stations in 1901 caught the eye of United States Department of Agriculture officials. Beverly T. Galloway, head of the Bureau of Plant Industry, had in mind a number of investigations dealing with the growing of various hays and grasses in the western states, with much of the work to be carried on in cooperation with experiment stations in the region. Since Spillman had the needed scientific background as well as an acquaintance with most of the station men involved, he seemed to be the ideal candidate. Resigning his post in Washington, Spillman hurried east, taking up his new duties January 1, 1902.[1]

As agrostologist in the Department of Agriculture's Office of Grass and Forage Plant Investigations, Spillman soon had under way cooperative programs with farmers and state experiment stations. The objectives in most cases were the better utilization of grasslands and the development of new grasses for improved pastures. Near Pekin, Illinois, Spillman's office joined with a landowner in an effort to discover grasses that could withstand occasional flooding; on a farm near Columbus, Texas, experiments dealt with the eradication of Johnson grass. In both instances, the owners supplied labor and machinery; Spillman's office provided plans, seeds, and certain specialized equipment. Elsewhere cooperators agreed to grow small experimental plots of various grasses, with the Department of Agriculture supplying the seed and dispatching an agent to visit the plots at least once during the summer. Such work was concentrated in the western states and usually involved only those farmers who applied for assistance in meeting their problems.[2]

[1] Bryan, *State College of Washington*, p. 532; B. T. Galloway to W. J. Spillman, December 9, 1901, and Spillman to Galloway, December 14, 1901, Bureau of Plant Industry Records; AAACES, *Proceedings*, 1931, p. 285; E. H. Thomson, "The Origin and Development of the Office of Farm Management in the United States Department of Agriculture," *Journal of Farm Economics*, XIV (January, 1932), pp. 10–11.

[2] W. J. Spillman to B. T. Galloway, January 14 and August 15, 1902, A. S. Hitchcock to Galloway, May 22 and June 1, 1904, Spillman to A. F. Woods, August 3, 1903, and Spillman to J. E. Jones, April 18, 1904, Bureau of Plant Industry Records.

Meanwhile, as early as June, 1902, Spillman had worked out co-operative projects with fifteen experiment stations, and the number increased in subsequent months. In some arrangements, Spillman's office provided only seed and assistance in planning the projects; in other instances, the Department of Agriculture carried practically all of the direct costs. Projects conducted with the Arizona station included experiments to determine the carrying-capacity of range country, the methods by which capacity might be increased, and the acreage needed to support a family unit. Other work dealt with the adaptability of alfalfa to Connecticut soils, the development of a cover crop for orchards in Delaware, and the examination of various forage crops in Missouri and their relative value in the production of beef cattle.[3]

Circumstances soon produced additional tasks for Spillman. His office had participated in the establishment of the five government farms that Seaman A. Knapp, as special agent for the promotion of agriculture in the South, had supervised in 1902–03. But when they failed and Knapp launched his highly successful Farmers' Cooperative Demonstration Work, Spillman continued to believe that government farms had a useful function. To aid in the struggle against the boll weevil, he now proposed that his office be allowed to establish across the cotton states a series of demonstration farms upon which the value and possibilities of diversification could be clearly shown. Spillman expected to illustrate how former cotton land could be used to produce forage and truck crops and livestock while rebuilding and retaining the fertility of the soil.[4]

When the project met with the approval of Spillman's superiors, the Missourian launched the work, establishing four farms in the

[3] C. R. Ball to A. F. Woods, February 12, 1902, W. J. Spillman to B. T. Galloway, January 21 and August 1, 1904, and January 13, 1905, Spillman to J. E. Jones, August 17, 1904, and D. A. Brodie to Galloway, August 13, 1906, Bureau of Plant Industry Records.

[4] W. J. Spillman to A. F. Woods, July 30, 1902, Bureau of Plant Industry Records; Spillman, Memo Pertaining to Demonstration Farms for the Purpose of Diversifying Agriculture in the South, n.d., *ibid.* Spillman's role in the broad program to improve southern agriculture is discussed in Bailey, *Knapp,* pp. 139–148, 161–162.

spring of 1904. Eighteen more came later in the year and twelve were added early in 1905. Of the total, South Carolina, Georgia, Alabama, and Mississippi had three, North Carolina one, Arkansas two, Louisiana seven, and Texas twelve.

The plan of operation was simple. After the general area for a farm was determined, Spillman's agents selected an alert, progressive farmer who was willing to follow explicitly all instructions given him. After careful study of all pertinent facts, experts worked out a cropping system covering every step to the marketing of the products. Stock raising received the most attention, but on some farms Spillman's men emphasized truck growing and other field crops. In drawing up the plans for an individual farm, the experts sought to develop a cropping system that would maintain soil fertility and otherwise promise economic stability over the long run. After a farm had been in operation for a period of time, Spillman hoped to produce a government bulletin that would summarize the results achieved and that would be useful in instruction of other farmers. Finally, he planned to hold institutes at the farms in the fall where nearby farmers might be contacted and influenced.[5]

Typical was a test farm established on a plantation near Uniontown, Alabama, in 1904. Some forty acres of land previously used in the production of cotton were placed under Spillman's supervision. Fertility had declined to the point that the land yielded only 170 pounds of cotton per acre. Spillman's investigators concluded that the land would grow alfalfa and corn and that the two crops might be used in the production of hogs. With those uses in mind, cropping and rotation systems were placed in operation. Records at the end of one year showed that $750 worth of pork was produced with a net return dramatically larger than might have been expected from cotton. A fall institute held on the grounds attracted almost a thousand interested farmers.[6]

[5] W. J. Spillman to B. T. Galloway, March 1, 1905, Bureau of Plant Industry Records; Spillman, Memo, November 16, 1905; *ibid.*; Galloway, "Work of the Bureau of Plant Industry," p. 503; *Southern Farm Gazette*, IX (October 1, 1904), p. 7; *World's Work*, IX (January, 1905), p. 5768.

[6] Henry B. Needham, "The Object-Lesson Farm," *World's Work*, XI (November, 1905), p. 6872; Galloway, "Work of the Bureau of Plant Industry," p. 504.

At first glance Spillman's program appeared to differ little from Knapp's demonstration work. Spillman, in fact, said that the purpose of his work was "to carry the results of scientific discovery to the farmer, and show him, by demonstration, that such methods were not only practical, but are the only methods by which the best results can be obtained."[7] That description would appear to be equally appropriate for Knapp's efforts. But there were significant differences. Recognizing them, Galloway directed that Spillman's plots be called diversification farms to distinguish them from those conducted by Knapp. The Department of Agriculture considered Farmers' Cooperative Demonstration Work to be primarily urgent in character, designed to show cotton farmers how they might survive in spite of the boll weevil. Spillman's project, on the other hand, was expected to break down the reliance on cotton and to produce a basic and long-term transformation of southern agriculture through diversification. Admittedly, Knapp's agents taught the value of crops other than cotton and discussed rotation, but in the early years the objective was only to provide cotton growers with some protection against the devastation of the boll weevil and to improve their chances of successfully producing cotton. Finally, Knapp's instructions tended to be significantly more elementary than Spillman's and to be aimed at the mass of southern cotton growers including all varieties of tenants. The diversification program looked to the most progressive farm operators; it utilized the latest information and employed ideas and techniques beyond the immediate ability of most farmers.[8]

In spite of Spillman's optimistic hopes, his diversification farms accomplished little. Well conceived in theory, they failed to take into account the nature and thinking of the farmers they were expected to instruct. They fell into the same trap that had destroyed the effectiveness of other so-called model farms: rural residents observed the results but quickly discounted them, saying that the techniques used

[7] Quoted in Needham, "Object-Lesson Farm," p. 6872.

[8] Spillman, Memo Pertaining to Demonstration Farms for the Purpose of Diversifying Agriculture in the South, Bureau of Plant Industry Records; S. A. Knapp to B. T. Galloway, January 4, 1905, and Galloway to Spillman, January 25, 1904, *ibid.*

were impractical on their own properties. Years later Spillman himself labeled the farms "dismal failures."[9]

Spillman, however, had diverse interests, and another of his projects in time would produce a new approach to the problem of educating the rural population. He had been impressed with the fact that in every part of the country some farmers prospered. Such men obviously had formulated methods that solved the problems common to their areas. If the Department of Agriculture could gather data concerning those methods, it would possess a valuable and useful body of information.

When Spillman went to Washington he hoped to approach the broad problems of farm management by studying the methods of those farmers who by their successes showed that they had found answers and were acting upon them. Soon his agents were seeking out such farmers, studying their practices, and publishing the findings in bulletin form. By July 1, 1905, the work had progressed to the point that an Office of Farm Management, with Spillman in charge, was established in the Bureau of Plant Industry.[10]

The establishment of the new office and the larger appropriations that came with it allowed Spillman to systematize and expand his project. Soon the nation was divided into districts consisting of two or more states, with an agent assigned to each. Spillman found it possible to work out cooperative agreements with the state experiment stations, an arrangement which allowed him to draw on the manpower and knowledge available on agricultural college campuses. Working in this manner, he was able over a period of years to build up detailed information concerning the operation of entire

[9] W. J. Spillman to B. T. Galloway, October 17, 1911, Bureau of Plant Industry Records; AAACES, *Proceedings*, 1912, pp. 145–146.

[10] C. B. Smith, "The Origin of Farm Economics Extension," *Journal of Farm Economics*, XIV (January, 1932), pp. 17–18; H. C. M. Case and D. B. Williams, *Fifty Years of Farm Management* (Urbana, Ill., 1957), pp. 16, 23; "History of Cooperative Extension Work in Michigan," p. 39; W. J. Spillman to B. T. Galloway, February 21, 1905, Bureau of Plant Industry Records; Spillman, Memo Concerning the Establishment of the Office of Farm Management, January 18, 1905, *ibid.*; *Breeder's Gazette*, LVI (December 1, 1909), p. 1164. For an example of the work of the Office of Farm Management, see C. L. Goodrich, "A Profitable Cotton Farm," U.S. Department of Agriculture, *Farmers' Bulletin* 364 (Washington, 1909).

farms, including such hitherto largely ignored matters as cost accounting, the optimum utilization of land and machinery, and the economic feasibility of various kinds of farm organization under different soil and climatic conditions. In a very real sense, Spillman was fusing the knowledge of the various bureaus of the Department of Agriculture and the agricultural colleges with the proven practices of successful farmers to provide a new kind of comprehensive data useful in the proper operation of any given farm anywhere in the United States.

Implicit in the authorization establishing the Office of Farm Management was the responsibility for disseminating among practicing farmers the information obtained. Bulletins detailing the operations of successful farmers constituted one step in that direction, but Spillman's agents soon adopted other techniques as well. Field trips to the farms of successful operators and institutes held on their properties came to be common. The government agents and their subordinates were often available for service as lecturers in institutes held by other agencies.[11]

Such teaching devices suffered from all of the weaknesses common to them. The fact that Spillman's bulletins contained a new type of practical hints on farm operation did not mean that farmers read them. His agents who served in institutes no doubt were better informed than most lecturers, but farmers continued to show their usual reluctance to apply the information presented to them. In short, Spillman had found a method for gathering and organizing a new type of potentially valuable information, but he had no technique for putting the information to use on the ordinary farms of the country.

Almost as soon as Spillman began his project, there were calls for another type of instruction. As early as 1907 congressmen were asking him to send into their districts men who might aid farmers. By 1910 some of Spillman's agents were answering the calls of farm-

[11] W. J. Spillman to B. T. Galloway, November 25, 1907, February 8, 1908, February 27 and December 18, 1909, and June 6, 1912, D. W. Brodie to Galloway, January 30, 1908, and Galloway to Spillman, March 7, 1907, Bureau of Plant Industry Records; AAACES, *Proceedings*, 1912, pp. 142, 145–146.

ers, going directly to their properties where they helped in working out comprehensive plans for the farms.[12] In reality, Spillman's office was slowly moving toward the concept of a man stationed in a locality, easily available to those desiring his services.

While Spillman was learning something about the nature of farmers and their needs, other men in the United States and Canada were coming to the same conclusion that Knapp had reached. It was becoming obvious everywhere that the man in the field offered the best hope for effectively instructing farmers. Some spokesmen had been calling for such a program for years. As early as 1900 John Hamilton, soon to be farmers' institute specialist in the Department of Agriculture, had proposed that Pennsylvania move beyond the institute stage and name agents who would go to private farms and in various ways aid farmers with their specific problems. James J. Hill, the railroad tycoon, was only one of many others who spoke out in a similar vein.[13]

An early example of the type of work possible came in Marathon County, Wisconsin. There Fred Rietbrock, a Milwaukee lawyer, had acquired several thousand acres of cutover land and had developed a large dairy farm as a model for others. When few neighbors followed his example, Rietbrock and local businessmen decided that they needed expert assistance. The Athens Advancement Association resolved to employ a trained dairyman who might aid nearby farmers to establish and maintain dairy herds. Rietbrock himself went to the University of Wisconsin where he located David O. Thompson, who agreed to begin his duties in July, 1905. Thompson spent three years in the area, promoting dairying and gradually extending his scope of operation until in reality he became a general expert, available for consultation on all agricultural problems.[14]

[12] B. T. Galloway to W. J. Spillman, May 7, 1907, Bureau of Plant Industry Records; *Progressive Farmer and Southern Farm Gazette*, XXVIII (March 22, 1913), p. 396.

[13] Fletcher, *Pennsylvania Agriculture and Country Life*, p. 446; *Wallaces' Farmer*, XXXI (September 28, 1906), p. 1124; *Southern Farm Gazette*, XIII (July 18, 1908), p. 3.

[14] D. O. Thompson and W. H. Glover, "A Pioneer Adventure in Agricultural Extension: A Contribution from the Wisconsin Cut-Over," *Agricultural History*,

Other pioneer efforts came in the same geographic area and elsewhere. In 1908 the Duluth Commercial Club hired A. B. Hostetter as "superintendent of farming developments" in Minnesota's St. Louis County. A year later three itinerant field agents were at work in the upper peninsula of Michigan. Appointed by the state college of agriculture, they apparently were paid at least in part from local sources. In a similar fashion and in the same year, Charles H. White, an employee of Massachusetts Agricultural College, began his duties in that state. Meanwhile in Canada, President C. G. Creelman of Ontario Agricultural College had adopted the Knapp technique. In 1907 he assigned six of his college graduates to as many counties where they functioned as teaching agents. Four years later twenty-two men were at work.[15]

The well-known Bedford County, Pennsylvania, experiment represented another step toward the creation of an agent system in the North. In 1907 A. B. Ross, a Cleveland lawyer forced by poor health to return to his home, began traveling about Bedford County, talking with farmers. In time he began providing them with information distilled from Department of Agriculture bulletins, and gradually he was able to induce some to undertake experiments or demonstrations with varieties of seed corn and legumes. Spillman learned of his work and decided to grant him the franking privilege and a limited fund for expenses. On March 1, 1910, Ross went on the payroll; he was directed to assist farmers in arranging cropping systems, to instruct them in the most improved methods of farm management, and to inaugurate such demonstration work as he might devise in Bedford and five adjacent counties. By the end of 1910 Ross was cooperating in one way or another with 716 farmers; two years later the number was more than 1,700.[16]

XXII (April, 1948), pp. 124–128; Helgeson, *Agricultural Settlement in Northern Wisconsin*, pp. 77, 102–103.

[15] *Hoard's Dairyman*, XXXIX (July 10, 1908), p. 652; "Fifty Years of Co-operative Extension in Wisconsin," p. 46; Cary, *University of Massachusetts*, p. 119; AAACES, *Proceedings*, 1911, pp. 206–207.

[16] Orville M. Kile, *The Farm Bureau through Three Decades* (Baltimore, 1948), pp. 29–32; W. A. Orton to A. B. Ross, January 29, 1909, C. B. Smith to A. Zappone, March 3, 1910, Smith to B. T. Galloway, August 31, 1912, and Ross to W. A.

Ross was the first agent appointed in the North by the Office of Farm Management for direct and general work with farmers. He was soon to be joined by others. Best known was John H. Barron in Broome County, New York.

Early in 1910 the Delaware, Lackawanna and Western Railroad, like other carriers, was deeply interested in techniques by which it might generate additional traffic. The Lackawanna had no easy task; agriculture in southern New York was characterized by eroded fields, declining fertility, and abandoned farms. Drawing upon precedents common in the industry, in January, 1910, the railroad approached the New York State Department of Agriculture and the agricultural college at Cornell, offering to purchase two farms in the area which it would turn over to the state for demonstration purposes.

Meanwhile, the Binghamton Chamber of Commerce, which had recently established a bureau of agriculture to carry on an educational campaign among neighborhood farmers, was considering the feasibility of operating a model farm. By September, 1910, the two agencies had combined their plans and agreed that the railroad would contribute the purchase price of a property, the bureau of agriculture would do much of the planning, and the state agricultural college would provide the technical knowledge needed to renovate and operate the farm.[17]

Before proceeding, George A. Cullen of the Lackawanna and Byers T. Gitchell of the Binghamton Chamber of Commerce visited the Department of Agriculture in search of additional advice and possible assistance. They met Spillman who suggested that the idea of a model farm be abandoned and that in its place the two agencies employ a single agent who would devote his full time to the problems of farmers in the Binghamton area. Spillman pledged the support of the Office of Farm Management. In line with these ideas, a memo-

Taylor, July 11, 1913, Bureau of Plant Industry Records; Fletcher, *Pennsylvania Agriculture and Country Life*, p. 471; *Breeder's Gazette*, LXVI (July 9, 1914), p. 42.

[17] *Experiment Station Record*, XXII, p. 598; G. A. Cullen to R. A. Pearson, January 26, 1910, L. H. Bailey to H. J. Webber, February 1, 1910, B. H. Gitchell to Bailey, November 29, 1910, and Gitchell to Cullen, September 6, 1910, Bailey Papers; Burritt, *County Agent*, p. 160.

randum of understanding was signed on March 20, 1911, and John H. Barron assumed his duties in the area within a fifty-mile radius of Binghamton.[18]

Much of the significance of the Broome County experiment arose from the fact that the project involved the participation of organized farmers in a joint arrangement with the United States Department of Agriculture and with the state agricultural college. When the Binghamton Chamber of Commerce created a farm bureau department it invited leading farmers in the community to participate in planning and implementing its program. A number of farmers did join the organization but many tended to be suspicious of a project in which businessmen had a large voice, so in 1913 the Broome County Farm Improvement Association was formed with an all-farmer membership. It assumed the role that the Binghamton farm bureau had previously played.[19]

Similar in many ways to the Broome County experiment, in so far as farmer participation was concerned, was a project launched in Pettis County, Missouri, in 1912. While a farmers' institute was in progress in Sedalia, members of the local boosters' club came to the conclusion that while institutes served a worthwhile purpose, neighborhood farmers needed an adviser who would be available throughout the year. They approached Samuel M. Jordan, a leading institute lecturer, with a proposition: Would he accept a permanent position in Pettis County?

Jordan, who was fully aware of the limitations of institutes and cognizant of the beginnings of county agent work elsewhere, gave a tentative acceptance. The boosters' club began to raise money for the project. The county scanned state laws and concluded that it could appropriate $1,500; the local school board agreed to pay

[18] B. H. Gitchell to L. H. Bailey, December 4, 1910, R. A. Pearson to Bailey, February 14, 1911, and Spillman to Bailey, February 27, 1911, Bailey Papers; N. A. Cobb to G. A. Cullen, March 28, 1911, Bureau of Plant Industry Records; Lloyd R. Simons, "New York State's Contribution to the Organization and Development of the County Agent-Farm Bureau Movement," Cornell Agricultural Extension, *Bulletin* 993 (Ithaca, N.Y., 1957), pp. 5–7.

[19] *Ibid.*; Bertels, "National Grange," pp. 161–162; Epsilon Sigma Phi, *Extension Work*, pp. 122–123.

$600 for a series of lectures at the high school; and local business-
men and farmers donated $900. A few days later the Crop Improve-
ment Committee of the American Grain Exchanges offered $1,000.
Assured of adequate financial support, Jordan formally accepted the
position at a salary of $250 a month and began his new duties April
15, 1912.[20]

Immediately upon taking the position, Jordan asked that an organization of farmers be established to assist him in carrying out
his tasks. The result was the formation of a bureau of agriculture.
Membership was open to all persons over sixteen who paid annual
dues of one dollar; officers were elected for one-year terms. Jordan
was given the title of manager. An advisory council took responsibility for planning and directing the general work of the organization
and its manager. Monthly meetings served as sessions in which policy
and programs were determined.[21]

Far more extensive was an arrangement worked out in North
Dakota. By 1910 Minneapolis and St. Paul business leaders had become concerned with deteriorating agricultural conditions in North
Dakota, and they resolved to improve farm practices. The Better
Farming Association of North Dakota began operations November
15, 1911. Farm implement manufacturers, lumber men, grain buyers, and bankers were among its supporters, but the project in its
original concept and early planning drew heavily on the ideas of
James J. Hill. He contributed at least fifteen thousand dollars in the
name of the Great Northern Railway Company, an amount that was
matched by the Northern Pacific and the Soo lines.[22]

The purpose of the organization was to take "personally to the

[20] *Missouri Ruralist*, August 17, 1912, p. 2; Missouri State Board of Agriculture,
Monthly Bulletin, XI (January, 1913), pp. 3, 11–12, 31; (Sedalia, Mo.) *Daily
Capital*, April 2 and 6, 1912; M. V. Carroll to D. F. Houston, June 4, 1913, Bureau
of Plant Industry Records.

[21] (Sedalia) *Daily Capital*, July 7, 1912, *ibid.*, July 17, 1912, p. 1; *ibid.*, July 21,
1912, p. 1; Missouri State Board of Agriculture, *Monthly Bulletin*, XI (January,
1913), pp. 25–29.

[22] Frank P. Stockbridge, "The North Dakota Man Crop," *World's Work*, XXV
(November, 1912), pp. 84–85; Better Farming Association of North Dakota, *Annual Report*, 1913 (n.p., 1913), p. 30.

individual operating his own farm the results of investigations as obtained by the Department of Agriculture, the Experiment Stations, and similar agencies." Agricultural agents of broad training and experience were to be placed in counties or districts of no more than 800 square miles to manage field and livestock demonstrations in conjunction with farmers and to carry on general farm advisory work. County officials, the agricultural college, and local groups of businessmen and farmers were invited to join with the association in forwarding the project.

Under the direction of Thomas P. Cooper, secretary and manager of the association, the first agent went to work in Bottineau County in January, 1912. By March, thirteen men were located in the eastern third of the state. In the course of the first year's operations, more than twelve thousand acres were farmed under the direct supervision of the agents, and 2,346 farmers had cooperated in various ways. In addition, the association sponsored a corn contest, held a boys' encampment at the state fair, issued a monthly educational bulletin, organized thirty-four farmers' clubs, and provided speakers for a multitude of rural meetings. In 1913 the number of agents increased to twenty-five; the number of cooperating farmers rose to 5,105. The association also expanded its operations to include the farm home and the farm wife. Spokesmen claimed that the association had contacted directly or indirectly 40 percent of the agricultural population in the state.[23]

Similar in nature and function to the Better Farming Association of North Dakota and obviously inspired by it were the West Central Development Association of Minnesota and the Better Farming Association of South Dakota. The former, established in 1912, took a leading role in placing agents in the western counties of Minnesota until it was absorbed by the University of Minnesota's extension program. The Better Farming Association of South Dakota, which from the outset had a nominal connection with the state agricultural college, was established in March, 1912. The group planned to limit

[23] *Ibid.*, 1912, pp. 6, 9–18; *ibid.*, 1913, p. 30; T. P. Cooper to B. T. Galloway, March 25, 1912, Bureau of Plant Industry Records; *Farm Implement News*, XXXIV (October 30, 1913), p. 16.

its first year's work to Brown County, but demand grew and before the end of the year three agents were at work in as many counties.[24]

Serving as a catalyst in the movement to place agents in agricultural counties was the Crop Improvement Committee of the Council of North American Grain Exchanges. The committee was the result of a meeting of the council in October, 1910 where members discussed the hesitancy of farmers to take up new methods. Other meetings followed, attended not only by grain dealers but by a diversity of other businessmen as well, and the committee was instructed to undertake an agricultural improvement program. Within a few months the committee was busily engaged in sponsoring seed improvement meetings in at least eight states and in the distribution of literature to farmers, but increasingly the group turned to the promotion of the county agent system as the best means of obtaining its goals.[25]

An important aspect of the committee's work came to be the administration of a large grant by Julius Rosenwald of Sears, Roebuck and Company. The Chicago mail-order magnate had long been interested in agricultural improvement, and on May 10, 1912, he offered to contribute $1,000 to any county which wanted to employ a county agent and which agreed to raise locally $2,000 to $5,000 and to retain the agent at least two years. Originally Rosenwald set aside $100,000 for the project but he indicated that, if the program succeeded, as much as $1,000,000 would be made available. The first counties to avail themselves of the offer were Kankakee County, Illinois, and Pettis County, Missouri. In time 110 counties received assistance under the program. One of the recipient counties was in Arkansas, but practically all the rest were in the Middle West.[26]

[24] McNelly, *County Agent Story*, pp. 19–20; *Kimball's Dairy Farmer*, XI (March 15, 1913), p. 175; *Breeder's Gazette*, LXII (November 27, 1912), p. 1170.

[25] Council of North American Grain Exchanges, Committee on Crop Improvement, *Proceedings of a Meeting in Chicago, February 8, 1911*, pp. 5, 9–15; Railway Industrial Association, *Proceedings*, 1911, pp. 5–8; McNelly, *County Agent Story*, p. 19.

[26] Werner, *Rosenwald*, pp. 62–63; *Breeder's Gazette*, LXI (May 15, 1912), p. 1150; *ibid.*, LXII (November 6, 1912), pp. 694–695; *Corn*, IV (January, 1915), p. 155; Evans, "Recollections of Extension History," p. 12; AAACES, *Proceedings*, 1912, p. 227.

267

Meanwhile, the General Education Board, which was playing such a large role in the South, turned its attention to New England. Seaman A. Knapp, on a trip just before his death, saw the signs of decaying agriculture in New England, and he suggested that the board undertake there a program roughly similar to its work in the South. Conferences between New Hampshire experiment station personnel and Wallace Buttrick, secretary of the board, led to a grant of $7,500 to launch demonstration work and boys' clubs in the state. One agent went to work in Grafton County in September, 1913, and by early 1914 two other men had been assigned to counties. A program for farm boys began in January, 1914, with 240 boys participating that year.

General Education Board activities began even earlier in Maine. In the fall of 1912, agents were placed in Cumberland, Kennebec, and Oxford counties; two others were added in 1913 and four in 1914. In all, by June 30, 1914, the board had spent over fifty thousand dollars in these two New England states.[27]

By 1911, with farmers, agricultural educators, and businessmen actively interested in the placing of agents in counties or larger districts, Spillman decided that the time had come to organize the work on a systematic basis. The widespread success and popularity of Farmers' Cooperative Demonstration Work and the fact that it had practically covered the South made the development of comparable programs in the North and the West almost a necessity.

Writing to Director J. L. Hills of the Vermont experiment station in May, 1911, Spillman outlined his tentative plans. At the outset he hoped to place in each congressional district an agent whose salary and expenses would be paid by the Department of Agriculture. These men would report to and be supervised by a state agent located at the state agricultural college. The state agent would be under the joint direction of the Department of Agriculture and the college and would receive his salary and expenses from those agencies, an arrangement that Spillman hoped would allay some of the growing

[27] Martin, *Demonstration Work*, p. 25; University of New Hampshire, *University of New Hampshire*, pp. 214–215; Fernald, *University of Maine*, pp. 264–265; General Education Board, *General Education Board*, p. 17.

suspicion in college circles of Department intentions. Between the state men and the Department of Agriculture would be representatives of Spillman's office who would supervise operations in three to five states.

The men assigned to congressional districts would devote themselves to careful study of the problems confronting farmers in their districts and to the discovery of solutions to those problems. Specifically, they would study the best types of farming, the most profitable crops, labor conditions, market facilities, and rotation systems. After the district men had been at work a few years, Spillman believed that they would make the best possible type of institute lecturers. In addition, Spillman hinted that after they had become well informed about conditions in their district, they might well devote a considerable part of their time to the regular extension work of the agricultural colleges in their states.

The work, said Spillman, would differ in a significant way from that being conducted in the South. Southern agriculture was carried on by small tenants; in the North the typical farmer was an owner or a tenant operating on a much larger scale with a dramatically greater investment. In short, agents in the North would be dealing with a different class of people with far more sophisticated needs. As a result, Spillman wanted as agents men of the highest quality, practical-minded college graduates in agriculture who had some years of experience as teachers, experimenters, or institute lecturers. "We want men who in a short while can come to be real help to the farmers in the districts where they work, men whom college graduates on the farm will be glad to have visit them."[28]

A preliminary but essential step was the establishment of a definite boundary between Farmers' Cooperative Demonstration Work and Spillman's new endeavors. Essentially, it was decided that Bradford Knapp should confine his activities to the cotton-growing states.

[28] Evans, "Recollections of Extension History," p. 22; W. J. Spillman to J. L. Hills, May 27, 1911, Office of Experiment Stations Records; W. J. Spillman to B. T. Galloway, October 17, 1911, Bureau of Plant Industry Records; AAACES, *Proceedings*, 1911, pp. 95–97; D. A. Brodie to L. H. Bailey, July 20, 1911, Bailey Papers.

Virginia was included but Maryland was not, and Knapp was directed to withdraw from Kentucky. West Virginia originally fell within Spillman's jurisdiction, but effective July 5, 1913, it was shifted to the office of Farmers' Cooperative Demonstration Work. The boys' corn club work and similar activities of Knapp's office were to be confined to the same area as the Farmers' Cooperative Demonstration Work. Spillman would be allowed to institute his plans throughout the remainder of the country, doing as much as time and funds permitted with boys' clubs and related efforts in the North and the West.[29]

Spillman's plans were modified even before they were placed in operation. By early 1912 Spillman had concluded that agents should be placed in counties or in no more than small groups of counties. These local agents would be supervised by district men and through them by the state agents. Spillman also decided that the United States Department of Agriculture could cooperate not only with the agricultural colleges but also with the experiment stations or such other state agencies as might be willing to join with it in the naming of state agents. But if cooperation was not forthcoming, Spillman was prepared to have the Department of Agriculture assume full responsibility for the implementation of his project.

More important than these changes, however, was Spillman's acceptance of the concept of local assistance in the maintenance of county agents. Communities in which such men were located were expected to provide one-half of the agents' salaries and expenses. These contributions might come from the farmers themselves, from rural organizations of one kind or another, from chambers of commerce, or from any other agency that might be interested in the development of agriculture in a given locality. As a general rule the Department of Agriculture expected to contribute the remaining one-half of the cost of maintaining agents.[30]

Since the plan involved the expenditure of increasing amounts of

[29] B. T. Galloway to W. J. Spillman and Bradford Knapp, February 29, 1912, and Galloway to Knapp, March 7, 1912, Bureau of Plant Industry Records; West Virginia Agricultural Experiment Station, *Report*, 1912–14 (n.p., n.d.), p. 91.

[30] W. J. Spillman to L. H. Bailey, April 6, 1912, Bailey Papers; Demonstration

money, the Department of Agriculture had to seek larger appropriations from Congress. Finally in August, 1912, influenced by rural demands, Congress granted $300,000 to "investigate and encourage the adoption of improved methods of farm management and farm practice, and for farm demonstration work . . . ." In later years southerners would recall with some bitterness that the appropriation was the first direct federal financial support for general demonstration work, despite the fact that it had been underway in the South for eight years.[31]

Spillman's project aroused considerable discussion and no little opposition among agricultural college leaders. Quite obviously his plans involved the Department of Agriculture in a general program of agricultural extension, an area into which the land-grant colleges were themselves moving. Nor was there an immediate emergency in the North and the West such as had been used to justify the launching of Knapp's work. Some academic men, such as Iowa's C. F. Curtiss, praised the Department of Agriculture's "generous desire to cooperate," but others were less complimentary. F. B. Mumford of Missouri and J. G. Lipman of New Jersey tended to think that the Department of Agriculture was attempting to preempt a field not rightfully belonging to it, thereby creating controversy and leading to a waste of time and money, a duplication of work, and a general lack of coordination. Moreover, according to Lipman, the department workers too often did not see fit "to consult with the local institutions. They have taught doctrines which, in view of their limited experience and knowledge of local conditions, they had no right to promulgate."[32]

The most vigorous criticism was mounted by Dean Eugene Daven-

Work of the Office of Farm Management and the General Plan for Its Conduct, May 17, 1912, Bureau of Plant Industry Records.

[31] *U.S. Statutes at Large*, XXXVII, part I, p. 277; Evans, "Recollections of Extension History," pp. 22–23; AAACES, *Proceedings*, 1912, p. 142.

[32] J. L. Hills to W. O. Thompson, June 1, 1911, Office of Experiment Stations Records; A. R. Hill to F. B. Mumford, July 31, 1912, Missouri Agricultural College Papers; AAACES, *Proceedings*, 1912, pp. 143–144; F. B. Mumford to B. T. Galloway, May 30, 1912, and W. J. Spillman to Galloway, June 6, 1912, Bureau of Plant Industry Records.

port of the University of Illinois. Spillman, according to Davenport, was exceeding his authority. The Office of Farm Management had been established for the primary purpose of gathering information concerning the principles of sound farm management. But Spillman had not been content with that "important and difficult field of economic research"; instead, he had set out to do in the North and the West what Knapp did in the South. To this end, Davenport felt, the Office of Farm Management had constructed a "gigantic scheme" that would ultimately either destroy the land-grant colleges or subvert them to federal will.[33]

Despite such criticism the county agent program went forward, with the Office of Farm Management in the forefront of the movement. In New York, the Broome County experiment set the pattern followed in subsequent expansion. There the program had the enthusiastic support of the agricultural college and of its influential dean, Liberty Hyde Bailey. Limitation of funds prevented the college from participating in the Binghamton enterprise except to provide technical advice. But Bailey welcomed the role of the Office of Farm Management; he recognized clearly the need for federal money although he believed that the program should be administered by state agencies. Consequently, by 1912 Cornell was deeply involved in the work, and as funds became available the institution began to contribute financially to the maintenance of agents.[34]

Additional support in New York came from the state department of agriculture, county boards of supervisors, the railroads, and other businessmen or groups. The 1912–13 session of the state legislature authorized the department of agriculture to grant $600 a year to counties employing agents, and by January, 1914, eighteen counties had taken advantage of the offer. Meanwhile, the counties began to contribute, while the railroads continued their assistance. In Jefferson County, for instance, the local contribution was $1,000, while the Office of Farm Management and the state gave $900 and $600

33 AAACES, *Proceedings,* 1913, p. 127.
34 L. H. Bailey to F. E. Dawley, May 25, 1911, Bailey to C. H. Tuck, November 27, 1911, and August 29, 1912, and Tuck to Bailey, August 14, 1912, Bailey Papers.

respectively. Railroad aid, substantial at the outset, fell as public support rose, and by the latter part of 1912 it usually took the form of free transportation for the agents. To give such assistance the railroads were forced to put the agents on their payrolls, usually at a rate of $5 a month. In some counties the Crop Improvement Committee donated $1,000. Other businessmen also added their support. The Borden Milk Company contributed to the support of the first agent in Oneida County.[35]

Elsewhere in the East, Pennsylvania was most active in launching county agent work prior to 1914. A. B. Ross continued his labors in Bedford County, but it was not until Congress increased the financial support of Spillman's office that other counties joined in the work. In November, 1912, Montgomery, Blair, Washington, Mercer, and Butler counties received agents. In each case the Office of Farm Management and local groups provided the funds to maintain the men; the agricultural college gave technical guidance. The next year the legislature gave the college $20,000 to aid in the work and at the same time authorized counties to appropriate additional funds. As in New York, railroads made significant contributions; in 1914 the Pennsylvania gave $900 to Bedford County.[36]

Little more than a start was made in other eastern states. In Maine and New Hampshire the General Education Board was inaugurating its program. In Sussex County, New Jersey, in March, 1912, the state experiment station, the Lackawanna Railroad, the Office of Farm Management, and a local group combined to employ an agent. Connecticut and Massachusetts had placed their first agents by 1914.[37]

Among northern states those in the Middle West were most active

[35] *Hoard's Dairyman,* XLIV (January 17, 1913), pp. 763–764; *Cornell Countryman,* XI (January, 1914), pp. 128–129; *Country Gentleman,* LXXVII (May 18, 1912), p. 11; *Farmer and Fruit Grower,* XXIV (February, 1912), p. 132; True, *Agricultural Extension Work,* p. 81.

[36] Fletcher, *Pennsylvania Agriculture and Country Life,* pp. 470–471; D. A. Brody to Mr. Bradley, April 29, 1912, Bureau of Plant Industry Records.

[37] B. T. Galloway to G. A. Cullen, March 22, 1912, W. A. Taylor to W. J. Spillman, December 12, 1913, and Lisle Morrison to J. E. Jones, April 27, 1914, Bureau of Plant Industry Records; *Breeder's Gazette,* LXIV (December 4, 1913), p. 1084; True, *Agricultural Extension Work,* p. 97.

in developing county agent work; all of them had made substantial progress before the enactment of the Smith-Lever Act. The greater success of the institute movement there no doubt contributed to the progress. In addition, midwestern farmers were typically more prosperous, better educated, and less apathetic than elsewhere. Their power in their respective state legislatures made those bodies more responsive to their wishes. Finally, businessmen in the Middle West were more active in attempting to promote the improvement of the countryside.

In Missouri the appointment of Samuel M. Jordan as agent in Pettis County caused the college of agriculture to formulate a plan for placing its representatives in other counties of the state. Authorities at the University of Missouri as elsewhere recognized that the greatest task in agricultural education lay in persuading "farmers to apply the knowledge we already have" and that the agent assigned to a small district offered the most promising method for achieving that goal. In March, 1912, after months of study, the college of agriculture proposed a plan that entailed placing a representative of the institution in each of four counties. The college expected to pay one-fourth of the agents' salaries, the balance to be raised in the communities in which they worked. Authorities in Cape Girardeau County promptly accepted the offer, and on July 1, 1912, C. W. McWilliams became the second county agent in the state.[38]

Like many of his colleagues, Dean F. B. Mumford believed that extension was clearly a function of the colleges. But he was a realist; he needed federal funds and he recognized the inevitability of federal participation. He therefore agreed to cooperate with the Office of Farm Management since "we do not think it would be a good thing to allow them to operate independently in Missouri," but he resolved to get the best possible terms. Bargaining began in May, 1912, and

[38] *Colman's Rural World*, LXV (April 17, 1912), p. 121; *ibid.* (June 25, 1912), p. 193; F. B. Mumford to Asbury Lever, January 31, 1912, Mumford to Executive Board, March 12, 1912, and Mumford to L. L. Allen, June 18, 1912, Missouri Agricultural College Papers; "Twenty-Five Years of Extension Work in Missouri," Missouri Agricultural Extension Service, *Circular* 420 (Columbia, Mo., 1940), p. 8; Vera B. Schuttler, *A History of the Missouri Farm Bureau Federation* (n.p., 1948), pp. 2–3.

continued until November when a memorandum of understanding was signed. The agreement provided for the placing of agents in every county as rapidly as conditions and funds permitted. The salaries and expenses of county men were to be shared by the Department of Agriculture, the college, and local authorities, with the national and state agencies each contributing 25 percent. A state leader, whose salary was shared by the two agencies, would supervise the county agents and, from his office on the campus, would report to both Spillman and Mumford.[39]

Under these terms the county agent system was firmly established in Missouri. D. Howard Doane, who had previously cooperated with Spillman in farm management studies, was appointed state leader. The position of agent in Pettis County was offered to Jordan who agreed to work under joint federal, state, and local direction. C. W. McWilliams in Cape Girardeau County entered the arrangement. Expansion into other counties followed.[40]

In Illinois the launching of county agent work followed closely the precedent set in Missouri's Pettis County. When the Rosenwald grant was announced, farmers and businessmen in Kankakee County formed the Kankakee County Soil and Crop Improvement Association, raised $10,000, and on June 1, 1912, employed John S. Collier at an annual salary of $2,500, thereby allowing the county to take advantage of the Rosenwald offer. In nearby DeKalb County William G. Eckhardt was employed under similar auspices.[41] The

[39] F. B. Mumford to K. L. Butterfield, December 23, 1912, Missouri Agricultural College Papers; B. T. Galloway to F. B. Mumford, May 22 and October 1 and 21, 1912, and Mumford to Galloway, May 30 and October 9, 1912, and December 23, 1913, Bureau of Plant Industry Records.

[40] E. A. Boeger to W. A. Taylor, December 23, 1912, D. H. Doane to W. J. Spillman, March 29, 1912, and B. T. Galloway to F. B. Mumford, January 28, 1913, Bureau of Plant Industry Records; F. B. Mumford to M. V. Carroll, November 18, 1912, W. J. Spillman to S. M. Jordan, November 27, 1912, Jordan to Mumford, December 17, 1912, and Carroll to Mumford, December 17, 1912, Missouri Agricultural College Papers; *Colman's Rural World*, LXVI (January 2, 1913), p. 15; "Twenty-Five Years of Extension Work in Missouri," p. 8; (Sedalia) *Daily Capital*, June 16, 1912, n.p.; F. B. Mumford, "History of the Missouri College of Agriculture," Missouri Agricultural Experiment Station, *Bulletin* 483 (Columbia, Mo., 1944), p. 118.

[41] *Breeder's Gazette*, LXI (January 31, 1912), p. 266; *ibid.*, LXII (August 21,

Office of Farm Management extended its aid to the two counties early the next year, and similar organizations and appointments came in Peoria, Tazewell, Iroquois, Bureau, Will, Champaign, and other counties. Reflecting Dean Davenport's continued opposition to Spillman's work, the University of Illinois refused to cooperate actively in the spread of the system.[42]

Purdue University took the first step in Indiana. Its extension activities had been sufficiently popular to cause the state legislature in an act of February 21, 1911, to authorize the establishment of an agricultural extension division within the university and to give it an annual appropriation of $30,000, beginning October 1. The funds were to be used for a variety of extension techniques including farmers' institutes, but the university soon began to use a portion of the appropriation for the maintenance of county agents. The first agent went to work in LaPorte County on October 1, 1912, supported jointly by the university, the Office of Farm Management, and a local group, the LaPorte County Crop Improvement Association.[43]

Formal beginnings of the county agent movement in Ohio came in 1912 when H. P. Miller was appointed in Portage County. He and other agents were employed by the agricultural experiment station in conjunction with county improvement associations and the Office of Farm Management. An act of May 3, 1913, provided additional state funds.[44]

1912), pp. 318–319; *ibid.* (November 6, 1912), pp. 964–965; *Country Gentleman,* LXXVII (November 23, 1912), p. 7; *Hoard's Dairyman,* LXVI (October 10, 1913), p. 289. Early developments in Illinois are discussed fully in John J. Lacey, *Farm Bureau in Illinois* (Bloomington, Ill., 1965), pp. 11–31.

[42] Bay, *County Agricultural Agents,* p. 10; B. T. Galloway to W. J. Spillman, January 10, 1913, and April 16, 1914, Lisle Morrison to J. E. Jones, April 4 and June 20, 1914, and W. A. Taylor to Spillman, August 26, 1913, Bureau of Plant Industry Records.

[43] True, *Agricultural Extension Work,* pp. 93–94; B. T. Galloway to W. J. Spillman, November 27, 1912, and Lisle Morrison to J. E. Jones, March 24, 1914, Bureau of Plant Industry Records; *Hoard's Dairyman,* XLVII (May 29, 1914), p. 665.

[44] Mendenhall, *Ohio State,* II, 159–160; *Breeder's Gazette,* LXIV (October 23, 1913), p. 780.

In Iowa Perry G. Holden was fully aware of the work being done in the South, and in September, 1911, he called a meeting in Des Moines where rural leaders discussed the establishment of a state-wide county agent system. Little came of that session, but a year later agricultural spokesmen were more enthusiastic, despite the tendency of some to denounce business contributions as "tainted money." The result was the usual type of cooperation between the agricultural college, the Office of Farm Management, and local groups. Fairly typical was an agreement that placed an agent in Muscatine County in February, 1913. It provided that the Office of Farm Management and a local crop improvement association would each contribute $1,200 a year toward an agent's salary; the local group would grant an additional $1,000 for his expenses. The county board of supervisors provided office space, while Iowa State College gave the agent "advice and encouragement." Under roughly similar arrangements M. L. Mosher, the first agent in Iowa, had taken his post in Clinton County September 1, 1912.[45]

In the Dakotas, expansion of the county agent system meant primarily the growth of projects begun by the better farming associations. Early in 1913 a memorandum of understanding was signed between the Better Farming Association of North Dakota and the Office of Farm Management, a step that was followed by the enactment of a state law authorizing the counties to levy a tax to help maintain agents. The South Dakota Better Farming Association and the state's agricultural college, which had placed H. N. Patterson in Brown County early in 1912, were joined a year later by the Office of Farm Management in maintaining agents in three counties. In each case Spillman's office contributed $1,200 annually to the salaries of the agents.[46]

[45] Morgan, *Extension Service of Iowa State College*, p. 37; *Wallaces' Farmer*, XXXVII (August 9, 1912), p. 6; Bay, *County Agricultural Agents*, p. 10; E. H. Bradley to D. A. Brodie, April 7, 1913, Bureau of Plant Industry Records.

[46] B. T. Galloway to W. J. Spillman, January 21, 1913, Bureau of Plant Industry Records; William A. Lloyd, "County Agricultural Agent Work under the Smith-Lever Act, 1914–1924," U.S. Department of Agriculture, *Miscellaneous Circular* 59 (Washington, 1925), p. 3; True, *Agricultural Extension Work*, p. 86; Powers, *South Dakota State College*, p. 93.

Merrick County was the first in Nebraska to employ an agent, naming V. S. Culver to the post in 1912. Funds for the pioneer effort came entirely from private sources. At the outset agricultural educators at the university hoped to expand the work to other counties by similar methods, appointing a representative who devoted his time to the organization of county groups that might support agents. Soon, however, the Office of Farm Management began to provide funds, and in 1913 the legislature authorized counties to use public money for the purpose. Gage, Seward, Thurston, Madison, and Dawes counties promptly took up the work and selected agents. Among the men named was Val Kuska in Madison County, later to be an agricultural development agent for the Burlington railroad.[47]

Kansas had less public money than did Nebraska; in fact, it had none until the passage of the Smith-Lever Act. The county agent movement in the state sprang exclusively from private contributions. A local group, the Progressive Agricultural Club of Leavenworth, employed the first agent August 1, 1912, and he was soon joined by men in Montgomery, Cowley, Harper, and Allen counties.[48]

Minnesota's first county agent was Frank Marshall, assigned to Traverse County September 1, 1912. Later in the year agents were appointed in Stevens, Pope, Grant, and Ottertail counties. The work was first launched by the West Central Development Association, a business group that proposed to raise money for the project. By the end of 1912 the Office of Farm Management, the grain exchanges' Crop Improvement Committee, and the Minnesota farmers' institute management were contributing substantial amounts. Meanwhile, the state legislature set aside $1,000 for each of twenty-five counties to extend the program in 1912–13, while the extension office of the university and the Office of Farm Management assumed overall direction of the work.[49]

[47] Crawford, *University of Nebraska,* pp. 140–141; C. Clyde Jones, "Val Kuska, Agricultural Development Agent," *Nebraska History,* XXXVIII (December, 1957), pp. 285–293.

[48] Willard, *Kansas State College,* pp. 221, 486; *Wallaces' Farmer,* XXXVIII (June 6, 1913), p. 920; Kansas State Board of Agriculture, *Biennial Report,* 1913–14 (Topeka, 1915), pp. 247–250.

[49] The Development of Agricultural Extension Work in Minnesota, manuscript

In Wisconsin in 1911 the state legislature authorized the employment of "agricultural representatives," a term taken from the program maintained by the Ontario Agricultural College. Such men were to be appointees of the board of regents of the university and were expected to teach in short courses and local high school classes, conduct demonstrations, and otherwise make themselves useful to farmers in their counties. To implement the program the state agreed to contribute 50 percent of the agents' salaries and expenses, provided each county raised the remainder. The state money was administered through the agricultural college. In November the Oneida County board of supervisors voted to institute the plan and in 1912 E. L. Luther was selected for the position. Expansion was slow, and only three men were at work by the end of 1913. A new state law called for the addition of ten more agents in 1914.[50]

Michigan's first agricultural agent was Harvey G. Smith, who took the position in Alpena County July 1, 1912. The Office of Farm Management gave $1,100 toward the cost of maintaining Smith during his first year, the county provided additional funds, and the Alpena Chamber of Commerce contributed to the cost of equipment and transportation. Within a year twelve counties had agents and the college of agriculture had joined with the Office of Farm Management in naming a state leader.[51]

The work got under way in the western states almost as soon as in the Middle West but failed to progress as rapidly. In California the agricultural college agreed to place agents in counties which pledged at least $2,000 a year and which created organizations containing at least 20 percent of the farmers in the county. When these

dated December 14, 1937; McNelly, *County Agent Story,* pp. 26–27; Minnesota Agricultural Experiment Station, *Annual Report,* 1912–13, pp. 65–66.

[50] "Fifty Years of Cooperative Extension in Wisconsin," p. 45; *Kimball's Dairy Farmer,* XI (January 15, 1913), p. 3; *Wallaces' Farmer,* XXXVIII (October 24, 1913), p. 1459; *ibid.,* XXXIX (February 20, 1914), p. 321; *Breeder's Gazette,* LXV (February 5, 1914), p. 282; Helgeson, *Agricultural Settlement in Northern Wisconsin,* pp. 102–103; Wisconsin Agricultural Experiment Station, *Bulletin* 228 (Madison, 1913), pp. 64–65.

[51] "History of Cooperative Extension Work in Michigan," pp. 6–7; D. A. Brodie to Mr. Bradley, April 29, 1912, Bureau of Plant Industry Records.

requirements were met, the college and the Office of Farm Management joined in appointing agents, the first being assigned to Humboldt County in July, 1913. Within a year three other men had taken their posts. One of these was George H. Hecke, assigned to Yola County. A wealthy horticulturist, Hecke accepted only a nominal salary from the college and the Office of Farm Management.[52]

In Colorado, county agent work commenced October 1, 1912, with the appointment of D. C. Bascom in Logan County. Cooperating with the Office of Farm Management in the project were the county commissioners, the governing board of the county high school, and the agricultural college. Two weeks later another agent went to work in El Paso County and a third man assumed his duties in Pueblo County in March, 1913.[53]

Beginning in 1913 the expansion of the county agent system in Colorado was aided by the Chicago, Burlington and Quincy Railroad. Early in the year the carrier announced its willingness to donate up to $150 a year to those counties served by its tracks that employed an agent. At least three Colorado counties—Morgan, Logan, and Boulder—benefitted from the program. In the case of Logan County the agent in the 1913–14 year received a salary of $1,800, of which $1,200 came from the Office of Farm Management, $300 from the county, and $150 apiece from the local high school and from the Burlington.[54]

Elsewhere in the West the pattern of development was similar to that in California and Colorado. The first agent in Washington went to work in Wahkiakum County in November, 1912, with the Office

[52] California Agricultural Experiment Station, *Circular* 112 (Berkeley, 1914), p. 2; True, *Agricultural Extension Work,* pp. 97–98; Ferrier, *University of California,* p. 617; Lisle Morrison to J. E. Jones, April 29, 1914, Bureau of Plant Industry Records.

[53] Steinel and Working, *Agriculture in Colorado,* pp. 615–616; G. A. Billings to B. T. Galloway, October 18, 1912, Bureau of Plant Industry Records; True, *Agricultural Extension Work,* p. 93.

[54] J. B. Lamson to F. B. Mumford, January 28, 1913, Missouri Agricultural College Papers; Jones, "Survey of the Agricultural Development Program of the Chicago, Burlington and Quincy Railroad," pp. 238–239; B. T. Galloway to W. J. Spillman, March 21, 1914, Lisle Morrison to J. E. Jones, July 1, 1914, and W. A. Taylor to Spillman, July 13, 1914, Bureau of Plant Industry Records.

of Farm Management providing two-thirds of the cost and the county grange the remainder. Early the next year state legislation authorized counties to levy taxes to support county agent work and established the Bureau of Farm Development, headed by the director of the experiment station, to administer the program. Seven counties had agents by June 30, 1914. Idaho's pioneer agent was employed in Bonner County on August 19, 1912, while Fremont and Sheridan counties in Wyoming appointed agents in May and July, 1913. In these cases the Burlington railroad made the same sort of contribution as it did in various counties in Colorado.[55]

The duties of the agents in the various states of the North and the West differed to some degree, depending upon the crops, soils, and climate in particular areas. Still, there was a basic uniformity; everywhere, the agents were expected to do what they could to help farmers meet their specific problems. The work of Samuel M. Jordan in Pettis County, Missouri was typical: After opening his office in Sedalia, he set aside two days a week as office days and invited farmers to bring their problems to him. Some 120 appeared during his first two months. He spoke at a variety of farm meetings and wrote short bulletins on agricultural subjects for publication in local newspapers. But most important Jordan visited farms, answering questions, giving advice, and generally providing the man in the field with the direct, personal assistance previously lacking. Calling only on those farmers who requested him to do so, he visited sixty-three farms, some several times, in the course of his first two months' work. As another method for contacting farmers, Jordan developed the out-on-the-farm institute, a meeting at which a progressive farmer demonstrated his methods to neighbors. Also through his efforts, the Missouri College of Agriculture held a short course at Sedalia.[56]

The establishment of formal relations with the Office of Farm

[55] True, *Agricultural Extension Work*, pp. 92–97; Washington Agricultural Experiment Station, *Bulletin* 120, pp. 5, 34–35; Wyoming Agricultural Experiment Station, *Annual Report*, 1913–14 (Laramie, Wyo., n.d.), pp. 129–130.

[56] Missouri State Board of Agriculture, *Monthly Bulletin,* XI (January, 1913), pp. 15–16; (Sedalia) *Daily Capital*, May 29, June 11 and 23, and July 19 and 31, 1912; A. J. Meyer to M. V. Carroll, November 27, 1912, and Carroll to Meyer, December 5, 1912, Missouri Agricultural College Papers.

Management and the college of agriculture did little to alter Jordan's operating habits. Visits to farms continued to be his most important duty. Farmers wanted advice concerning such diverse matters as the planting of alfalfa, the installation of water systems, and the treatment of diseased hogs. Occasionally he assisted farmers wanting to purchase livestock and at times he was called upon to settle controversies between neighbors. While the Missouri State Fair was in progress he maintained a tent on the grounds, discussing agriculture with those who visited it.

Nor did Jordan ignore the women of Pettis County. He urged them to follow the example of their husbands by establishing an organization in which they might discuss common problems. Consequently, when in December, 1912, an institute lecturer broached the subject of organization to a group of Pettis County women, they accepted Jordan's suggestion and formed the Pettis County Homemakers' Club, the first such group in the state. By early 1915 the club claimed 300 members. It functioned as the ladies' auxiliary to the Pettis County Bureau of Agriculture.

While engaged in the multifarious duties of a county agent, Jordan promoted a number of definite programs designed to improve Pettis County agriculture. Foremost among these were the expansion of alfalfa and cowpea acreages, the introduction of soybeans, and the establishment of Pettis County as a center of the purebred Holstein industry in the state. In his contacts with farmers, both individually and collectively, Jordan emphasized the value of purebred sires and encouraged the planting of soilbuilding crops. Partly to promote the growing of alfalfa, cowpeas, and soybeans, he spent considerable time in testing soils for farmers who wanted to experiment with the crops.[57]

[57] Weekly Report, Pettis County, April 5, April 12, September 20, March 15, May 3, October 4, 25, 1913; Monthly Report, Pettis County, January 3, 1914, Records of Missouri Agricultural Extension Service (Missouri College of Agriculture, Columbia, Mo.); *University Missourian,* February 5, 1913; Missouri State Board of Agriculture, *Monthly Bulletin,* XII (January, 1914), p. 21; *Colman's Rural World,* LXVI (March 13, 1913), p. 168; "Ten Years of Extension Work

Elsewhere the work differed only in kind. Men placed through the cooperation of the Office of Farm Management were specifically directed to investigate problems of farm management in their counties and to disseminate among farmers the best principles discovered. Demonstrations of various kinds were expected to be the primary method. In actual practice, however, agents and their immediate supervisors were left relatively free to develop such programs as circumstances warranted and to seek to promote them through such methods as appeared feasible.

In Kankakee County, Illinois, for instance, John S. Collier took as his first step the platting of every farm in the county. The task was laborious, but among other benefits it provided an accurate soil map of each farm so that Collier was better able to advise the operator when the need arose.[58]

Probably the introduction of new crops or new varieties of existing crops occupied more of the agents' time than any other single project. In Clinton County, Iowa, the agent used the demonstration technique to show farmers how to grow alfalfa, while in Clay County the agent organized an alfalfa club. The introduction of soybeans received attention in Jefferson County, New York, and elsewhere. Almost as common were efforts to improve established crops through the use of better seed. Methods in use were many and varied. G. R. Bliss, working in Scott County, Iowa, in the fall of 1912, conducted a seed corn campaign in the rural schools, demonstrating to pupils and parents alike the value and methods of properly selecting and storing seed. In many areas winter corn shows were used to good effect.[59]

Livestock improvement also received attention. In New Hampshire the first agent concentrated his efforts on dairy herd improve-

in Missouri," Missouri Agricultural Extension Service, *Project Announcement* 16 (Columbia, Mo., 1923), pp. 83–84.

[58] California Agricultural Experiment Station, *Circular* 112, pp. 2–3; *Breeder's Gazette*, LXII (August 21, 1912), pp. 318–319.

[59] *Hoard's Dairyman*, XLV (February 7, 1913), p. 45; *Wallaces' Farmer*, XXXIX (June 5, 1914), p. 878; *Country Gentleman*, LXXVII (November 2, 1912), p. 10; *Corn*, I (December, 1912), p. 207.

ment, teaching farmers to keep records of production and helping them to weed out unprofitable cows, raise better producers, and care for their stock more economically and efficiently. Minnesota agents stressed the value and methods of building silos, urged farmers to form cow testing associations, and induced dairymen to build suitable milk cooling facilities. Throughout the Middle West hog cholera vaccination campaigns were common. In Portage County, Ohio, the improvement association that contributed to the support of an agent acquired under his guidance a number of pedigree sires—boars, bulls, and work studs—to be used cooperatively by the members. Early agents in Wyoming devoted considerable attention to urging farmers and ranchers to finish out their cattle, using home-grown feeds.[60]

Agents were generally expected to cooperate with other rural educational agencies. Quite often they participated in institutes, spoke before farmers' clubs, conducted boys' and girls' work of various kinds, and attempted to stimulate the teaching of agriculture in the country schools. The rural press and local newspapers constituted mediums through which agents reached farmers who could not be contacted in other ways. Finally, agents often participated in traveling short courses held in their counties; in many instances, in fact, they were prime movers in arranging for such courses.[61]

In several areas cooperative economic activity appeared very early. In Illinois the members of improvement groups, encouraged by their agents, undertook the cooperative buying of alfalfa seed, linseed oil meal, tankage, and limestone and phosphate rock, producing substantial savings in the process. There were isolated examples of cooperative selling. In other instances bureaus and similar groups became exchange offices, facilitating the sale and purchase among members of seed, livestock, and implements. They also

[60] University of New Hampshire, *University of New Hampshire*, p. 214; *Wallaces' Farmer*, XXXVIII (August 22, 1913), p. 1163; *ibid.*, XXXIX (June 5, 1914), p. 878; *Hoard's Dairyman*, XLV (February 7, 1913), p. 45; *Breeder's Gazette*, LXV (January 22, 1914), pp. 176–177; Wyoming Agricultural Experiment Station, *Annual Report*, 1913–14, pp. 129–130.

[61] *Hoard's Dairyman*, XLVII (May 29, 1914), p. 665; *ibid.* (June 12, 1914), p. 723; *Wallaces' Farmer*, XXXVIII (August 22, 1913), p. 1163.

functioned as employment agencies, telling farmers where they might find hired men.[62]

Illustrative of the scope of the work performed by typical agents was a report submitted by the agent in Clay County, Iowa, for the year ending March 10, 1914. He had organized 626 boys and girls in clubs, held numerous meetings to encourage the planting of alfalfa, formed an alfalfa club, vaccinated over 500 herds of hogs against cholera, and organized branches of the county improvement association in ten of the sixteen townships. He had arranged or participated in a county picnic, a county institute, a three-day school of agriculture and home economics, and a winter corn show. Finally, he had made a study of the common weeds in the county and of the means to eradicate them.[63]

Like many of his contemporaries, Samuel M. Jordan in Missouri's Pettis County reported that it was "next to impossible to get folks to do just as they are told," but the results of his labors were soon obvious. During Jordan's first two years in the county, alfalfa acreage increased 1,000 percent, and soybean and cowpea acreage gained sevenfold. In three years of work, the number of purebred Holstein cattle increased from 30 to 400. Other gains directly attributable to Jordan's efforts were the expansion of the gravel road mileage from 40 to 110, the establishment of four consolidated rural schools employing vocational agriculture teachers, and the introduction of general agricultural courses in a number of one-room schools.[64]

Despite such accomplishments through the North and the West, farmers remained far from unanimous in their reactions to the appearance of county agents among them. The old rural suspicion and apathy were still present. All too often farmers wanted to know why they should follow instructions given by men whom many considered to be little more than college boys. In other instances farmers made it plain that when they wanted advice they would ask for it.

---

[62] *Breeder's Gazette*, LXV (June 22, 1914), p. 1287; *ibid.*, LXII (August 21, 1912), p. 313; *ibid.*, LXV (January 15, 1914), p. 122.

[63] *Wallaces' Farmer*, XXXIX (June 5, 1914), p. 878.

[64] Railway Development Association, *Proceedings,* 1915, pp. 53–54; "Ten Years of Extension Work in Missouri," p. 84.

Conservatives bitterly objected to being taxed by their counties to raise funds for the maintenance of an agent.

Some of the opposition arose from a misconception of the agents' nature and duties. In a few localities the agent was known as a farm advisor, an unfortunate term, since it implied that the agent was an authority on all agricultural matters. Too often farmers failed to grasp the fact that agents were merely the mediums through which information could be transmitted, not the sources of information themselves.[65]

A fair part of the opposition during the formative years arose from the role of businessmen in the movement. The town-country antagonisms were still strong in many areas, and it was easy for opponents to arouse hostility with statements expressing sentiment that would not have been out of place in the populist period. John H. Barron in Broome County, New York, was only one of the early agents who found farmers looking for ulterior motives in the efforts of businessmen; he might well have failed had he not been a member of the Grange in close personal relation with some of its local officers. The contributions of the grain exchanges, railroads, and Sears, Roebuck and Company touched off waves of suspicion that were never completely dissipated. The opposition of some college officials to Spillman and his plans did little to allay the hostility of farmers.[66]

Nevertheless, by the middle of 1914, the county agent system was well established in the North and the West, and its development to that date provided a firm foundation for subsequent expansion under the terms of the Smith-Lever Act. In all there were 236 agents at work in 234 counties. Of the thirty-three states in the North and the West, only seven had made no start in the placing of agents. In twenty-two states a state leader had been appointed. The work was most vigorously pushed in the Middle West. There twelve states had 151 agents. By contrast the ten states of the northeastern quadrant of the nation had only 47 men at work, while eleven western states

[65] Erickson, *Rural Youth*, p. 72; *Hoard's Dairyman*, XLV (June 6, 1913), p. 681.

[66] Erickson, *Rural Youth*, p. 72; McNelly, *County Agent Story*, pp. 38–39; Burritt, *County Agent*, pp. 161–162; AAACES, *Proceedings*, 1913, p. 265.

employed but 38 agents. Among the states Indiana was the leader, with 28 agents assigned to as many counties. New York, Minnesota, and North Dakota each had 24. Pennsylvania, Illinois, Michigan, Iowa, and Missouri had 10 or more.

Impressive though the county agent movement was in the northern and western states, it compared rather poorly with accomplishments in the South. The fifteen states in which demonstration work was conducted under the general plans developed by Seaman A. Knapp had 859 county agents by June 30, 1914, or a per-state average of over eight times as many as in the North and the West.[67]

Still, by any measure, the rural reception of the county agent system suggested that farmers would demand its expansion to all parts of the United States. The advisor in the fields using the demonstration method had proven to be the teaching technique for the future. But difficult questions remained: From where might come the financial support for its universal adoption and how could the new technique be reconciled with older ones?

[67] Lisle Morrison to J. E. Jones, July 9, 1914, Bureau of Plant Industry Records; McNelly, *County Agent Story*, pp. 41–42.

XI. THE SMITH-LEVER ACT:
THE CULMINATION

SPEAKING BEFORE THE 1904 MEETING of the Association of American Agricultural Colleges and Experiment Stations, Kenyon L. Butterfield said, "To carry out the functions of the agricultural college we need . . . a vast enlargement of extension work among farmers." This work should "rank as a distinct department . . . with a faculty of men whose chief business is to teach the people who cannot come to the college." Such departments should manage farmers' institutes, cooperative experiments, reading courses, and extension lectures, he continued. They might also undertake the formation of young people's clubs, the development of agricultural education in the public schools, the distribution of various types of literature, and the handling of the multitude of inquiries from farmers that burdened experiment station personnel.[1]

Within six years a majority of the nation's land-grant colleges had acted upon Butterfield's suggestions. Extension departments had come into existence, and they had undertaken the types of off-campus instruction that the prominent educator had suggested. Farmers' institutes continued to be of great importance and in fact seemed to be growing rapidly in popularity.

Meanwhile, in the South and at scattered points elsewhere agricul-

[1] AAACES, *Proceedings,* 1904, p. 60.

tural extension had taken different forms: Demonstration agents had appeared in the fields, instructing farmers and consulting with them concerning specific problems. Youth work had been systematized, and dedicated women had begun to move from the pioneer tomato clubs toward the more sophisticated and useful concept of home demonstration work.

In fact, many authorities had concluded that the county agent and his female counterpart using the demonstration technique constituted the final answer in the long search for an effective teaching device. But there remained serious problems concerning the inauguration, organization, and financing of the work on a national basis. Some important educators, especially those at the northern colleges, were slow to grasp the significance of the type of demonstrations used by Knapp and Spillman. Moreover, arrangements would have to be worked out to unite effectively the various agencies and groups engaged in agricultural improvement without arousing dangerous jealousies and suspicions. Funds in sizeable amounts would have to be found for inauguration of a plan that envisioned one agent in every agricultural county. These issues were finally settled in the Smith-Lever Act.

By 1910, at least, there was general recognition that the establishment of a nationwide agricultural extension system would require the outlay of federal funds. In the main, agricultural educators favored such aid and, in support of their position, professed to believe that a well-financed extension program would solve most of the social and economic as well as educational problems in the countryside. Other rural spokesmen displayed little reluctance to turn to the federal government for money.[2] Meanwhile, the United States Department of Agriculture was moving ahead with its demonstration work, in both the South and the North, at such a rate as to suggest that the establishment of a national system under its auspices was but a matter of time.

[2] For instance, see statements of Dean Charles F. Curtiss of Iowa and Director Whitman H. Jordan of New York in "Agricultural Colleges and Experiment Stations," U.S. Senate Committee on Agriculture and Forestry, *Hearings on S. 4676* (1910), p. 13.

Business elements were prepared to embrace enthusiastically federal aid to agricultural extension if it included the use of teaching techniques that promised to reach every farmer in an effective way. To promote their cause, in 1911 businessmen formed the National Soil Fertility League, an organization that listed on its advisory committee such prestigious figures as William Howard Taft, William Jennings Bryan, Champ Clark, James J. Hill, and Samuel Gompers. Under the leadership of Howard H. Gross, a Chicagoan, the league became a powerful exponent of the Knapp demonstration technique. Its primary purpose came to be to "secure State and Federal legislation that will enable the agricultural colleges to place and maintain . . . experts in the respective counties to the end that demonstrations shall be carried on . . . for building up the soil." The league proposed to prepare and have introduced in Congress a bill appropriating for the program an amount increasing from $1 million to $4.5 million, those sums to be prorated among the agricultural colleges and to be matched by state appropriations. It then planned to launch an energetic campaign to insure that members of Congress and state legislatures would "hear from home." An early success came when President Taft publicly endorsed federal aid to agricultural extension at a speech in Kansas City late in 1911. Theodore Roosevelt had expressed his general approval earlier.[3]

Lesser politicians also were finding it advantageous to obtain a place on the bandwagon. In the South, especially, the popularity of Knapp's work was such that few politicians opposed it; in fact, many supported agricultural extension with an enthusiasm that transcended their knowledge of farming. One avid friend of agriculture, in the course of a speech praising Knapp and his work, talked of the time when as a farm boy he had "helped to plough the wheat and mow the corn."[4] Northern politicians were only slightly less en-

[3] AAACES, *Proceedings*, 1911, p. 205; *ibid.*, 1912, p. 102; Howard H. Gross, *The National Soil Fertility League* (Chicago, 1911), pp. 3, 11–12; Bailey, *Knapp,* p. 244.

[4] Alpha Gamma Chapter of Epsilon Sigma Phi, *Extension Work in Virginia, 1907–1940: A Brief History* (Blacksburg, Va., 1941), p. 12.

thusiastic in proclaiming their willingness to support agricultural extension. In their platform of 1912 the Democrats urged federal "appropriations for . . . extension teaching in agriculture in cooperation with the several States," while the Progressive party pledged to promote the development of "agricultural college extension."[5]

In outlining plans for federal financial assistance the Association of American Agricultural Colleges and Experiment Stations was among the first in the field. As early as 1904 a standing committee on extension had been appointed and in 1909 a section on extension had been formed within the organization giving the new function of the colleges equal status with the older resident instruction and experiment station work. The association's committee on extension, under the chairmanship of Kenyon L. Butterfield, led a struggle to place agricultural extension on a firm basis. Influenced especially by the thinking of B. H. Rawl, then a dairy expert in the United States Department of Agriculture, the committee soon concluded that the magnitude of the extension task required federal financial support. "The chief means of stimulating the proper recognition and adequate organization of extension work in agriculture . . . is a Federal appropriation for the work," said Butterfield's committee in 1908.[6]

This view squared well with that of the influential Commission on Country Life. Created by President Roosevelt in August, 1908, it included such national figures as Liberty Hyde Bailey, Henry Wallace, Kenyon L. Butterfield, Gifford Pinchot, and Walter Hines Page. The commission was charged with the task of reporting to the president "upon the present condition of country life, upon what means are now available for supplying the deficiencies which exist, and upon the best methods of organizing permanent effort" to improve rural living. The commission issued its report in January, 1909. The group found that the "incubus of ignorance and inertia

[5] Kirk H. Porter and Donald B. Johnson, *National Party Platforms, 1840–1956* (Urbana, Ill., 1956), pp. 171, 177.

[6] *Hoard's Dairyman*, XLVIII (August 21, 1914), pp. 82–83; AAACES, *Proceedings*, 1908, p. 40.

is so heavy and widespread as to constitute a national danger" and that conditions could be improved by "the establishment of a nation-wide extension work."[7]

Soon the extension committee of the Association of American Agricultural Colleges and Experiment Stations spelled out in detail the terms of the desired federal assistance. A satisfactory extension law, the committee thought, would appropriate annually to each state and territory a given amount and would offer an additional sum on a matching basis. These funds would be made available only to the agricultural colleges; state departments of agriculture and similar agencies need have no extension role. Each school receiving federal money would be required to organize and staff a separate extension department or division, giving it a status comparable to teaching and research divisions. Extension publications would have the franking privilege. Such a program, said the committee, would complete "the circle of national aid," stimulate even the poor states to begin systematic extension work, and allow all states to progress at a rate suited to their needs and capabilities.

At the 1909 meeting the association as a whole finally took positive action. Adopting the recommendations of its committee on extension, it went on record as favoring a federal appropriation and assigned to its executive committee the task of producing a draft of a bill and obtaining its introduction into Congress.[8]

As a result of these actions, Representative James C. McLaughlin of Michigan introduced an extension bill into Congress on December 15, 1909. Drafted by John Hamilton of Pennsylvania and Kenyon L. Butterfield, by then president of the Massachusetts agricultural college, the bill provided for an annual appropriation of $10,000 to each state and territory and, after two years, an additional sum not to exceed one cent per capita of its population. Such additional funds were to be available only to those states that agreed to con-

---

[7] The quotations may be found in Epsilon Sigma Phi, *Extension Work*, pp. 88, 89, 90. The commission and its work are discussed in Clayton S. Ellsworth, "Theodore Roosevelt's Country Life Commission," *Agricultural History*, XXXIV (October, 1960), pp. 155–172.

[8] AAACES, *Proceedings*, 1909, pp. 34, 37–38, 45; *ibid.*, 1910, p. 95.

tribute an equal amount. In those states having two agricultural colleges, state authorities might divide the federal appropriation or assign the entire sum to one institution. The franking privilege would be extended to the colleges for all extension materials.

Essentially, the McLaughlin bill was designed simply to encourage the establishment of extension departments in the land-grant colleges and to provide better support for existing extension departments, much as the Hatch Act of 1887 had encouraged the establishment and better support of experiment stations. The colleges receiving funds would be left free of any real control from Washington to develop and implement extension programs deemed suitable to their particular conditions, using such teaching techniques as their officers chose.[9]

The McLaughlin bill progressed no further than the House agricultural committee. Still, its reception had not been hostile and supporters believed that no serious barriers lay in the way of the ultimate passage of such a bill.

Unfortunately, the effort to obtain federal aid for agricultural extension was caught up in a broader movement to secure federal assistance for vocational education. As early as 1906 Representative Ernest M. Pollard and Senator E. J. Burkett of Nebraska introduced bills providing for the appropriation of funds to normal schools for the training of teachers in agriculture, home economics, mechanical arts, and related subjects. A year later Representative Charles R. Davis of Minnesota presented a bill that would allocate federal money to district agricultural high schools for instruction in agriculture and home economics and to the states for the establishment of branch experiment stations. A 1909 version of the bill added a provision that would help the state normal schools offer instruction in agriculture and home economics. Despite support from such diverse groups as the National Grange and the National League for Industrial Education, none of these bills was enacted into law.[10]

---

[9] *Congressional Record*, December 15, 1909, p. 180; *Breeder's Gazette*, LVII (April 6, 1910), p. 874.

[10] Roy W. Roberts, *Vocational and Practical Arts Education: History, Development, and Principles* (New York, 1965), pp. 127–128.

Supporters were not discouraged, and in 1910 vocational education found a vigorous champion in Senator Jonathan P. Dolliver of Iowa. On January 5 he introduced two bills. One was similar to the McLaughlin bill, the other was a vocational education measure. Both were referred to the Senate Committee on Agriculture and Forestry of which Dolliver was chairman.[11]

After hearings the Senate committee directed Dolliver to present a new bill combining the features of both of his proposals and reported the new measure with a recommendation favoring passage. The combined Dolliver bill provided $5 million for instruction in trades and industries, home economics, and agriculture in city high schools; $5 million for district agricultural high schools and branch experiment stations; $1.5 million for agricultural extension departments in the state agricultural colleges; $1 million for the preparation by state normal schools of vocational teachers; and $70,000 for administration. These were lump sum grants; to obtain its share each state had only to appropriate an equal amount. The Secretary of the Interior, who was given the task of administering the various programs, would exercise little effective control over the expenditure of the funds.

According to advocates the Dolliver bill would go far toward meeting existing problems in education. It would enable "the cities to establish several hundred splendidly equipped schools for the trades and industries" and would allow "the States to establish several hundred well-equipped agricultural schools" such as those at Crookston, Minnesota, and at Madison, Georgia. Moreover, the measure would help the states bring "rapidly accumulating new truths and processes" to farmers through such mediums as the "farmers' institute, the itinerant school, the demonstration farm, and the correspondence school."

With its multiple purposes the Dolliver bill did generate wide popular support. Among the groups endorsing it were the National Grange, National Farmers' Congress, Farmers' Union, American

[11] *Congressional Record,* January 5, 1910, p. 311; *Wallaces' Farmer,* XXXV (May 20, 1910), p. 798; "Agricultural Colleges and Experiment Stations," U.S. Senate Committee on Agriculture and Forestry, *Hearings on S. 4676.*

Society of Equity, American Federation of Labor, American Association of Home Economics, and National Education Association. But Congress remained unconvinced and the bill failed to win approval.[12]

Later, in October, 1910, the cause lost a warm friend with the death of Senator Dolliver. Leadership in the struggle for vocational education now fell upon Carroll S. Page of Vermont who took up the task of pushing bills similar in form to those offered by the Iowan.

Page's bills and comparable proposals were continuously before Congress from 1911 to 1913. All were copies of the Dolliver bill and all sought to unite the advocates of vocational education and of agricultural extension in support of one major piece of legislation. A version of a Page bill introduced June 14, 1912, was amended and reported favorably by the Senate Committee on Agriculture and Forestry. It provided $5 million for vocational instruction in urban high schools; $4 million for agricultural high schools; $1 million for the maintenance of "branch feed test and breeding stations" in connection with those agricultural high schools; $70,000 for administration; and $1.5 million for the training of vocational teachers in the public normal schools. In contrast to the Dolliver bill, which had provided a flat grant of $1.5 million for agricultural extension departments, the Page proposal gave $500,000 but increased that amount by $300,000 a year until the outlay reached $2.9 million annually. The matching arrangement was continued. Administration of the extension fund was shifted to the Department of Agriculture.[13]

The proposal linking vocational education with agricultural ex-

[12] *Congressional Record,* June 22, 1910, p. 8713; "Cooperation with the States in Providing Vocational Education," U.S. Senate *Report* 902 (1910), pp. 1–5. A biography of Dolliver, not very useful in this study, is Thomas R. Ross, *Jonathan Prentiss Dolliver: A Study in Political Integrity and Independence* (Iowa City, 1958).

[13] *Congressional Record,* March 3, 1911, p. 4033; *Ibid.,* April 6, 1911, p. 101; *ibid.,* April 7, 1913, p. 51; "Vocational Education," U.S. Senate *Report* 405 (1912), pp. 1–14; *Breeder's Gazette,* LXI (June 19, 1912), p. 1370. For a defense and explanation of his measure, see Carroll S. Page, "Vocational Education," U.S. Senate *Document* 845 (1912), pp. 3–66.

tension failed partly because of the bitter hostility of the Association of American Agricultural Colleges and Experiment Stations. Although some members of that group, including Dean Charles F. Curtiss of Iowa and Minnesota's Willet M. Hays (then assistant secretary of agriculture), praised the Dolliver bill and suggested that progress in vocational education was inevitable, a majority of the association's members believed that proposals for federal aid to agricultural extension and to vocational education should be separated. Speaking at the 1910 conclave of the association, Director Whitman H. Jordan of New York's Geneva station complained that the association "left in the paternal care of Congress what is known as the McLaughlin bill. It was a minor when we left it . . . and we were very much astonished later to find that it had been married without the consent of its parents. Indeed, it was a polygamous marriage . . . and I now appeal for a divorcement, not only on the grounds of illegality, but of incompatibility."

Other members of the association pointed to more specific objections to the Dolliver and Page bills: A. M. Soule of Georgia contended that federal aid to secondary schools would in due course lead to federal control of public education, Illinois' Eugene Davenport feared that the multisided Dolliver bill would divert funds from the agricultural colleges, and H. L. Russell of Wisconsin complained that such bills were not sufficiently flexible to meet the needs in all areas of the country. Several spokesmen contended that branch experiment stations were of limited value and might well curtail more important work at the main stations.

Accordingly, the association resolved to concentrate its efforts on obtaining a federal appropriation for agricultural extension. The whole matter of federal aid to secondary education was so complex that in 1910 the group did "not feel prepared to commit itself either for or against the proposition."[14] A year later the association acknowledged the need for an expansion of vocational education and agreed that in all probability a federal appropriation was essential. Still, the educators contended that federal assistance to

[14] AAACES, *Proceedings*, 1910, pp. 95–99, 100, 102–105, 120.

agricultural extension was of primary concern and they were certain that the two propositions should not be united in the same bill.[15]

The separation of agricultural extension and vocational education proposals was finally accomplished early in 1914. Sentiment for doing so was strong in Congress, and friends of vocational education agreed to abandon their demands for the prompt passage of a national education bill if Congress by joint resolution would establish a commission to study the whole subject. Congress accepted that suggestion, and on January 20, 1914, President Wilson approved the resolution that created the Commission on National Aid to Vocational Education. The settlement cleared the way for favorable consideration of an extension bill.[16]

Even while the struggle with the advocates of vocational instruction went on, a number of bills similar to the original McLaughlin proposal appeared in Congress. Among those introducing such bills in one or both sessions of the Sixty-second Congress in addition to McLaughlin were Representatives John G. McHenry of Pennsylvania, Dick T. Morgan of Oklahoma, and Henry D. Flood of Virginia and Senator Robert S. Owen of Oklahoma. In the first session of the Sixty-third Congress, Congressman Flood reintroduced his bill, while Representatives Samuel J. Tribble of Georgia and John A. Adair of Indiana presented their versions.[17] Unique among this group of bills that progressed no further than congressional committees was one first introduced on August 10, 1911, by William B. McKinley of the nineteenth district of Illinois. Drafted by a committee of the Illinois Bankers' Association headed by Benjamin F. Harris of Champaign, the measure specifically provided for the expansion of Seaman A. Knapp's work throughout the North and the West, with control of the program in the hands of the state agricultural colleges.[18]

[15] *Ibid.*, 1912, p. 131; Evans, "Recollections of Extension History," pp. 23–24.
[16] Roberts, *Vocational and Practical Arts Education*, pp. 129–130.
[17] *Congressional Record*, May 30, 1911, p. 1647; *ibid.*, February 16, 1912, p. 2146; *ibid.*, August 1, 1912, p. 10043; *ibid.*, January 12, 1912, p. 1296; *ibid.*, April 7, 1913, p. 83; *ibid.*, December 4, 1913, p. 231; *ibid.*, December 12, 1913, p. 817.
[18] *Ibid.*, August 10, 1911, p. 3819; *ibid.*, December 7, 1911, p. 106; *Breeder's Gazette*, LX (October 25, 1911), pp. 828–829.

But it was Asbury F. Lever of South Carolina who was destined to place his name on the law that would provide for a nationwide county agent system. Born in 1875 on a farm near Lexington, South Carolina, Lever was raised under less than promising circumstances. Still, he found the means to complete his education, graduating from Newberry College in 1895 and later obtaining a degree in law. After teaching school for two years, he served as secretary to Congressman J. William Stokes. When Stokes died, Lever was elected in his place, taking the seat in 1901. Lever served in Congress for eighteen years. A steady friend of agriculture, he fought the farmers' battles in Washington seeking always to improve the rural conditions that he remembered so well. He regarded the Smith-Lever Act "as the most significant and pregnant event of my Congressional career."[19]

As a new congressman, Lever learned of the boll weevil and of its impact on southern agriculture. At a meeting of the House Committee on Agriculture, of which Lever was a junior member, Representative Albert S. Burleson of Texas displayed a bottle filled with alcohol and weevils and "in his usual dramatic fashion told the Committee something of the dangers he saw lurking in these little insects in their migration across the cotton belt."

Burleson's performance produced an appropriation for the study of the pest; it also turned Lever's attention to the general area in which he would make his greatest contribution. A few years later Lever met Knapp and learned that Knapp "had not only discovered a method of growing cotton notwithstanding the presence of the boll weevil, but that . . . he was laying the foundations firmly and cautiously for a new agriculture and a new rural life in the South." Later, Lever wrote, "I became his devoted disciple. I embraced his teachings and philosophies without reserve and with the ardor and enthusiasm of youth."[20]

[19] Mrs. A. F. Lever, "Biography of A. Frank Lever," pp. 67, 102, Lever Papers; Napier, *South Carolina Agricultural Extension Workers*, p. 14.

[20] Mrs. A. F. Lever, "Biography of A. Frank Lever," pp. 96, 105, Lever Papers; A. Frank Lever, "Origins and Objects of Agricultural Extension Work." South Carolina Agricultural Extension Service, *Circular* 143 (Clemson, S.C., 1935), n.p.;

Lever introduced his first extension bill on June 12, 1911. Entitled "A bill to establish agricultural extension departments in connection with the agricultural colleges and experiment stations ...," it was similar to McLaughlin's bill and received the same treatment, dying quietly in committee.[21]

Nevertheless, Lever's increasing importance in the Democratic party and in the House, which went Democratic in the election of 1910 making him the second ranking member of the Committee on Agriculture, suggested to the friends of agricultural extension that the South Carolinian might well be their most powerful supporter. Accordingly, when representatives of the Association of American Agricultural Colleges and Experiment Stations, the United States Department of Agriculture, and the National Soil Fertility League met in the fall of 1911 to draft a bill that presumably all could support, they invited Lever to participate in their deliberations. The result was a bill, resting heavily upon the proposals of the National Soil Fertility League, that Lever introduced into Congress January 17, 1912. Senator Hoke Smith of Georgia, having been enlisted in the cause, presented the same bill in the upper house a day earlier.[22]

Reintroduced in the House April 4, 1912, the Lever bill received favorable attention. The Committee on Agriculture returned it with only minor amendments and with the recommendation that it pass. In the form approved in committee, the bill authorized agricultural colleges to establish extension departments, permitted the states to designate the institutions to administer the program, and defined the duty of the extension departments as the giving of "instruction and practical demonstrations in agriculture and home economics

Lloyd, "County Agricultural Agent Work under the Smith-Lever Act, 1914–1924," p. 2.

[21] *Congressional Record,* June 12, 1911, p. 1948.

[22] Mrs. A. F. Lever, "Biography of A. Frank Lever," p. 64, Lever Papers; *Congressional Record,* January 17, 1912, p. 1052; *ibid.,* January 16, 1912, p. 961; *Breeder's Gazette,* LXI (January 17, 1912), p. 130; F. B. Mumford to Asbury Lever, January 31, 1912, Missouri Agricultural College Papers. For the views of educators and others on the Senate version, see "To Establish Agricultural Extension Departments," U.S. Senate Committee on Agriculture and Forestry, *Hearings on S. 4563* (1912).

through field demonstrations, publications, and otherwise." The bill further provided for an annual grant of $10,000 to each state and the appropriation of an additional fund, increasing from $300,000 to $3,000,000 over a period of ten years. Again the additional funds were to be made on a matching basis and were to be allocated among the states "in the proportion which their rural population bears to the total population of the United States. . . ."

In recommending the bill for passage, the members of the agricultural committee indicated very clearly their concept of what agricultural extension should entail. "Your committee," they wrote, "is informed and believes that [Knapp's] system brings home to the actual farmer upon his actual farm the best methods of agriculture. . . . The proposed legislation intends to do this same kind of work on a bigger, broader, and better scale, under the direction of State rather than Federal authorities. . . ." A significant amendment added in committee provided that "75 percent of all moneys available under this act shall be expended . . . for field instruction and demonstrations."[23]

The bill was debated in the House in August, 1912. There southern congressmen added an amendment designed to insure that passage of the bill would not interfere with the existing Farmers' Cooperative Demonstration Work, and on August 23 the lower house approved it and sent it to the Senate.[24]

That the bill had wide support among rural educators and country spokesmen was unquestioned. Influential farm papers praised it, and agricultural college officials urged its passage. Writing to one of his senators, Frederick B. Mumford of the University of Missouri pointed out that the "College of Agriculture is overwhelmed with demands for demonstration work, lectures at Farmers' Institutes,

[23] *Congressional Record*, April 4, 1912, p. 4318; *ibid.*, April 13, 1912, p. 4764; "Establishment of Agricultural Extension Departments," U.S. House *Report* 546 (1912), pp. 1–7. The hearings in the House are in "Agricultural Extension Departments," U.S. House of Representatives, Committee on Agriculture, *Hearings on H. R. 18160* (1912).

[24] *Congressional Record*, August 23, 1912, p. 11743; *ibid.*, August 24, 1912, p. 11770; "Establishment of Agricultural Extension Departments," U.S. Senate *Report* 1072 (1912), pp. 1–7.

publications and other means of extending agricultural knowledge to the farmers. If the Agricultural Extension Bill is approved . . . it will provide the means for us to meet this demand." Meanwhile, in a memorial to Congress dated November 14, 1912, the Association of American Agricultural Colleges and Experiment Stations gave its wholehearted support.[25]

Unfortunately, the Lever bill floundered in the Senate on the rock of the vocational education issue. On January 29, 1913, Senator Page induced the upper house by a vote of 31 to 30 to substitute his measure for the Lever bill. In conference the House refused to accept the multisided Page bill, with its much larger appropriations, and hopes died for an extension bill in the Sixty-second Congress.[26]

A renewed effort was certain in the Sixty-third Congress. Supporters of agricultural extension were justifiably encouraged by events. The Lever bill had passed the House by a large margin, suggesting the wide popular support for the general concept of extension. Moreover, there was a general feeling that the vocational education forces had damaged their chances by their tactics in the Sixty-second Congress. Senator Page and his followers had "tried to use extension as a horse to get themselves across the stream" and had broken "the horse's back," causing several representatives and senators who had supported both propositions to conclude that extension should be considered first. Finally, in the new Congress the Senate as well as the House was Democratic, and it was felt that the upper house might be more willing to consider extension and vocational education separately, since both bills would now carry the names of Democrats.[27]

[25] *Breeder's Gazette*, LXI (May 22, 1912), pp. 1200, 1202; *Country Gentleman*, LXXVII (November 30, 1912), p. 10; F. B. Mumford to J. A. Reed, December 17 and 23, 1912, Missouri Agricultural College Papers; AAACES, *Proceedings*, 1912, pp. 99–101.

[26] *Country Gentleman*, LXXVIII (February 22, 1913), p. 261; *Congressional Record*, January 29, 1913, p. 2228. While Georgia's Hoke Smith was a powerful advocate of agricultural extension in the Senate, he played only a minor role in shaping the various bills in Congress. See Dewey W. Grantham, Jr., *Hoke Smith and the Politics of the New South* (Baton Rouge, 1958), pp. 256–264.

[27] *Country Gentleman*, LXXVIII (March 29, 1913), p. 493; *ibid.* (April 9, 1913), p. 537.

Prospects appeared especially bright in the spring of 1913. Asbury F. Lever had become chairman of the House Committee on Agriculture and Hoke Smith held a position on the Senate Committee on Agriculture and Forestry as well as the chairmanship of the Committee on Education and Labor. It seemed certain that agricultural extension would receive an early and favorable hearing. On April 7, 1913, in fact, Lever and Smith introduced bills in the House and Senate that were practically identical with the one which had passed the House in August, 1912.[28]

But problems of a serious nature remained even at that late date. By 1913 Farmers' Cooperative Demonstration Work had become a fixture in southern agriculture. Meanwhile, Spillman's work in the North was receiving wide attention and favorable comment from farmers, businessmen, and others. In both the North and the South, observers were aware that these popular programs had been launched and carried forward by the Department of Agriculture, often with little encouragement and less understanding from the agricultural colleges. Increasingly, those who favored the type of work carried on in the South by Bradford Knapp and in the North by Spillman feared that passage of the Smith-Lever bills as they were originally drawn might well mean the end of those programs or their subordination to other types of extension teaching conducted by the colleges. At the least, the enactment of the bills in their original form might lead to the creation of two extension systems, one managed by the states, the other by the Department of Agriculture, with resulting confusion, inefficiency, and waste.

At the bottom of the problem lay a long-standing disagreement between the Department of Agriculture and the agricultural colleges. That controversy stemmed from a number of factors including jealousy, fear of federal domination, and a difference of opinion concerning the place and role of the county agent in an extension program.

The latter was perhaps most important; certainly, it contributed

[28] *Congressional Record,* April 7, 1913, pp. 52, 85; AAACES, *Proceedings,* 1915, p. 42; *Progressive Farmer and Southern Farm Gazette,* XXVIII (July 26, 1913), p. 810.

to academic resentment of the Department of Agriculture and fear of control from Washington. On the other hand, the Department of Agriculture was convinced that the colleges' extension programs, using institutes, publications, itinerant lecturers, and similar techniques, were totally ineffectual and that they stood in the way of the inauguration of a nationwide county agent system. To men in Washington, Knapp had proven beyond any doubt the value of the demonstration and county agent systems; Spillman was in the process of proving it in the North and the West.

It was perfectly obvious that the colleges, particularly the northern institutions, were slow to develop an interest in the county agent system as a teaching device. The Association of American Agricultural Colleges and Experiment Stations ignored Knapp's work during its early years; Dean Charles F. Curtiss of Iowa claimed that he was unaware of its existence until 1910, and two years later Frederick B. Mumford of Missouri could profess to believe that the Department of Agriculture had no significant extension programs in operation. Moreover, the association consistently indicated that it believed that the more traditional forms of instruction held out the greatest promise. In 1910 the association's extension committee stated flatly that emphasis should be placed on those types of work that represented "systematic or formal teaching," including lecture courses, reading and correspondence courses, movable schools, college-managed demonstration plots or farms, and study clubs. "In our judgement this is the great permanent work of the extension department," proclaimed the academicians. According to their view of the matter, less formal methods, such as conventions and conferences, institutes, traveling advisors or field agents, and publications, had their place, primarily to reach a wider group of farmers and to arouse interest to a point at which the "thoroughly scientific and pedagogically sound" methods might be used effectively.[29]

Nor did the association hasten to change its views, despite the

[29] "Agricultural Colleges and Experiment Stations," U.S. Senate Committee on Agriculture and Forestry, *Hearings on S. 4676*, p. 5; AAACES, *Proceedings*, 1910,

growing popularity of county agent work. In 1911 its spokesmen complained that there was "an increasing number of requests for personal inspection of farms by experts" and that if "the calls are answered, after a time the burden of such service will become enormous." The association's extension committee could only suggest "the principle that the individual is to be helped as a part of the help rendered to the community" while proposing rather vaguely that the "efforts of permanent field agents" might be necessary to reach the "backward farmer." The next year the committee observed that "our concept of extension" was still evolving, but the county agent had little place in its scheme of things. Instead, the committee thought that a model extension department might consist of a central office and a staff of specialists to act somewhat "in the capacity of a flying squadron and assist in the organization and supervision of such lines of work as movable schools and boys' and girls' clubs." Meanwhile, the association's executive committee cautiously expressed the view that the county agent technique might be of some value, assuming that it was used with "certain well-defined limitations."

Even a year before the Smith-Lever Act became law, academic men remained unclear concerning the nature and significance of county agent work. The fact that it was most in demand among farmers was now penetrating the thinking of the scholars, and the association's committee on extension admitted that in the end the demonstration technique might well be "most permanent and beneficial, if properly safeguarded while the work is new." But still some traditional educators considered county agent teaching as superficial, while others continued to confuse demonstrations with controlled experiments. Finally and perhaps most important, many agricultural educators failed to understand that through a system of agents in the field using demonstrations they could reach the mass of the nation's farmers, rather than the few who constituted a form of rural elite. In fact, as late as 1911, Dean Eugene Davenport of

pp. 83–84; F. B. Mumford to Asbury Lever, January 31, 1912, Missouri Agricultural College Papers.

Illinois stated flatly that "I do not agree with the proposition that the colleges or any other public agency should now, or ever, take the message to every individual. . . ."[30]

If the academic men had a narrow view of the importance of county agent work, they displayed no hesitancy in proclaiming the colleges as the primary agencies in extension work and in downgrading the roles of other institutions. They were convinced that state boards and departments of agriculture should limit themselves to regulatory and administrative duties. "The farmer's institute . . . will always have its place" but in every case, they felt, these meetings should be managed by the colleges as a part of their extension programs. Neither the state normal schools nor the agricultural high schools should have an extension role, except in those instances in which such institutions might serve as local centers for college-managed extension programs. So far as the United States Department of Agriculture was concerned, the college men were certain that the federal agency should abandon all extension teaching, except for the distribution of what they labeled as "popular literature." Nor did many academicians acknowledge the right of the Department of Agriculture to supervise the expenditure of funds appropriated by Congress for agriculture. In the final analysis, wrote Cornell's Liberty Hyde Bailey, such money "belongs to the people, so there is nothing improper about other agencies, state or local, spending it."[31]

The emergence of the Office of Farm Management as an extension agency especially challenged the northern academicians' view of the proper function of the Department of Agriculture. With few exceptions, the college men praised the investigative work of the Office and joined willingly with it in cooperative projects of that type. But when Spillman formulated his plans to place agents in the countryside to demonstrate improved methods to farmers, north-

[30] AAACES, *Proceedings*, 1911, pp. 36, 78; *ibid.*, 1912, pp. 60, 64; *ibid.*, 1913, pp. 96–99, 152, 265; Lloyd, "County Agricultural Agent Work under the Smith-Lever Act, 1914–1924," p. 1.

[31] AAACES, *Proceedings*, 1911, pp. 34–35; L. H. Bailey to Eugene Davenport, September 29, 1911, Bailey Papers.

ern college men were almost unanimous in denouncing the program as an invasion of a field belonging rightfully to them. Such a plan, they maintained, was certain to force an unhealthy uniformity upon the different states, might well harm the colleges by rendering them less responsive to local needs or by confusing the people as to the sources of aid, and could open the door to political control of extension. Some of the more outspoken college leaders were prepared to accuse Spillman of empire building and to suggest that his plans for the North and the West stemmed from little more than his personal ambition. The fact that Congress voted sharp increases in appropriations for the support of Spillman's work in fiscal 1913 and 1914 did nothing to settle the minds of nervous college leaders.[32]

By the spring of 1913, officials in the Department of Agriculture were convinced that these issues would have to be settled before a nationwide extension program could be instituted. The type of extension bill that had appeared in Congress during the preceding three years was now deemed unsatisfactory. Essentially, these proposals had provided only that extension departments would be established in conjunction with the agricultural colleges. Although federal money was to be used in substantial amounts, the character of the teaching to be performed by these departments was left to the colleges themselves. Admittedly, several of the bills had included or had been amended to include provisions specifying that a percentage of all federal appropriations would be used for demonstrations and that existing Department of Agriculture extension programs would be continued, but ultimate control of the work would still rest with the colleges. Federal officials now declined to accept this, since they were far from convinced that the academicians were dedicated to the use of those techniques that experience clearly showed were most effective.

On May 16, 1913, Secretary of Agriculture David F. Houston met with the executive committee of the Association of American Agricultural Colleges and Experiment Stations to iron out differ-

[32] AAACES, *Proceedings*, 1912, pp. 135–140, 142; *ibid.*, 1913, p. 127; *U.S. Statutes at Large*, XXXVII, part I, pp. 277, 836; *Country Gentleman*, LXXVIII (February 22, 1913), p. 261.

ences. Houston was particularly well qualified to arrange a compromise. A former agricultural college president, he well understood
the fears and prejudices of the academicians. Both sides agreed that
the nation should have only one extension system. Houston contended that his department had not sought an extension role; the
task had been thrust upon it by obvious need and circumstances.
Quite clearly, he said, the department had a responsibility to the
taxpayers to see that the information gathered by its researchers was
utilized by the people, suggesting that some interest in extension was
inescapable. Still, he recognized that the colleges were the logical
agencies to manage extension programs. They were themselves res-
ervoirs of agricultural knowledge and they were closer to local problems and to actual farmers than the department could hope to be.
To Houston, the answer was clear; extension should be a cooperative
enterprise.

Working on these premises, the conclave produced agreement on
a number of basic issues. First, the absolute necessity of a federal
appropriation was reiterated and accepted by all concerned. The
secretary agreed that agricultural extension should be administered
under the immediate direction of the agricultural college. The college men accepted the principle that those extension projects financed by federal funds would be mutually agreed to by the college
involved and by the Department of Agriculture. Houston agreed
that no cooperative arrangement for maintaining extension work
would be made by the department with any corporation or commercial body, except those that might wish to donate funds to be
used in projects administered by the colleges in consultation with
the department. In summary, the college men were offered a federal
appropriation, which all of them recognized they had to have, in
return for their accepting a federal role in extension that would
insure the continuation and universal adoption of demonstration as
the basic teaching technique.[33]

To meet the terms of this compromise, Lever introduced another
bill on September 6, 1913. A bill "to provide for cooperative agri-

[33] "Cooperative Agricultural Extension Work," U.S. House *Report* 110 (1913),
pp. 9–10; *Hoard's Dairyman,* XLVI (October 31, 1913), p. 389.

cultural extension work between the agricultural colleges and the United States Department of Agriculture,"[34] it differed from the previous proposal only in the particulars necessary to satisfy those who feared for the future of the demonstration technique. "Cooperative" was the key word. In hearings on the bill, Assistant Secretary of Agriculture Galloway pointed out that all work conducted under the terms of the measure would be mutually agreed to by both parties. There would be no effort to bind the colleges to any specific type of work, but the implication was clear that the Department of Agriculture would require that the major emphasis be placed on demonstration. Moreover, Galloway assured those in attendance that the existing programs of the department would be fitted into the new scheme of things.

When Secretary Houston appeared before the committee, he pointed to the characteristics of farmers that made demonstration the essential teaching technique and praised the proposed legislation as the best possible arrangement. "A farmer," he said, "is rather prejudiced; he is conservative and rather hard-headed. He is a man of sense and wants to be shown, and he is skeptical until he is shown." The bill would provide an educational system that would overcome these handicaps to learning, eliminate much duplication of effort, and give all bureaus of the Department of Agriculture, not just the Bureau of Plant Industry, a means of reaching farmers. He rejected the proposal of some that at least a portion of the extension funds be made available to the state departments or boards of agriculture, saying that only the colleges were equipped to manage the work.[35]

The bill met little opposition in the House. On December 8, 1913, the Committee on Agriculture presented its favorable report. It summarized again those conditions that made passage of the measure imperative. According to the report, sufficient "information has been gathered and is awaiting distribution to revolutionize rural conditions . . . in the next ten years, but it is dead information until it becomes vitalized by the service to which the farmer puts it." Point-

[34] *Congressional Record,* September 6, 1913, p. 4414.

[35] "Lever Agricultural Extension Bill," U.S. House Committee on Agriculture, *Hearings on H. R. 7951* (1913), pp. 3–43. The quotation is from p. 33.

ing out that over the years various teaching techniques had been tried, the committee's statement noted that because "the farmer is naturally conservative" he could be "taught newer methods only by personal appeal and ocular demonstration." The bill, said the report, would provide machinery to carry "to the farm the approved methods and practices of the agricultural colleges, experiment stations, the Department of Agriculture, and the best farmers . . ." and to demonstrate "their value under the immediate environment of the farm itself." In short, the extension system envisioned by the bill would be the "connecting link between the sources of information . . . and the people. . . ."[36]

The full House debated the Lever bill on January 19, 1914, and quickly passed it by a vote of 177 to 9. The proposal of some members that the funds appropriated should be distributed on the basis of total rather than rural population of each state was rejected. Similar treatment awaited the suggestion of a few congressmen that the individual states could very well support their own extension systems. Members who admired the work of the Bureau of Plant Industry were sufficiently numerous to amend the bill to provide that "nothing in this Act should be construed to discontinue either the farm management work or the farmers' cooperative demonstration work . . . ," although Lever contended that such a stipulation was no longer needed.[37]

Meanwhile, the Smith bill was progressing through the legislative machinery in the Senate. In introducing his bill, Hoke Smith stated that its object was "to bring more completely into harmony the Department of Agriculture and the colleges . . . for performing demonstration work." The Committee on Agriculture and Forestry reported the bill favorably, but after the House passed the Lever bill Smith moved and the Senate agreed to drop his measure and to substitute the House version.[38]

[36] "Cooperative Agricultural Extension Work," U.S. House *Report* 110, pp. 1–3.

[37] *Congressional Record,* January 19, 1914, pp. 1932–1947; *American Fertilizer,* XL (January 24, 1914), p. 29; *Progressive Farmer and Southern Farm Gazette,* XXIX (February 21, 1914), p. 244.

[38] *Congressional Record,* September 6, 1913, p. 4330; *ibid.,* January 28, 1914,

The Senate considered the Lever bill between January 28 and February 7. The race issue entered the debate when some northern senators, reacting in part to resolutions from the National Association for the Advancement of Colored People, sought unsuccessfully to amend the bill to provide for a division of the funds in southern states between the Negro and white colleges. The southern position essentially was that Negro farmers could best be instructed by white agents because of the shortage of trained Negro agriculturists and because ordinary Negro farmers were more likely to accept directions from white agents. Moreover, they pointed out correctly that from its inception Knapp's work had involved and benefitted Negroes and, in effect, had functioned without obvious discrimination. Finally, southerners maintained flatly that if the funds were divided, their states would reject the program in its entirety.

Another controversy involved the method of allocating the funds among the states. Senator Cummins of Iowa was the spokesman for a group of senators from the Middle West who asked that the distribution be on the basis of improved farm land, an arrangement that would materially increase the share of the total outlay allocated to well-developed states. In the course of a speech by Cummins on that subject, Senator Vardaman of Mississippi observed that the proposed amendment "would make it a matter of fertilizing acres instead of enhancing the mentality of men." That change also was rejected.[39]

Passage in the Senate came February 7. Since the Senate version carried a number of minor amendments, the measure was in conference until early in May when Congress accepted the bill in its final form. President Wilson signed it on May 8, 1914.[40]

As approved by the president, the Smith-Lever Act differed but

---

pp. 2426–2427; "Providing for Cooperative Agricultural Extension Work," U.S. Senate *Report* 139 (1913), p. 1.

[39] *Congressional Record*, January 30, 1914, pp. 2748–2750; *ibid.*, February 5, 1914, pp. 3063, 3068; *ibid.*, February 6, 1914, p. 3146; *ibid.*, February 7, 1914, p. 3242; *ibid.*, February 2, 1914, pp. 2850–2861, 2900.

[40] *Ibid.*, February 7, p. 3130; *ibid.*, May 4, 1914 pp. 7658, 7691; *ibid.*, May 16, 1914, p. 8719; Mrs. A. F. Lever, "Biography of A. Frank Lever," p. 114, Lever Papers.

little from the bills that had been introduced eight months earlier. Its purpose, the law stated, was to "aid in diffusing among the people . . . useful and practical information on subjects relating to agriculture and home economics and to encourage the application of the same. . . ." The measure stated clearly that the work was to be cooperative in nature between the Department of Agriculture and the colleges and that no federal money would be "available to any college" until "plans for the work to be carried on . . . shall be . . . approved by the Secretary of Agriculture." In those states having more than one land-grant college, the legislatures were empowered to designate the institution to manage the program. The federal contribution was set at $10,000 for each state, plus an additional amount increasing over a period of seven years from $600,000 to $4,100,000, that sum to be distributed among the states on the basis of their rural population and to be matched by "State, county, college, local authority, or individual contributions from within the State. . . ."

The new law described in a general way the types of instruction to be offered and listed a number of teaching techniques that could not be maintained with federal funds. The work, said the law, "shall consist of the giving of instruction and practical demonstrations in agriculture and home economics . . . through field demonstrations, publications, and otherwise. . . ." Pending inauguration of the plan envisioned by the law, the act specifically provided that both Farmers' Cooperative Demonstration Work and farm management work would be continued. No portion of the federal funds could be expended for the promotion of educational trains "or for any purpose not specified" and no more than 5 percent could be used for publications and their distribution.[41]

The language of the Smith-Lever Act left a number of important issues to interpretation and to experience. Titles of extension workers had to be changed, and for years there was controversy concerning the proper title for local agents. A multitude of other details remained to be worked out. Relationships between the Department

[41] *U.S. Statutes at Large*, XXXVIII, part II, pp. 372–374; Houston, *Eight Years with Wilson's Cabinet*, I, 204.

of Agriculture, the colleges, and those groups that in time would become the American Farm Bureau Federation had to be clarified. The first step was the drawing up of a memorandum of understanding between each agricultural college and the Department of Agriculture. The organization of the department had to be modified to provide for more efficient administration. Soon the States' Relation Service came into existence to manage the department's extension programs. Until 1921 it consisted of two offices, one for the North and the West and the other for the South, reflecting the different origins, standards, and characteristics of the work conducted in those regions.[42]

Some old institutions, techniques, and roles were lost as the new system was inaugurated. Farmers' institutes tended to disappear in the new scheme of things. The Department of Agriculture indicated its general belief that they were ineffective, an opinion that was widely shared, and with few exceptions they had faded from the scene by the World War years. Agricultural trains enjoyed a resurgence in the 1920s, despite the prohibition against the use of federal money in their operation.[43] Meanwhile, the concentration of extension activities at the colleges hastened the decline of the educational role of the state boards or departments of agriculture; by the 1920s they were almost exclusively regulatory and administrative agencies.[44]

Establishment of a nationwide extension system with federal support brought to an end the role of the General Education Board, thus terminating certainly one of the most important programs in the history of American philanthropy. By 1914 the board had been under attack for several months. Some agricultural educators always

[42]. A. C. True to P. S. Spence, December 1, 1914, Office of Experiment Stations Records; McNelly, *County Agent Story*, p. 28; Baker, *County Agent*, pp. 38–39; *Hoard's Dairyman*, XLVII (June 19, 1914), p. 757.

[43] AAACES, *Proceedings*, 1913, p. 264; *Breeder's Gazette*, LXVI (November 19, 1914), pp. 880–881; A. C. True to D. F. Houston, November 5, 1914, Office of Experiment Stations Records; C. Clyde Jones, "The Burlington Railroad and Agricultural Policy in the 1920's," *Agricultural History*, XXXI (October, 1957), pp. 67–74.

[44] See, for instance, Scott, "Career of Samuel M. Jordan," pp. 256–257.

resented its work, and that resentment grew among northerners when the board began its demonstration program in New Hampshire and Maine. Moreover, in 1913 the board had made available $37,-000 for a program to study and improve rural marketing procedures. The project proved to be poorly managed, and it produced widespread misunderstanding in some government and academic circles. Finally, the inept handling of the Colorado Fuel and Iron Company strike engendered additional hostility to Rockefeller interests. The provision in the Smith-Lever Act that limited private financial support of extension to "contributions from within the State" was aimed squarely at the General Education Board and effectively ended its participation in agricultural extension as of June 30, 1914.[45] By that time, of course, the board had achieved the primary objective for which it and other pioneers in the field had labored for years.

Almost three decades after 1914, one educator said that the Smith-Lever Act was the most statesmanlike piece of legislation ever drawn for the benefit of agriculture. Some might argue that such an evaluation was an exaggeration; but there could be no doubt that the measure marked the end of an era and the beginning of a new age in American agriculture and economic development. For more than a century men had sought methods for educating the mass of farmers; by trial and error they had now found it.

[45] Fosdick, *General Education Board*, p. 57; "Rockefeller Foundation," U.S. Senate *Document* 538 (1914), pp. 1–5.

BIBLIOGRAPHY

I. Primary Materials

A. MANUSCRIPTS

Liberty Hyde Bailey Papers (Cornell University Library, Ithaca, N. Y.).

Bureau of Plant Industry Records (National Archives, Washington).

Champaign Grange, Proceedings (Illinois Historical Survey, Champaign).

Clemson University History Collection (Clemson University Library, Clemson, S. C.).

The Development of Agricultural Extension Work in Minnesota, manuscript dated December 14, 1937 (University of Minnesota Institute of Agriculture Archives, St. Paul).

Executive Committee of the University of Minnesota Board of Regents, Minutes (University of Minnesota Archives, Minneapolis).

Extension Service Records (National Archives, Washington).

William Watts Folwell Papers (University of Minnesota Archives, Minneapolis).

J. C. Hardy Correspondence (Mississippi State University Library, State College).

P. G. Holden Memoirs (Michigan State University Library, East Lansing).

P. G. Holden Papers (Michigan State University Library, East Lansing).

Robert R. Hudelson and Anna C. Glover, History of Agricultural Education of Less than College Grade: Contributions of the University of Illinois, manuscript (University of Illinois Library, Urbana).

R. C. King Letterbooks (Mississippi State University Library, State College).

Asbury F. Lever Papers (Clemson University Library, Clemson, S.C.).

Mississippi Agricultural and Mechanical College, President's Report to the Board of Trustees, June 3, 1907, manuscript (Department of History, Mississippi State University, State College).

Mississippi Extension Collection (Mississippi State University Library, State College).

Missouri Agricultural College Papers (University of Missouri Library, Columbia).

Frederick B. Mumford Papers (University of Missouri Library, Columbia).

Office of Experiment Stations Records (National Archives, Washington).

Records of the Missouri Agricultural Extension Service (Missouri College of Agriculture, Columbia).

Records of the Office of the Secretary of Agriculture (National Archives, Washington).

M. L. Sanford Papers (University of Minnesota Archives, Minneapolis).

Juniata L. Shepperd Papers (University of Minnesota Archives, Minneapolis).

Hoke Smith Papers (University of Georgia Library, Athens).

W. C. Stubbs Papers (Louisiana State University Library, Baton Rouge).

James E. Tanner Papers (Mississippi State University Library, State College).

Jackson E. Towne, Perry Greeley Holden, 1865–1959, paper read before Michigan Historical Society, October 8, 1960. Copy provided by the author.

B. GOVERNMENT DOCUMENTS

"Agricultural Colleges and Experiment Stations," U.S. Senate Committee on Agriculture and Forestry, Hearings on S. 4676 (1910).

"Agricultural Extension Departments," U.S. House of Representatives, Committee on Agriculture, Hearings on H. R. 18160 (1912).

Alabama Agricultural Experiment Station, Circular 12 (Auburn, 1911).

Aull, G. H., "The South Carolina Agricultural Experiment Station: A

316

Brief History, 1887–1930," South Carolina Agricultural Experiment Station, *Circular* 44 (Clemson, S. C., 1931).

Bailey, Liberty Hyde, "Farmers' Institutes: History and Status in the United States and Canada," United States Office of Experiment Stations, *Bulletin* 79 (Washington, 1900).

———, "Farmers' Reading Courses," United States Office of Experiment Stations, *Bulletin* 72 (Washington, 1899).

California Agricultural Experiment Station, *Circular* 112 (Berkeley, 1914).

Call, Leland E., "Agricultural Research at Kansas State Agricultural College before the Enactment of the Hatch Act," Kansas Agricultural Experiment Station, *Bulletin* 441 (Manhattan, 1961).

Congressional Record, 1885–1914.

"Cooperation with the States in Providing Vocational Education," U.S. Senate *Report* 902 (1910).

"Cooperative Agricultural Extension Work," U.S. House *Report* 110 (1913).

Cornell University Agricultural Experiment Station, *Bulletin* 110, 159 (Ithaca, N.Y., 1896, 1899).

Crosby, Dick J., "Boys' Agricultural Clubs," United States Department of Agriculture, *Yearbook,* 1904 (Washington, 1905), pp. 489–496.

Edwards, Everett E., "Jefferson and Agriculture," United States Department of Agriculture, *Agricultural History Series* 7 (Washington, 1943).

"To Establish Agricultural Extension Departments," U.S. Senate Committee on Agriculture and Forestry, *Hearings on S. 4563* (1912).

"Establishment of Agricultural Extension Departments," U.S. House *Report* 546 (1912).

"Establishment of Agricultural Extension Departments," U.S. Senate *Report* 1072 (1912).

Evans, J. A., "Recollections of Extension History," North Carolina Extension Service, *Circular* 224 (Raleigh, 1938).

Experiment Station Record.

"Farmers' Institute Bulletin, 1902, 1906, 1907 and 1908," Mississippi Agricultural Experiment Station, *Bulletin* 80, 100, 120 (Agricultural College, 1903, 1906, 1908).

"Farmers' Institutes in Georgia," University of Georgia, *Bulletin* IX (Athens, 1909).

317

"Fifty Years of Cooperative Extension in Wisconsin, 1912–1962," Wisconsin Agricultural Extension Service, *Circular* 602 (Madison, 1962).

Galloway, Beverly T., "Work of the Bureau of Plant Industry in Meeting the Ravages of the Boll Weevil and Some Diseases of Cotton," United States Department of Agriculture, *Yearbook*, 1904 (Washington, 1905), pp. 497–508.

"General Education Board," U.S. Senate *Document* 453 (1910).

Goodrich, C. L., "A Profitable Cotton Farm," United States Department of Agriculture, *Farmers' Bulletin* 364 (Washington, 1909).

Hamilton, John, ed., "College Extension in Agriculture: Discussions before the Graduate School of Agriculture, at the Iowa State College, Ames, Iowa, July 4–27, 1910," United States Office of Experiment Stations, *Bulletin* 231 (Washington, 1910).

———, "Farmers' Institutes and Agricultural Extension Work in the United States in 1913," United States Department of Agriculture, *Bulletin* 83 (Washington, 1914).

———, "Farmers' Institutes for Women," United States Office of Experiment Stations, *Circular* 85 (Washington, 1909).

———, "History of the Farmers' Institutes in the United States," United States Office of Experiment Stations, *Bulletin* 174 (Washington, 1906).

———, "Progress in Agricultural Education Extension," United States Office of Experiment Stations, *Circular* 98 (Washington, 1910).

———, and J. M. Stedman, "Farmers' Institutes for Young People," United States Office of Experiment Stations, *Circular* 99 (Washington, 1910).

Historical Statistics of the United States, 1789–1945 (Washington, 1949).

"History of Cooperative Extension Work in Michigan, 1914–1939," Michigan State College, Extension Division, *Extension Bulletin* 229 (East Lansing, 1941).

Howe, F. W., "Boys' and Girls' Agricultural Clubs," United States Department of Agriculture, *Farmers' Bulletin* 385 (Washington, 1910).

Howe, John D., "The New York Agricultural Society—Its History and Objects," New York State Department of Farms and Markets, *Bulletin* 161 (Albany, 1924).

Idaho State Farmers' Institutes, *Year Book*, 1901–02 (Moscow, n.d.).

318

Illinois Farmers' Institute, *Annual Report*, 1897–1914 (Springfield, 1897–1914).

Illinois Industrial University, *Annual Report of the Board of Trustees*, 1868–76 (Springfield, 1868–77).

Illinois State Board of Agriculture, *Transactions*, 1872 (Springfield, 1873).

State of Indiana, *Report on Farmers' Institutes for the Year 1896* (Lafayette, 1896).

Iowa State Agricultural College, *Fourth Biennial Report*, 1871 (Des Moines, 1872).

Kansas State Agricultural College, *Biennial Report*, 1868–94 (Topeka, 1868–95).

Kansas State Board of Agriculture, *Biennial Report*, 1913–14 (Topeka, 1915).

Knapp, Seaman A., "Demonstration Work in Cooperation with Southern Farmers," United States Department of Agriculture, *Farmers' Bulletin* 319 (Washington, 1908).

————, "Demonstration Work on Southern Farms," United States Department of Agriculture, *Farmers' Bulletin* 422 (Washington, 1910).

————, "The Farmers' Cooperative Demonstration Work," United States Department of Agriculture, *Yearbook,* 1909 (Washington, 1910), pp. 153–160.

————, "Farmers' Cooperative Demonstration Work in Its Relation to Rural Improvement," United States Bureau of Plant Industry, *Circular* 21 (Washington, 1908).

————, "The Work of the Community Demonstration Farm at Terrell, Texas," United States Bureau of Plant Industry, *Bulletin* 51 (Washington, 1905).

Lever, A. Frank, "Origins and Objectives of Agricultural Extension Work," South Carolina Agricultural Extension Service, *Circular* 143 (Clemson, 1935).

"Lever Agricultural Extension Bill," U.S. House Committee on Agriculture, *Hearings on H. R. 7951* (1913).

Lloyd, William A., "County Agricultural Agent Work under the Smith-Lever Act, 1914–1924," United States Department of Agriculture, *Miscellaneous Circular* 59 (Washington, 1925).

Louisiana Farmers' Institute, *Bulletin* 1–9 (Baton Rouge, 1898–1906).

Louisiana State Agricultural Society, *Proceedings*, 1897–1909 (Baton Rouge, 1898–1910).

Louisiana State Board of Agriculture and Immigration, *Annual Report,* 1903, 1908–09 (Baton Rouge, 1904–10).

Mercier, William B., "Extension Work among Negroes, 1920," United States Department of Agriculture, *Department Circular* 190 (Washington, 1921).

———, "Status and Results of Extension Work in the Southern States, 1903–1921," United States Department of Agriculture, *Department Circular* 248 (Washington, 1922).

Michigan Agricultural College, *President's Report,* 1898 (Lansing, 1898).

Michigan State Board of Agriculture, *Fourteenth Annual Report,* 1875 (Lansing, 1876).

Michigan State Farmers' Institutes, *Institute Bulletin* 12 (Agricultural College, 1906).

Minnesota Agricultural Experiment Station, *Annual Report,* 1907–08, 1912–13 (n.p., 1908, 1914).

University of Minnesota Board of Regents, *Fourth Biennial Report, Supplement I* (St. Paul, 1887).

Minnesota Farmers' Institutes, *Annual* 1–30 (St. Paul, 1888–1917).

Minnesota, *General Laws,* 1887 (Minneapolis, 1887).

Minnesota State Horticultural Society, *Annual Report,* 1888–1914 (St. Paul, 1889–1914).

Mississippi Agricultural and Mechanical College, *Annual Catalogue,* 1880–1914 (Jackson, 1880–1914).

Mississippi Agricultural and Mechanical College, *Report,* 1883–1915 (Jackson, 1883–1915).

Missouri State Board of Agriculture, *Annual Report,* 1869–1914 (Jefferson City, 1870–1914).

———, *Monthly Bulletin,* I–XII (1901–14).

Montana Farmers' Institutes, *Annual Report,* 1902–14 (Helena, 1902–14).

Mumford, F. B., "History of the Missouri College of Agriculture," Missouri Agricultural Experiment Station, *Bulletin* 483 (Columbia, 1944).

Napier, J. M., *Guide and Suggestions for South Carolina Agricultural Extension Workers* (Clemson, S. C., 1950).

Nebraska Farmers' Institute, *First Report* (Lincoln, 1906).

New York Bureau of Farmers' Institutes, *Report,* 1908–09 (Albany, 1910).

New York Farmers' Institute and Normal Institutes, *Report*, 1906 (Albany, 1907).

North Dakota Agricultural Experiment Station, *Seventh Annual Report*, 1907 (Fargo, 1907).

————, *Bulletin* 104 (Fargo, 1913).

North Dakota Farmers' Institute, *Annual*, 1900–05 (Fargo, 1900–05).

Page, Carroll S., "Vocational Education," U.S. Senate *Document* 845 (1912).

Pennsylvania Agricultural College, *The Agricultural College of Pennsylvania* (Philadelphia, 1862).

Pennsylvania Department of Agriculture, *Annual Report*, 1904 (Harrisburg, 1905).

Pennsylvania Farmers' Annual Normal Institute, *Proceedings*, 1913 (Harrisburg, 1913).

"Providing for Cooperative Agricultural Extension Work," U.S. Senate *Report* 139 (1913).

Purdue University, *Twenty-Second Report*, 1896 (Indianapolis, 1897).

"Rockefeller Foundation," U.S. Senate *Document* 538 (1914).

Simons, Lloyd R., "New York State's Contribution to the Organization and Development of the County Agent-Farm Bureau Movement," Cornell Agricultural Extension, *Bulletin* 993 (Ithaca, N. Y., 1957).

Smith, C. Beaman, and K. H. Atwood, "The Relation of Agricultural Extension Agencies to Farm Practices," United States Bureau of Plant Industry, *Circular* 117 (Washington, 1913).

Snowden, Mason, "Demonstration Work in Louisiana," United States Bureau of Education, *Bulletin* 30, 1913 (Washington, 1913).

Stedman, John M., "Farmers' Institute Work in the United States in 1914, and Notes on Agricultural Extension Work in Foreign Countries," United States Department of Agriculture, *Bulletin* 269 (Washington, 1915).

Strauss, Frederick, and Louis H. Bean, "Gross Farm Income and Indices of Farm Production and Prices in the United States, 1869–1937," United States Department of Agriculture, *Technical Bulletin* 703 (Washington, 1940).

Stuntz, Stephen C., "List of the Agricultural Periodicals of the United States and Canada Published during the Century July, 1810, to July, 1910," United States Department of Agriculture, *Miscellaneous Publication* 398 (Washington, 1941).

"Ten Years of Extension Work in Missouri," Missouri Agricultural Extension Service, *Project Announcement* 16 (Columbia, 1923).

Texas Agricultural Extension Service, *Silver Anniversary: Cooperative Demonstration Work, 1903–1928* (College Station, n.d.).

True, Alfred C., *A History of Agricultural Education in the United States, 1785–1923* (Washington, 1929).

————, *A History of Agricultural Experimentation and Research in the United States, 1607–1925* (Washington, 1937).

————, *A History of Agricultural Extension Work in the United States, 1785–1923* (Washington, 1928).

————, "Popular Education for the Farmer in the United States," United States Department of Agriculture, *Yearbook*, 1897 (Washington, 1898), pp. 279–290.

————, and Dick J. Crosby, "The American System of Agricultural Education," United States Office of Experiment Stations, *Circular* 106 (Washington, 1911).

"Twenty-Five Years of Extension Work in Missouri," Missouri Agricultural Extension Service, *Circular* 420 (Columbia, 1940).

United States Commissioner of Agriculture, *Report*, 1862–71 (Washington, 1863–72).

United States Commissioner of Education, *Report*, 1895–1914 (Washington, 1897–1915).

United States Commissioner of Patents, *Report*, 1851–58 (Washington, 1852–59).

United States Department of Agriculture, *Report on Agricultural Experiment Stations and Cooperative Extension Work in the United States*, 1915 (Washington, 1916).

United States Department of Agriculture, *Yearbook*, 1894–1940 (Washington, 1895–1940).

United States Office of Experiment Stations, *Annual Report*, 1902–14 (Washington, 1903–15).

United States Office of Experiment Stations, *Circular* 9 (Washington, n.d.). *United States Statutes at Large.*

United States *Thirteenth Census: Population; Agriculture.*

Utah Farmers' Institutes, *First Annual Report*, 1897 (n.p., n.d.).

"Vocational Education," U.S. Senate *Report* 405 (1912).

Washington Agricultural Experiment Station, *Bulletin* 2, 3, 120 (Pullman, Wash., 1892, 1915).

West Virginia Agricultural Experiment Station, *Report*, 1912–14 (n.p., n.d.).

Wisconsin Agricultural Experiment Station, *Bulletin* 228 (Madison, 1913).

Wisconsin State Agricultural Society, *Transactions*, XXI (Madison, 1884).

Witter, D. P., "A History of Farmers' Institutes in New York," New York Department of Farms and Markets, *Bulletin* 109 (Albany, 1918).

Woodward, John A., "Farmers' Institutes in Other States," Pennsylvania Department of Agriculture, *Annual Report*, 1895 (n.p., 1896).

"The Work of the College of Agriculture and Experiment Stations," California Agricultural Experiment Station, *Bulletin* 111 (Berkeley, 1896).

Wyoming Agricultural Experiment Station, *Annual Report*, 1913–14 (Laramie, n.d.).

C. BOOKS AND ARTICLES

"Agricultural Demonstration Trains of the University of California," *University of California Chronicle*, XI (April, 1909), pp. 186–187.

Agricultural Extension Committee of the National Implement and Vehicle Association, *Agricultural Extension* (Chicago, 1916).

Alford, George H., *Autobiography* (Progress, Miss., 1949).

———, *Diversified Farming in the Cotton Belt* (Chicago, n.d.).

"American Agricultural Societies," in Liberty Hyde Bailey, ed., *Cyclopedia of American Agriculture* (4 vols., New York, 1909–1910), IV, 341–354.

American Association of Farmers' Institute Managers, *Report of the Meetings Held at Watertown, Wisconsin, March 13, and at Chicago, October 14, 1896* (Lincoln, Nebr., 1897).

———, *Proceedings*, 1897 (Lincoln, Nebr., 1898).

———, "Proceedings, 1899," in New York State Agricultural Society, *Annual Report*, 1899 (New York, 1899), pp. 627–716.

American Association of Farmers' Institute Workers, *Proceedings*, 1901–17 (Washington, 1902–18).

American Bankers' Association, *Proceedings*, 1911–13 (New York, 1911–13).

Association of American Agricultural Colleges and Experiment Stations, *Proceedings*, 1887–1931 (Washington, 1888–1932).

Atkeson, Thomas C., *Semi-Centennial History of the Patrons of Husbandry* (New York, 1916).

————, and Mary W. Atkeson, *Pioneering in Agriculture: One Hundred Years of American Farming and Farm Leadership* (New York, 1937).

N. W. Ayer and Son's American Newspaper Annual, 1912 (Philadelphia, 1912).

Bailey, Liberty Hyde, *Agricultural Education in New York State* (Ithaca, 1904).

————, *Annals of Horticulture*, 1891 (New York, 1892).

————, *The Country Life Movement in the United States* (New York, 1911).

————, ed., *Cyclopedia of American Agriculture* (4 vols., New York, 1909–1910).

————, "Newer Ideas in Agricultural Education," *Educational Review*, XX (November, 1900), pp. 377–382.

————, "The Revolution in Farming," *World's Work*, II (July, 1901), pp. 945–948.

————, *The State and the Farmer* (New York, 1908).

————, *The Training of Farmers* (New York, 1909).

Ball, Walter S., "Farming by Mail," *Independent*, LXXI (May 16, 1912), pp. 1052–1057.

Barrett, Charles S., *The Mission, History and Times of the Farmers' Union* (Nashville, Tenn., 1909).

Bell, Arthur J., *The Grout Farm Encampment* (Urbana, Ill., 1906).

Better Farming Association of North Dakota, *Annual Report*, 1912–13 (n.p., n.d.).

Boston and Albany Railroad, *Souvenir of the Better Farming Special* (n.p., n.d.).

"Boys' and Girls' Clubs in Agriculture and Home Economics in Massachusetts," *The School Review*, XXIV (December, 1916), pp. 765–766.

Buell, Jennie, "The Educational Value of the Grange," *Business America*, XIII (January, 1913), pp. 50–54.

Butterfield, Kenyon L., *Chapters in Rural Progress* (Chicago, 1908).

————, "Farmers' Institutes," *The Chautauquan*, XXXIV (March, 1902), pp. 638–641.

————, "Farmers' Social Organizations," in Liberty Hyde Bailey, ed.,

Cyclopedia of American Agriculture (4 vols., New York, 1909–1910), IV, 289–297.

————, "The Grange," *Forum*, XXXI (April, 1901), pp. 231–242.

————, "A Significant Factor in Agricultural Education," *Educational Review*, XXI (March, 1901), pp. 301–306.

————, "An Untilled Field in American Agricultural Education," *Popular Science Monthly*, LXIII (July, 1903), pp. 257–261.

Buttrick, Wallace, "Seaman A. Knapp's Work as an Agricultural Statesman," *Review of Reviews*, XLIII (June, 1911), pp. 683–685.

Calvin, Henrietta, "Extension Work," *Journal of Home Economics*, IX (December, 1917), pp. 565–566.

Campbell, Thomas M., *The Movable School Goes to the Negro* (Tuskegee, Ala., 1936).

Coleman, W. A., *Farmers' Institute and Reference Book for Page County, Iowa* (Des Moines, 1900).

Council of North American Grain Exchanges, Committee on Crop Improvement, *Proceedings of a Meeting in Chicago, February 8, 1911* (n.p., n.d.).

Craig, John, "Teaching Farmers at Home," *World's Work*, II (June, 1901), pp. 810–812.

Davis, Benjamin M., "Agricultural Education: Boys' Agricultural Clubs," *Elementary School Teacher*, XI (March, 1911), pp. 371–380.

————, "Agricultural Education: State Organizations for Agriculture and Farmers' Institutes," *Elementary School Teacher*, XI (November, 1910), pp. 136–145.

Drew, Frank M., "The Present Farmers' Movement," *Political Science Quarterly*, VI (June, 1891), pp. 282–310.

Dunning, Nelson A., ed., *Farmers' Alliance History and Agricultural Digest* (Washington, 1891).

Erickson, T. A., *My Sixty Years with Rural Youth* (Minneapolis, 1956).

Farmers' Alliance of South Carolina, *Proceedings*, 1893 (Spartanburg, 1893).

Farmers' Alliance of Virginia, *Constitution Adopted August 18–20, 1891* (Richmond, 1891).

Farmers' Mutual Benefit Association, *General Charter, Declaration of Purpose, and Constitution and By-Laws* (Mt. Vernon, Ill., 1890).

Flagg, Willard C., "Report on the Agricultural, Horticultural, and Other Industrial Interests of the County of Madison," Illinois State Agricultural Society, *Transactions*, VIII (1869–70), pp. 203–243.

Fullerton, Edith L., *The Lure of the Land* (New York, 1906).

General Education Board, *The General Education Board: An Account of Its Activities, 1902–1914* (New York, 1915).

Graham, A. B., "Boys' and Girls' Agricultural Clubs," *Agricultural History*, XV (April, 1941), pp. 65–68.

Gross, Howard H., *The National Soil Fertility League* (Chicago, 1911).

Halstead, Byron D., "A New Factor in American Education," *The Chautauquan*, XVI (December, 1892), pp. 284–288.

Haney, J. G., *IHC Demonstration Farms in the North* (Chicago, 1917).

Harris, Mrs. Frank V., ed., "The Autobiography of Benjamin Franklin Harris," Illinois State Historical Society, *Transactions*, 1923 (Springfield, 1923), pp. 72–101.

Hilgard, Eugene W., "Progress in Agriculture by Education and Government Aid," *Atlantic Monthly*, XLIX (April, 1882), pp. 531–541.

Houston, David F., *Eight Years with Wilson's Cabinet* (2 vols., Garden City, N.Y., 1926).

Illinois Central Railroad, *Mississippi, A Wonderful Agricultural State* (Chicago, n.d.).

———, *Organization and Traffic of the Illinois Central System* (Chicago, 1938).

Illinois State Grange, *Proceedings*, 1890 (Peoria, 1891).

International Harvester Company, *Annual Report*, 1912 (n.p., n.d.)

———, *Every Farm a Factory* (Chicago, 1916).

———, *For Better Crops in the South* (Chicago, 1913).

———, *The Golden Stream* (Chicago, 1910).

———, *IHC Demonstration Farms in the South* (Chicago, 1917).

Kelley, Oliver H., *Origin and Progress of the Order of the Patrons of Husbandry in the United States: A History from 1866–1873* (Philadelphia, 1875).

Kern, O. J., *Among Country Schools* (Boston, 1906).

———, " 'Learning by Doing' for the Farmer Boy," *Review of Reviews*, XXVIII (October, 1903), pp. 456–461.

Kile, Orville M., *The Farm Bureau through Three Decades* (Baltimore, 1948).

Kirkwood, W. P., "The Man Who Roused the Farmer," *Northwestern Miller*, LXXXVIII (December 13, 1911), pp. 649–650.

Knapp, Bradford, "Education through Farm Demonstration," The American Academy of Political and Social Science, *The Annals*, LXVII (September, 1916), pp. 224–240.

Knapp, Seaman A., "An Agricultural Revolution," *World's Work*, XII (July, 1906), pp. 7733–7738.

———, "Farmers' Co-Operative Demonstration Work and Its Results," *Southern Educational Review*, III (October, 1906), pp. 49–65.

Latta, William C., *Indiana Farmers' Institutes from Their Origin in 1882, to 1904* (Lafayette, Ind., 1904).

———, *Outline History of Indiana Agriculture* (Lafayette, Ind., 1938).

Luther, Ernest L., "Farmers' Institutes in Wisconsin, 1885–1933," *Wisconsin Magazine of History*, XXX (September, 1946), pp. 59–68.

McKimmon, Jane S., *When We're Green We Grow* (Chapel Hill, N. C., 1945).

"Making Rural Life Profitable," *World's Work*, XVI (May, 1908), pp. 10178–10180.

Martin, Edward W., *History of the Grange Movement* (Chicago, 1874).

Martin, Oscar B., "Boys' and Girls' Clubs," Conference for Education in the South, *Proceedings*, 1912 (Washington, n.d.), pp. 206–215.

———, *The Demonstration Work: Dr. Knapp's Contribution to Civilization* (San Antonio, 1941).

Mayo, Earl, "The Good Roads Train," *World's Work*, II (July, 1901), pp. 956–960.

Minnesota State Farmers' Alliance, *Constitution and By-Laws*, 1890 (St. Paul, 1890).

———, *Resolutions Adopted at Minneapolis, February 25, 1886* (n.p., 1886).

Minnesota State Grange, *Proceedings*, 1892 (Minneapolis, 1893).

Mississippi Patrons of Husbandry, *The State Grange and the A. and M. College* (n.p., n.d.).

Morgan, W. Scott, *History of the Wheel and Alliance and the Impending Revolution* (New York, 1968).

Mosher, Martin L., *Early Iowa Corn Field Tests and Related Later Programs* (Ames, Iowa, 1962).

National Farmers' Alliance and Industrial Union, *Proceedings of the Supreme Council*, 1890 (Washington, 1891).

National Grange of the Patrons of Husbandry, *Proceedings*, 1883 (Philadelphia, 1883).

Needham, Henry B., "The Object-Lesson Farm," *World's Work*, XI (November, 1905), pp. 6871–6874.

Ohio State Board of Agriculture, *A Brief History of the State Board of*

Agriculture, the State Fair, District and Agricultural Societies and Farmers' Institutes in Ohio (Columbus, Ohio, 1899).

Periam, Jonathan, *The Groundswell: A History of the Origins, Aims, and Progress of the Farmers' Movement* (Cincinnati, 1874).

Pyle, Joseph G., "A Farm Revolution That Began in a Greenhouse," *World's Work*, XXV (April, 1913), pp. 665–671.

Railway Development Association, *Proceedings*, 1913–15 (Roanoke, Va., n.d.).

Railway Industrial Association, *Proceedings*, 1911 (St. Louis, n.d.).

Roberts, Isaac P., *Autobiography of a Farm Boy* (Albany, N.Y., 1916).

Schoffelmayer, Victor H., *Southwest Trails to New Horizons* (San Antonio, 1960).

Smith, Everett W., "Raising a Crop of Men," *The Outlook*, LXXXIX (July 18, 1908), pp. 603–608.

Snyder, A. H., "Extension Work in the West," *Addresses Delivered at the University of Virginia Summer School in Connection with the Conference for the Study of the Problems of Rural Life*, 1909 (n.p., n.d.), pp. 28–31.

———, "Traveling Schools," *Addresses Delivered at the University of Virginia Summer School in Connection with the Conference for the Study of the Problems of Rural Life*, 1909 (n.p., n.d.), pp. 48–54.

Southern States Association of Commissioners of Agriculture and Other Agricultural Workers, *Proceedings*, 1907 (Raleigh, 1908).

"State Organizations for Agriculture," in Liberty Hyde Bailey, ed., *Cyclopedia of American Agriculture* (4 vols., New York, 1909–1910), IV, 328–341.

Stockbridge, Frank P., "The North Dakota Man Crop," *World's Work*, XXV (November, 1912), pp. 84–93.

True, Alfred C., "University Extension in Agriculture," *Forum*, XXVIII (February, 1900), pp. 701–707.

Van Rensselaer, Martha, "Home Economics at the New York State College of Agriculture," *Cornell Countryman*, XI (May, 1914), pp. 260–263.

Waggener, O. O., *Western Agriculture and the Burlington* (Chicago, 1938).

Walker, Charles S., "The Farmers' Movement," American Academy of Political and Social Science, *Annals*, IV (March, 1894), pp. 790–798.

Wallace, Henry, *Uncle Henry's Own Story of His Life* (3 vols., Des Moines, 1917–19).

Washington, Booker T., "Farmers' College on Wheels," *World's Work*, XIII (December, 1906), pp. 8352–8354.

Waters, H. J., "The Duty of the Agricultural College," *Science,* n.s. XXX (December 3, 1909), pp. 777–789.

Witham, James W., *Fifty Years on the Firing Line* (Chicago, 1924).

"Women's Institutes in North Carolina," *Journal of Home Economics,* I (April, 1909), pp. 161–163.

D. FARM JOURNALS AND MISCELLANEOUS NEWSPAPERS

(Abbeville, La.) *Meridional*
Albion (Ill.) *Journal*
Alton (Ill.) *Sentinel-Democrat*
American Agriculturist (New York)
American Bankers' Association, *Journal* (New York)
American Fertilizer (Philadelphia)
Ariel (Minneapolis)
Arkansas Farmer and Homestead (Little Rock)
Banker-Farmer (Champaign, Ill.)
Bloomington (Ill.) *Daily Pantagraph*
Breeder's Gazette (Chicago)
Centralia (Ill.) *Daily Sentinel*
Chicago Tribune
College Farmer (Columbia, Mo.)
Colman's Rural World (St. Louis)
Corn (Waterloo, Iowa)
Cornell Countryman (Ithaca, N. Y.)
Cotton Plant (Columbia, S. C.)
Country Gentleman (Philadelphia)
Cultivator and Country Gentleman (Albany, N. Y.)
Decatur (Ill.) *Daily Republican*
Farm Implement News (Chicago)
Farm, Stock, and Home (Minneapolis)
Farm Students' Review (St. Anthony Park, Minn.)
The Farmer (St. Paul)
Farmers' Voice (Chicago)
Farmington (Mo.) *Times and Herald*

Fruit Grower and Farmer (St. Joseph, Mo.)
The Golden Era (McLeansboro, Ill.)
Hoard's Dairyman (Fort Atkison, Wis.)
Hopkins (Mo.) *Journal*
Illinois Central Magazine (Chicago)
(Springfield) *Illinois State Journal*
(Springfield) *Illinois State Register*
Jefferson City (Mo.) *Daily Tribune*
Kimball's Dairy Farmer (Waterloo, Iowa)
Lafayette (La.) *Advertiser*
Literary Digest (New York)
Marseilles (Ill.) *Plaindealer*
(Maryville, Mo.) *Weekly Democratic Forum*
Mississippi Agricultural Student (Agricultural College)
Mississippi School Journal (Jackson)
Missouri Agricultural College Farmer (Columbia)
Missouri Farmer (Columbia)
Missouri Ruralist (Kansas City)
National Economist (Washington, D. C.)
National Rural (Chicago)
(Nevada, Mo.) *Southwest Mail*
Northwestern Agriculturist (Minneapolis)
Northwestern Farmer (St. Paul)
(Opelousas, La.) *Courier*
Ottawa (Ill.) *Free Trader*
The Outlook (New York)
Pacific Rural Press (San Francisco)
Prairie Farmer (Chicago)
Progressive Farmer and Southern Farm Gazette (Raleigh, N. C.)
Railway Age (Chicago)
Railway Age Gazette (Chicago)
The Record (East Lansing, Mich.)
Review of Reviews (New York)
(St. Paul) *Pioneer Press*
(Sedalia, Mo.) *Daily Capital*
Southern Farm Gazette (Starkville, Miss.)
Southern Farm Magazine (Baltimore)
Southern Live-Stock Journal (Starkville, Miss.)
Southern Planter (Richmond, Va.)

Stanberry (Mo.) *Owl*
University Missourian (Columbia, Mo.)
Wallaces' Farmer (Des Moines)
Western Fruit Grower (St. Joseph, Mo.)
Western Rural (Chicago)
White Hall (Ill.) *Register*
World's Work (New York)

E. INTERVIEWS

C. A. Cobb
M. F. Miller
J. E. Sides
P. E. Spinks

II. Secondary Materials

A. BOOKS AND ARTICLES

Agnew, Ella G., "Agricultural Fairs—Yesterday and Today," *Southern Planter*, XVI (August 15, 1929), pp. 5, 18.

Allmond, C. M., ed., "The Agricultural Memorandums of Samuel H. Black: 1815–1820," *Agricultural History*, XXXII (January, 1958), pp. 56–61.

Alpha Gamma Chapter of Epsilon Sigma Phi, *Extension Work in Virginia, 1907–1940: A Brief History* (Blacksburg, Va., 1941).

Anderson, W. A., "The Granger Movement in the Middle West with Special Reference to Iowa," *Iowa Journal of History and Politics*, XXII (January, 1924), pp. 3–51.

Arnett, Alex M., *The Populist Movement in Georgia* (New York, 1922).

Bailey, Joseph C., *Seaman A. Knapp: Schoolmaster of American Agriculture* (New York, 1945).

Baker, Gladys, *The County Agent* (Chicago, 1939).

Bardolph, Richard, *Agricultural Literature and the Early Illinois Farmer* (Urbana, Ill., 1948).

Barnett, Claribel R., *"The American Museum*: An Early American Agricultural Periodical," *Agricultural History*, II (April, 1928), pp. 99–102.

Barns, William D., "The Influence of the West Virginia Grange upon Public Agricultural Education of College Grade," *West Virginia History*, IX (January, 1948), pp. 128–157.

331

Bay, Edwin, *The History of the National Association of County Agricultural Agents, 1915–1960* (Springfield, Ill., 1961).

Beal, W. J., *History of the Michigan Agricultural College* (East Lansing, 1915).

Bettersworth, John K., *Mississippi: A History* (Austin, Texas, 1959).

———, *People's College: A History of Mississippi State* (University, Ala., 1953).

Bidwell, Percy W., and John I. Falconer, *History of Agriculture in the Northern United States, 1620–1860* (Washington, 1925).

Bliss, Ralph K., *History of Cooperative Agriculture and Home Economics Extension in Iowa: The First Fifty Years* (Ames, 1960).

Bogue, Allan G., *From Prairie to Corn Belt: Farming on the Illinois and Iowa Prairies in the Nineteenth Century* (Chicago, 1963).

———. "Pioneer Farmers and Innovation," *Iowa Journal of History*, LVI (January, 1958), pp. 1–36.

Bonner, James C., "Advancing Trends in Southern Agriculture," *Agricultural History*, XXII (October, 1948), pp. 248–259.

———, *A History of Georgia Agriculture, 1732–1860* (Athens, 1964).

Boss, Andrew, *The Early History and Background of the School of Agriculture at University Farm, St. Paul* (St. Paul, 1941).

Briggs, John E., "The Sioux City Corn Palaces," *Palimpsest,* III (October, 1922), pp. 313–326.

Bronson, Walter C., *The History of Brown University, 1764–1914* (Providence, R. I., 1914).

Brown, Ralph M., "Agricultural Science and Education in Virginia before 1860," *William and Mary College Quarterly History Magazine,* XIX (April, 1939), pp. 197–213.

Bryan, Enoch A., *Historical Sketch of the State College of Washington* (Spokane, 1928).

Buck, Solon J., *The Granger Movement: A Study of Agricultural Organization and Its Political, Economic and Social Manifestations, 1870–1880* (Cambridge, Mass., 1913).

Burkett, C. W., *History of Ohio Agriculture* (Concord, N. H., 1900).

Burritt, Maurice C., *The County Agent and the Farm Bureau* (New York, 1922).

Callcott, George H., *A History of the University of Maryland* (Baltimore, 1966).

Campbell, J. Phil, "Action Programs in Agriculture," *Agricultural History*, XV (April, 1941), pp. 68–71.

Carman, Harry J., "Jesse Buel, Early Nineteenth Century Agricultural Reformer," *Agricultural History*, XVII (January, 1943), pp. 1–13.

Carriel, Mary Turner, *The Life of Jonathan Baldwin Turner* (Urbana, Ill., 1961).

Carrier, Lyman, "The United States Agricultural Society, 1852–1860," *Agricultural History*, XI (October, 1937), pp. 278–288.

Carstensen, Vernon, "The Genesis of an Agricultural Experiment Station," *Agricultural History*, XXXIV (January, 1960), pp. 13–20.

Cary, Harold W., *The University of Massachusetts: A History of One Hundred Years* (Amherst, Mass., 1962).

Case, H. C. M., and D. B. Williams, *Fifty Years of Farm Management* (Urbana, Ill., 1957).

Cassidy, Lawrence A., *Kentucky Fairs, 1816–1959: A Brief History* (Lyndon, Ky., 1960).

Caswell, Lilley B., *Brief History of the Massachusetts Agricultural College* (Springfield, Mass., 1917).

Cathey, Cornelius O., *Agricultural Developments in North Carolina, 1783–1860* (Chapel Hill, 1956).

Cavanaugh, Helen M., *Seed, Soil, and Science: The Story of Eugene D. Funk* (Chicago, 1959).

Clark, Ira G., *Then Came the Railroads: From Steam to Diesel in the Southwest* (Norman, Okla., 1958).

Clevenger, Homer, "The Teaching Techniques of the Farmers' Alliance: An Experiment in Adult Education," *Journal of Southern History*, XI (November 1945), pp. 505–518.

Cline, Rodney, *The Life and Work of Seaman A. Knapp* (Nashville, Tenn., 1936).

Colman, Gould P., *Education and Agriculture: A History of the College of Agriculture at Cornell University* (Ithaca, N. Y., 1963).

———, "Innovation and Diffusion in Agriculture," *Agricultural History*, XLII (July, 1968), pp. 173–187.

Commons, John R., and others, eds., *A Documentary History of American Industrial Society* (10 vols., Cleveland, 1910).

Conover, Milton, *The Office of Experiment Stations: Its History, Activities and Organization* (Baltimore, 1924).

Cooley, Edgar W., "Service, the Duty of Citizenship," *Farming*, XIX (January, 1921), pp. 210–211.

Corliss, Carlton J., *Main Line of Mid-America: The Story of the Illinois Central* (New York, 1950).

Coulter, E. Merton, *James Monroe Smith; Georgia Planter, before Death and after* (Athens, Ga., 1961).

Craven, Avery O., "The Agricultural Reformers in the Ante-Bellum South," *American Historical Review*, XXXIII (January, 1928), pp. 302–314.

——, *Edmund Ruffin, Southerner: A Study in Secession* (New York, 1932).

Crawford, Robert P., *These Fifty Years: A History of the College of Agriculture of the University of Nebraska* (Lincoln, 1925).

Day, Clarence A., *Farming in Maine, 1860–1940* (Orono, Maine, 1963).

——, *A History of Maine Agriculture, 1604–1860* (Orono, Maine, 1954).

Demaree, A. L., "The Farm Journals, Their Editors, and Their Public, 1830–1860," *Agricultural History*, XV (October, 1941), pp. 182–188.

Destler, Chester M., "David Dickson's 'System of Farming' and the Agricultural Revolution in the Deep South, 1850–1885," *Agricultural History*, XXXI (July, 1957), pp. 30–39.

"Development of Wisconsin Fairs," *Wisconsin Agriculturist and Farmer*, LVIII (August 17, 1929), pp. 3, 21.

Dorf, Phillip, *Liberty Hyde Bailey: An Informal Biography* (Ithaca, N.Y., 1956).

Drache, Hiram M., *The Day of the Bonanza: A History of Bonanza Farming in the Red River Valley of the North* (Fargo, N. D., 1964).

Dunaway, Wayland F., *History of the Pennsylvania State College* (n.p., 1946).

Duncan, Clyde C., *Straight Furrows: A Story of 4-H Club Work* (Albuquerque, 1954).

Eddy, Edward D., *Colleges for Our Land and Time: The Land-Grant Idea in American Education* (New York, 1957).

Ellsworth, Clayton S., "Theodore Roosevelt's Country Life Commission," *Agricultural History*, XXXIV (October, 1960), pp. 155–172.

Epsilon Sigma Phi, *The Spirit and Philosophy of Extension Work* (Washington, 1952).

Eschenbacher, Herman F., *The University of Rhode Island: A History of Land-Grant Education in Rhode Island* (New York, 1967).

Etheridge, William C., "The College of Agriculture," in Jonas Villes, ed., *The University of Missouri: A Centennial History* (Columbia, Mo., 1939).

"Famous Hybrid Corn Scientist Dies at 94," *Michigan Extension News*, XXX (October–December, 1959), pp. 1, 4.

Ferguson, James S., "The Grange and Farmer Education in Mississippi," *Journal of Southern History*, VIII (November, 1942), pp. 497–512.

Fernald, Merritt C., *History of Maine State College and the University of Maine* (Orono, Maine, 1916).

Ferrier, William W., *Origin and Development of the University of California* (Berkeley, 1930).

Fletcher, Stevenson W., *Pennsylvania Agriculture and Country Life, 1640–1840* (Harrisburg, 1950).

————, *Pennsylvania Agriculture and Country Life, 1840–1940* (Harrisburg, 1955).

Flick, Hugh M., "Elkanah Watson's Activities on Behalf of Agriculture," *Agricultural History*, XXI (October, 1947), pp. 193–198.

Fosdick, Raymond B., *Adventure in Giving: The Story of the General Education Board* (New York, 1962).

Gardner, Charles M., *The Grange, Friend of the Farmer* (Washington, 1949).

Gates, Paul W., *The Farmer's Age: Agriculture, 1815–1860* (Vol. III of *The Economic History of the United States*, Henry David and others, eds., New York, 1960).

————, *The Illinois Central Railroad and Its Colonization Work* (Cambridge, Mass., 1934).

Glover, Wilbur H., "The Agricultural College Crisis of 1885," *Wisconsin Magazine of History*, XXXII (September, 1948), pp. 17–25.

————, "The Agricultural College Lands in Wisconsin," *Wisconsin Magazine of History*, XXX (March, 1947), pp. 261–272.

————, *Farm and College: The College of Agriculture of the University of Wisconsin* (Madison, 1952).

Good, H. G., "Early Attempts to Teach Agriculture in Old Virginia," *Virginia Magazine of History and Biography*, XLVIII (October, 1940), pp. 341–351.

Graham, I. D., "The Kansas State Board of Agriculture: Some High Lights of History," Kansas State Historical Society, *Collections*, XVII (1928), pp. 788–813.

Grantham, Dewey W., Jr., *Hoke Smith and the Politics of the New South* (Baton Rouge, 1958).

335

Gray, James, *The University of Minnesota, 1851–1951* (Minneapolis, 1951).

Gray, L. C., *History of Agriculture in the Southern United States to 1860* (2 vols., Washington, 1933).

Hale, Harrison, *University of Arkansas, 1871–1948* (Fayetteville, 1948).

Hall, Darwin S., and R. I. Holcombe, *History of the Minnesota State Agricultural Society* (St. Paul, 1910).

Hargreaves, Mary W. M., "Hardy W. Campbell (1850–1937)," *Agricultural History*, XXXII (January, 1958), pp. 62–65.

Hayes, Herbert K., *A Professor's Story of Hybrid Corn* (Minneapolis, 1963).

Hedrick, W. O., "The Tutored Farmer," *Scientific Monthly*, VII (August, 1918), pp. 158–165.

Helgeson, Arlan, *Farms in the Cutover: Agricultural Settlement in Northern Wisconsin* (Madison, 1962).

Hepburn, William M., and L. M. Sears, *Purdue University: Fifty Years of Progress* (Indianapolis, 1925).

Hicks, John D., *The Populist Revolt: A History of the Farmers' Alliance and the People's Party* (Lincoln, Nebr., 1961).

———, "The Origin and Early History of the Farmers' Alliance in Minnesota," *Mississippi Valley Historical Review*, IX (December, 1922), pp. 203–226.

Hill, Kate A., *Home Demonstration Work in Texas* (San Antonio, 1958).

Hofstadter, Richard, *The Age of Reform from Bryan to F. D. R.* (New York, 1960).

Holden, W. C., "Experimental Agriculture on the Spur Ranch, 1885–1909," *Southwestern Social Science Quarterly*, XIII (1932), pp. 16–23.

Hunt, Robert L., *A History of Farmers' Movements in the Southwest* (n.p., n.d.).

Hunter, William C., *Beacon across the Prairie: North Dakota's Land-Grant College* (Fargo, 1961).

Hutchison, Claude B., ed., *California Agriculture* (Berkeley, 1946).

Ivins, Lester S., and A. E. Winship, *Fifty Famous Farmers* (New York, 1925).

Jarchow, Merrill E., *The Earth Brought Forth: A History of Minnesota Agriculture to 1885* (St. Paul, 1949).

Jesness, Oscar B., and others, *Andrew Boss: Agricultural Pioneer and Builder* (St. Paul, 1950).

Jones, C. Clyde, "The Burlington Railroad and Agricultural Policy in the 1920's," *Agricultural History*, XXXI (October, 1957), pp. 67–74.

———, "A Survey of the Agricultural Development Program of the Chicago, Burlington and Quincy Railroad," *Nebraska History*, XXX (September, 1949), pp. 226–256.

———, "Val Kuska, Agricultural Development Agent," *Nebraska History*, XXXVIII (December, 1957), pp. 285–293.

Jones, Lewis W., "The South's Negro Farm Agent," *Journal of Negro Education*, XXII (Winter, 1953), pp. 38–45.

Jordan, Weymouth T., "Agricultural Societies in Ante-Bellum Alabama," *Alabama Review*, IV (October, 1951), pp. 241–253.

———, "Noah B. Cloud and the *American Cotton Planter*," *Agricultural History*, XXXI (October 1957), pp. 44–49.

Keener, Orrin L., *Struggle for Equal Opportunity: Dirt Farmers and the American Country Life Association* (New York, 1962).

Kniffen, Fred, "The American Agricultural Fair: The Pattern," Association of American Geographers, *Annals*, XXXIX (December, 1949), pp. 264–282.

Kuhn, Madison, *Michigan State: The First Hundred Years, 1855–1955* (East Lansing, 1955).

Lacey, John J., *Farm Bureau in Illinois* (Bloomington, Ill., 1965).

Larson, Vernon C., "The Development of Short Courses at the Land Grant Institutions," *Agricultural History*, XXXI (April, 1957), pp. 31–35.

Lemly, James H., *The Gulf, Mobile and Ohio* (Homewood, Ill., 1953).

Lemmer, George F., "The Early Agricultural Fairs in Missouri," *Agricultural History*, XVII (July, 1943), pp. 145–152.

———, "Early Agricultural Editors and Their Farm Philosophies," *Agricultural History*, XXXI (October, 1957), pp. 3–22.

Lindsay, Julian I., *Tradition Looks Forward; the University of Vermont: A History, 1791–1904* (Burlington, 1954).

Lockmiller, David A., *History of the North Carolina State College* (Raleigh, 1939).

Loehr, Rodney C., "The Influence of English Agriculture on American Agriculture, 1775–1825," *Agricultural History*, XI (January, 1937), pp. 3–15.

Lord, Russell, *The Agrarian Revival: A Study of Agricultural Extension* (New York, 1939).

———, *The Care of the Earth: A History of Husbandry* (New York, 1962).

McConnell, Grant, *The Decline of Agrarian Democracy* (Berkeley and Los Angeles, 1953).

McFarlane, D. C., " 'Corn Club' Smith," *Collier's*, LI (May 17, 1913), pp. 19, 24–26.

McGiffert, Michael, *Higher Learning in Colorado: An Historical Study, 1860–1940* (Denver, 1964).

McMillen, Wheeler, *Land of Plenty: The American Farm Story* (New York, 1961).

McNelly, C. L., *The County Agent Story: The Impact of Extension Work on Farming and Country Life* (Anoka, Minn., 1960).

Madill, A. J., *History of Agricultural Education in Canada* (Toronto, 1930).

Mahan, Bruce E., "The Blue-Grass Palace," *Palimpsest*, III (October, 1922), pp. 327–335.

Mairs, Thomas I., *Some Pennsylvania Pioneers in Agricultural Science* (State College, Pa., 1928).

Massachusetts Horticultural Society, *History of the Massachusetts Horticultural Society, 1829–1878* (Boston, 1880).

"The Massachusetts Society for Promoting Agriculture," *Garden and Forest*, V (August 3, 1892), pp. 371–372.

Mendenhall, Thomas C., ed., *History of the Ohio State University* (3 vols., Columbus, Ohio, 1920–26).

Merrison, J. G., "Extension of Agriculture to the American Farmer," *Journal of the Board of Agriculture*, XXVI (December, 1919), pp. 881–891.

Metcalf, Henry H., *New Hampshire Agriculture. Personal and Farm Sketches* (Concord, N. H., 1897).

Miller, August C., "Jefferson as an Agriculturist," *Agricultural History*, XVI (April, 1942), pp. 65–78.

Moore, John H., *Agriculture in Ante-Bellum Mississippi* (New York, 1958).

Morgan, Barton, *A History of the Extension Service of Iowa State College* (Ames, Iowa, 1934).

Neely, Wayne C., *The Agricultural Fair* (New York, 1935).

Nettels, Curtis P., *The Emergence of a National Economy, 1775–1815* (Vol. II of *The Economic History of the United States*, Henry David and others, eds., New York, 1962).

Nevins, Allan, *Illinois* (New York, 1917).

————, *The Origins of the Land-Grant Colleges and State Universities* (Washington, 1962).

Noblin, Stuart, *Leonidas LaFayette Polk; Agrarian Crusader* (Chapel Hill, N. C., 1949).

Noffsinger, John S., *Correspondence Schools, Lyceums, Chautauquas* (New York, 1926).

Ousley, Clarence, *History of the Agricultural and Mechanical College of Texas* (College Station, Texas, 1935).

Overton, Richard C., *Burlington Route: A History of Burlington Lines* (New York, 1965).

————, *Burlington West: A Colonization History of the Burlington Railroad* (Cambridge, Mass., 1941).

Paine, A. E., *The Granger Movement in Illinois* (Urbana, Ill., 1904).

Parker, William B., *The Life and Public Service of Justin Smith Morrill* (New York, 1924).

Partin, Robert, "Black's Bend Grange, 1873–77: A Case Study of a Subordinate Grange of the Deep South," *Agricultural History*, XXXI (July, 1957), pp. 49–59.

Pinkett, Harold T., "The *American Farmer*, a Pioneer Agricultural Journal," *Agricultural History*, XXIV (July, 1950), pp. 146–151.

Porter, Kirk H., and Donald B. Johnson, *National Party Platforms, 1840–1956* (Urbana, Ill., 1956).

Powell, Burt E., *The Movement for Industrial Education and the Establishment of the University, 1840–1870* (Urbana, Ill., 1918).

Powell, Fred W., *The Bureau of Plant Industry: Its History, Activities and Organization* (Baltimore, 1927).

Powers, William H., ed., *A History of South Dakota State College* (Brookings, S. D., 1931).

Proctor, Samuel, "The Early Years of the Florida Experiment Station," *Agricultural History*, XXXVI (October, 1962), pp. 213–221.

Pyle, Joseph G., *The Life of James J. Hill* (2 vols., New York, 1917).

Range, Willard, *A Century of Georgia Agriculture, 1850–1950* (Athens, 1954).

Reck, Franklin M., *The 4-H Story: A History of 4-H Club Work* (Ames, Iowa, 1951).

Ricks, Joel E., *The Utah State Agricultural College: A History of Fifty Years, 1888–1938* (Salt Lake City, 1938).

339

Roberts, Roy W., *Vocational and Practical Arts Education: History, Development, and Principles* (New York, 1965).

Rogers, Andrew D., *Liberty Hyde Bailey: A Story of American Plant Sciences* (Princeton, N. J., 1949).

Rogers, William W., "The Alabama State Fair, 1865–1900," *Alabama Review*, XI (April, 1958), pp. 100–116.

———, "The Alabama State Grange," *Alabama Review*, VIII (April, 1955), pp. 105–118.

———, "Reuben F. Kolb: Agricultural Leader of the New South," *Agricultural History*, XXXII (April, 1958), pp. 109–119.

Ross, Earl D., *Democracy's College: The Land-Grant Movement in the Formative Stage* (Ames, Iowa, 1942).

———, "The Evolution of the Agricultural Fair in the Northwest," *Iowa Journal of History and Politics*, XXIV (July, 1926), pp. 445–480.

———, "The 'Father' of the Land-Grant College," *Agricultural History*, XII (April, 1938), pp. 151–186.

———, *A History of the Iowa State College* (Ames, Iowa, 1942).

———, "A Neglected Source of Corn Belt History: Prime's Model Farms," *Agricultural History*, XXIV (April, 1950), pp. 108–112.

———, "The New Agriculture," *Iowa Journal of History*, XLVII (April, 1949), pp. 119–139.

Ross, Thomas R., *Jonathan Prentiss Dolliver: A Study in Political Integrity and Independence* (Iowa City, 1958).

Saloutos, Theodore, and John D. Hicks, *Agricultural Discontent in the Middle West* (Madison, Wis., 1951).

Schlebecker, John T., "Dairy Journalism: Studies in Successful Farm Journalism," *Agricultural History*, XXXI (October, 1957), pp. 23–33.

Schuttler, Vera B., *A History of the Missouri Farm Bureau Federation* (n.p., 1948).

Scott, Roy V., *The Agrarian Movement in Illinois, 1880–1896* (Urbana, Ill., 1962).

———, "American Railroads and Agricultural Extension, 1900–1914: A Study in Railway Developmental Techniques," *Business History Review*, XXXIX (Spring, 1965), pp. 74–98.

———, "The Career of Samuel M. Jordan: A Study in the Evolution of Agricultural Education," *Missouri Historical Society, Bulletin,* XVIII (April, 1962), pp. 239–259.

————, "Early Agricultural Education in Minnesota: The Institute Phase," *Agricultural History*, XXXVII (January, 1963), pp. 21–34.

————, "Farmers' Institutes in Louisiana, 1897–1906," *Journal of Southern History*, XXV (February, 1959), pp. 73–90.

————, "John Patterson Stelle: Agrarian Crusader from Southern Illinois," Illinois State Historical Society, *Journal*, LV (Autumn, 1962), pp. 229–249.

————, "Pioneering in Agricultural Education: Oren C. Gregg and Farmers' Institutes," *Minnesota History*, XXXVII (March, 1960), pp. 19–29.

————, "Railroads and Farmers: Educational Trains in Missouri, 1902–1914," *Agricultural History*, XXXVI (January, 1962), pp. 3–15.

————, "The Rise of the Farmers' Mutual Benefit Association in Illinois, 1883–1891," *Agricultural History*, XXXII (January, 1958), pp. 44–55.

————, and Jane B. Scott, "A Forgotten Phase of Agricultural Education: The Institutes in Missouri," *Mississippi Quarterly*, XIV (Fall, 1961), pp. 169–182.

Seal, Albert G., "John Carmichael Jenkins, Scientific Planter of the Old Natchez District," *Journal of Mississippi History*, I (January, 1939), pp. 14–28.

Shannon, Fred A., *The Farmer's Last Frontier: Agriculture, 1860–1897* (Vol. V of *The Economic History of the United States*, Henry David and others, eds., New York, 1945).

Sheldon, William D., *Populism in the Old Dominion: Virgina Farm Politics, 1885–1900* (Princeton, N. J., 1935).

Silver, James W., "C. P. J. Mooney of the Memphis *Commercial Appeal*, Crusader for Diversification," *Agricultural History*, XVII (April, 1943), pp. 81–89.

Slay, Ronald J., *The Development of the Teaching of Agriculture in Mississippi* (New York, 1928).

Smith, C. B., "The Origin of Farm Economics Extension," *Journal of Farm Economics*, XIV (January, 1932), pp. 17–22.

Smith, George W., "The Old Illinois Agricultural College," Illinois State Historical Society, *Journal*, V (January, 1913), pp. 475–480.

Smith, Ralph A., "The Contribution of the Grangers to Education in Texas," *Southwestern Social Science Quarterly*, XXI (March, 1941), pp. 312–324.

Smith, Ruby G., *The People's Colleges: A History of the New York State Extension Service in Cornell University and the State, 1876–1948* (Ithaca, N. Y., 1949).

Steinel, Alvin T., and D. W. Working, *History of Agriculture in Colorado* (Fort Collins, 1926).

Stemmons, Walter, *Connecticut Agricultural College: A History* (Storrs, Conn., 1931).

Stephens, Frank F., *A History of the University of Missouri* (Columbia, Mo., 1962).

Stevens, Neil E., "America's First Agricultural School," *Scientific Monthly*, XIII (December, 1921), pp. 531–540.

Swisher, Jacob A., "The Corn Gospel Trains," *Palimpsest*, XXVIII (November, 1947), pp. 321–333.

Taylor, Carl C., *The Farmers' Movement* (New York, 1953).

Thompson, D. O., and W. H. Glover, "A Pioneer Adventure in Agricultural Extension: A Contribution from the Wisconsin Cut-Over," *Agricultural History*, XXII (April, 1948), pp. 124–128.

Thomson, E. H., "The Origin and Development of the Office of Farm Management in the United States Department of Agriculture," *Journal of Farm Economics*, XIV (January, 1932), pp. 10–16.

Tontz, Robert L., "Membership of General Farmers' Organizations, United States, 1874–1960," *Agricultural History*, XXXVIII (July, 1964), pp. 143–156.

Trump, Fred, *The Grange in Michigan* (Grand Rapids, 1963).

University of New Hampshire, *History of the University of New Hampshire* (Durham, N. H., 1941).

Villes, Jones, ed., *The University of Missouri: A Centennial History* (Columbia, Mo., 1939).

Wallace, Wesley H., "North Carolina's Agricultural Journals, 1838–1861: A Crusading Press," *North Carolina Historical Review*, XXXVI (July, 1959), pp. 275–306.

Werner, M. R., *Julius Rosenwald: The Life of a Practical Humanitarian* (New York, 1939).

Wheeler, John T., *Two Hundred Years of Agricultural Education in Georgia* (Danville, Ill., 1948).

Wiest, Edward, *Agricultural Organization in the United States* (Lexington, Ky., 1923).

Willard, Julius T., *History of the Kansas State College of Agriculture and Applied Science* (Manhattan, Kans, 1940).

Willers, Diedrich, *The New York State Agricultural College at Ovid, N. Y. and Higher Agricultural Education* (Geneva, N. Y., 1907).

Williamson, Frederick W., *Origin and Growth of Agricultural Extension in Louisiana, 1860–1948* (Baton Rouge, 1951).

————, *Yesterday and Today in Louisiana Agriculture* (Baton Rouge, 1940).

Wing, Andrew S., " 'Alfalfa Joe' Wing, Exponent of Beardless Barley, Brome Grass, and Other Forage Crops," *Agricultural History*, XXXVIII (April, 1964), pp. 96–101.

Woodall, Clyde E., and George H. Aull, "The Pendleton Farmer's Society," *Agricultural History*, XXXI (April, 1957), pp. 36–37.

————, and William H. Faver, "Famous South Carolina Farmers," *Agricultural History*, XXXIII (July, 1959), pp. 138–141.

Woodward, Carl R., *The Development of Agriculture in New Jersey, 1640–1880* (New Brunswick, 1927).

————, and Ingrid N. Waller, *New Jersey's Agricultural Experiment Station, 1880–1930* (New Brunswick, N. J., 1932).

B. THESES

Aldous, Lois G., "The Grange in Kansas since 1895" (M.A. Thesis, University of Kansas, 1941).

Armstrong, Lindsey O., "The Development of Agricultural Education in North Carolina" (M.A. Thesis, North Carolina State College, n.d.).

Bardolph, Richard, "Agricultural Education in Illinois to 1870: The Press" (Ph.D. Thesis, University of Illinois, 1944).

Beck, Oscar A., "The Agricultural Press and Southern Rural Development, 1900–1940" (Ph.D. Thesis, George Peabody College, 1952).

Bertels, Sister M. Thomas More, "The National Grange: Progressives in the Land, 1900–1930" (Ph.D. Thesis, Catholic University, 1962).

Brown, Donald R., "The Educational Contributions of Jonathan Baldwin Turner" (M.A. Thesis, University of Illinois, 1954).

Burt, Jesse C., "History of the Nashville, Chattanooga and St. Louis Railway, 1873–1916" (Ph.D. Thesis, Vanderbilt University, 1950).

Cochran, John P., "The Virginia Agricultural and Mechanical College" (Ph.D. Thesis, University of Alabama, 1961).

Colman, Gould P., "A History of Agricultural Education at Cornell University" (Ph.D. Thesis, Cornell University, 1962).

Elkins, Francis C., "The Agricultural Wheel in Arkansas, 1882–1890" (Ph.D. Thesis, Syracuse University, 1953).

Evans, Samuel L., "Texas Agriculture, 1880–1930" (Ph.D. Thesis, University of Texas, 1960).

Ferguson, James S., "Agrarianism in Mississippi, 1871–1900: A Study in Nonconformity" (Ph.D. Thesis, University of North Carolina, 1952).

McDaniel, Curtis E., "Educational and Social Interests of the Grange in Texas" (M.A. Thesis, University of Texas, 1938).

Nordin, Dennis S., "The Educational Contributions of the Patrons of Husbandry, 1867–1900" (M.A. Thesis, Mississippi State University, 1965).

Pope, George J., "Agricultural Extension in Mississippi Prior to 1914" (M.A. Thesis, Mississippi State University, 1963).

Rich, Mary A., "Railroads and Agricultural Interests in Kansas, 1865–1915" (M.A. Thesis, University of Virginia, 1960).

Ritland, Everett G., "The Educational Activities of P. G. Holden in Iowa" (M.A. Thesis, Iowa State College, 1941).

Robson, George L., Jr., "The Farmers' Union in Mississippi" (M.A. Thesis, Mississippi State University, 1963).

Turner, Fred H., "The Illinois Industrial University" (Ph.D. Thesis, University of Illinois, 1931).

Watts, Stanley G., "Knowledge for the Tennessee Farmer: Agricultural Extension in Tennessee Prior to 1914" (M.A. Thesis, Mississippi State University, 1967).

INDEX

Aberdeen, S.D.: demonstration farm at, 194
Adair, John A.: introduces extension bill, 297
Adams, Charles K., 73
Adams, C. G., 238
Adams Act, 112
Adams Co., Miss.: county agent in, 228
Adger, J. E., 213
Agnew, Ella G., 249
Agrarian crusade: role of, 37; history of, 39-41; and the agrarian myth, 62
Agrarian myth, 62
Agricultural colleges, 24–27. *See also* land-grant colleges
Agricultural experiment stations: failure to meet needs, 4; development of, 32–33
Agricultural extension, early calls for, 34–36
Agricultural fairs: failure to meet needs, 4; development of, 14–17; geographical patterns of, 16; importance of, 16–17; supported by Grange and Alliance groups, 48–49; land-grant college personnel at, 162–163; and railroads, 175
Agricultural innovators: effect of, 4; in pre–Civil War South, 4–6; contributions of, 4–9; in the North, 6–7; in the West, 7–8; among ordinary farmers, 8; in post–Civil War South, 9
Agricultural instruction, early attempts at, 22–24
Agricultural journalism: development of, 17–22; circulation of, 19–20; sources of information for, 20 21; rural attitudes toward, 21; impact of, 21
Agricultural lectures: at Oberlin College, 66; in Illinois, 68–69; in Kansas, 69–70; in Minnesota, 71
Agricultural mechanization, effect of, 38–39
The Agricultural Museum, 18
Agricultural production in late nineteenth century, 38
Agricultural professors: teaching loads of, 32; hostility of farmers toward, 47–48
Agricultural representatives in Wisconsin, 279
Agricultural societies: failure to meet needs, 4; development of, 9-14; and ordinary farmers, 10; in different states, 11; and agricultural fairs, 14
Agricultural Students' Union: and youth work in Ohio, 125; and cooperative experiments, 151

Charleston, Ill.: short course at, 159

Chesapeake and Ohio railroad, alfalfa clubs, 176

Chicago and Northwestern railroad: distributes educational literature, 173; leases land for alfalfa growing, 176

Chicago, Burlington and Quincy railroad: operates educational train, 178–179; alfalfa campaign of, 196–197; aids in placing county agents, 280–281. *See also* Burlington railroad

Chicago, University of, 141

Chubbuck, Levi, 101

Cincinnati, Ohio: agricultural college at, 25

Civil War, 7

Clair, J. C., 114

Clark, Champ: and the National Soil Fertility League, 290

Clark County, Ohio: youth work in, 125–126

Clay County, Iowa: work of county agent in, 285

Clayton, W. D., 228

Clemson College: controversy at, 29–30; and early institutes, 88; extension work centered at, 229

Cleveland, Grover, 17

Cleveland, Ohio: agricultural college at, 25

Clinton County, Iowa: county agent in, 277; work of county agent, 283

Cobb, Cully A., 242–243

Cocke, Philip St. George, 23

Coeducation: attitude of farm groups toward, 57

Collier, John S.: appointed county agent in Kankakee Co., Ill., 275; work of, 283

Colman, Henry, 65

Colman, Norman J.: speaks at agricultural lectures in Illinois, 68; and early institutes in Missouri, 78; on the value of institutes, 102

Colorado: farmers' institutes in, 71

Colorado Fuel and Iron Company strike, 313

Columbia College, 22

Columbian Agricultural Society fair, 15

Columbian Correspondence College, 144

Commission on National Aid to Vocational Education, 297

Connecticut: farmers' institutes in, 67

Connecticut, University of, 53

Connecticut Agricultural College, 143

Cooking schools: in Middle West, 118–119

Cooper, Thomas P., 266

Cooperative experiments: at Cornell, 148; limitations of, 151–152; in Connecticut, Ohio, Iowa, and other states, 151–153

Corn clubs: origins of, 122, 238–240; organized in Holmes County, Miss., 239; expansion and changes in, 240–246; prizes for winners in, 244; for Negroes, 245; impact of, 246–247. *See also* boys' clubs

Cornell Countryman, 62

Cornell University: develops institute program, 73; instruction by correspondence at, 142–143; agricultural extension programs at, 145–148; aids county agents, 272

Correspondence: of land-grant colleges, 140–141

Correspondence College of Agriculture, 144

Correspondence courses: and institutes, 111

Correspondence school agriculture courses, 141–145

Corsa, W. S., 132

Council of North American Grain Exchanges, establishes National Crop Improvement Committee, 202

County agents: early calls for, 169, 260; work of in the South, 230–231; expansion in the North and West, 272–281; duties in the North, 281–285; attitude of farmers toward, 285–286; status in the North in 1914, 286–287; resistance of educators to, 289, 303–304

Cowley Co., Kans.: county agent in, 278

Craig, John, 152

Crane, Fred A.: speaks at boys' encampment, 132; employed by Great Northern railroad, 187

Creelman, C. G., 262

Cromer, Marie, 248

Crop lien system, 7, 206–207

Cullen, George A., 263

Cultivator, 7

New York Central railroad, 184
New York State Agricultural College, 25
New York State Agricultural Society, 65
New York Sun, 22
New York Tribune, 22
Nicholas, W. C., 5
Nixon, S. F., 145
Norfolk and Southern railroad, and demonstration farm, 184
Norfolk and Western railroad, demonstration farm, 184
Normal institutes, 113
North Carolina: farmers' institutes in, 89
North Carolina, University of: agricultural education considered at, 22–23; attacked by grangers, 53
North Carolina Farmers' Association, 40
North Carolina State College, 53
North Dakota Agricultural College: and farmers' institutes, 83–84; and youth work, 130; short courses at, 154; extension publications at, 161
Northern Pacific railroad, 174
Northwestern Dairymen's Association, 82
Noxubee Co., Miss.: agricultural campaign in, 196
Nutt, Joel B., 65

Oberlin College, 66
Ohio: farmers' institutes in, 75–76
Ohio Agricultural College, 25
Ohio State Board of Agriculture, 65
Ohio State University: and early rural youth work, 125; establishes extension department, 126, 166; short course at, 153–154; extension literature issued by, 161; youth work by, 161–162; maintains exhibit at fairs, 163
Oktibbeha Co., Miss.: pig clubs in, 245
Olmstead Co., Minn.: rural youth clubs in, 127
Oneida Co., Wis.: county agent in, 279
Ontario Agricultural College: begins instruction by correspondence, 141; and origins of educational trains, 176; employs agricultural agents, 262
Oregon: farmers' institutes in, 87
Oregon Railroad and Navigation Co. demonstration plots, 185–186

Orton, Edward, 75
Ottertail Co., Minn.: county agent in, 278
Otwell, Will T., 123–124
Otwell's Farmer Boy, 123–124
Ovid, N.Y.: agricultural college in, 25
Oxford Co., Maine: county agent in, 268
Owen, Robert S.: introduces extension bill, 297

Page, Carroll S., 295
Page, Jesse, 132
Page bills, 295
Page Co., Iowa: youth work in, 128; boys' encampment in, 132
Palmyra, N.Y.: grange program at, 48
Patrons of Husbandry, 39
Patrons of Industry, 40
Pendleton Farmer's Society, 5
Pennsylvania: establishes agricultural college, 26; farmers' institutes in, 74–75
Pennsylvania Agricultural College: John Hamilton at, 134; correspondence instruction at, 142
Pennsylvania Department of Agriculture, 74
Pennsylvania Horticultural Society, 13
Pennsylvania Railroad: operates educational train, 180; establishes demonstration farm, 184; aids county agent work, 273
People's College, 25
Peoria Co., Ill.: county agent in, 276
Perry Co., Ala.: agricultural campaign in, 196
Pettis Co., Mo.: county agent in, 264–265; Bureau of Agriculture in, 282; homemakers' club in, 282
Philadelphia Society for Promoting Agriculture, 9, 65
Pierce, J. B., 234
Pillsbury, John S., 82
Pinckney, Thomas, 4
Planters, Mississippi Delta, 115
Plough Boy, 19
Politicians support agricultural extension, 290–291
Polk, Leonidas L., 45
Pollard, Ernest M.: introduces vocational education bill, 293

357

extension programs by academic men, 302–303
U.S. Office of Experiment Stations: gathers information on farmers' institutes, 133; publishes proceedings of farmers' institute organization, 134
U.S. Office of Farm Management: established, 259; dissemination of information by, 260; begins county agent work in Bedford Co., Pa., and Broome Co., N.Y., 263–264; helps to place agricultural agents in counties, 272–273, 276–281, 283; criticized by academic men, 305–306
U.S. Office of Farmers' Institute Specialist, 134–135
U.S. Office of Grass and Forage Plant Investigations, 255–256
U.S. States' Relation Service, 312
Utah: farmers' institutes in, 87

Vanderbilt, Cornelius, 17
Van Rensselaer, Martha, 120
Van Rensselaer, Stephen: establishes school at Troy, N.Y., 24; employs itinerant lecturer, 65
Vardaman, James K., 310
Vermilion Parish, La.: farmers' club in, 110
Vermont: farmers' institutes in, 67
Vincennes University, 254
Virginia: farmers' institutes in, 89
Virginia, University of, 23
Virginia Farmers' Alliance, 45
Virginia Polytechnic Institute, 158–160
Virginia State Board of Agriculture, 5
Vocational education: movement for, 293; separated from agricultural extension in Congress, 297

Waggoner, J. E., 192
Wahkiakum Co., Wash.: county agent in, 280
Wailes, Benjamin L. C., 23
Wallace, Henry: speaks at farmers' institute, 95; and establishment of extension department at Iowa State College, 164–165; serves on educational train, 177; mentioned, 208
Wallaces' Farmer, 129
Washington, Booker T., 233

Washington, George: interested in agricultural improvement, 4; proposes establishment of national agricultural society, 13; urges holding of agricultural fairs, 14; proposes a national university, 34
Washington: farmers' institutes in, 86
Washington Co., Pa.: county agent in, 273
Washington, D.C.: trips to, for corn club winners, 244
Washington State Agricultural College, 86
Washington State College, 158, 160
Watson, Elkanah: establishes Berkshire Agricultural Society, 10; and origin of agricultural fairs, 15
Watts, Frederick, 36
Waycross and Western railroad, 185
Weaver, James B., 41
Webster, Daniel, 14, 65
Welch, A. S., 70, 99
Wemple, J. E., 213
West Baton Rouge Parish, La.: Negro farmers' club in, 110
West Central Development Association: program of, 266–267; aids in county agent work, 278
Western New York Horticultural Society, 73
Western Rural, 50–51
West Virginia: farmers' institutes in, 90
West Virginia Grange, 50
West Virginia University, 55
Wheatland, N.Y.: agricultural school at, 24
Wheelock, Eleazer, 24
White, Charles H., 262
White, F. S., 173
Whitehall, Ill., 132
Whitewater, Wis., *Register*, 22
Whitten, J. C., 98
Wilder, Marshall P., 14
Wiley, David, 18
Will Co., Ill.: agricultural agent in, 276
Williams Co., Ohio: International Harvester campaign in, 195
Wilson, Archie D.: appointed head of institutes in Minnesota, 113; on farmers' clubs, 162
Wilson, James: and agricultural students at Iowa agricultural college, 32; and

Roy V. Scott is professor of history at Mississippi State University. He received his B.S. from Iowa State University in 1952 and his Ph.D. from the University of Illinois in 1957. He won the Everett Eugene Edwards Award in Agricultural History in 1958. His first book was *The Agrarian Movement in Illinois, 1880–1896* (1963).

UNIVERSITY OF ILLINOIS PRESS